The Symbolic Species Evolved

BIOSEMIOTICS

VOLUME 6

Aims and Scope of the Series

Combining research approaches from biology, philosophy and linguistics, the emerging field of biosemiotics proposes that animals, plants and single cells all engage insemiosis – the conversion of physical signals into conventional signs. This has important implications and applications for issues ranging from natural selection to animal behaviour and human psychology, leaving biosemiotics at the cutting edge of the research on the fundamentals of life.

The Springer book series *Biosemiotics* draws together contributions from leading players in international biosemiotics, producing an unparalleled series that will appeal to all those interested in the origins and evolution of life, including molecular and evolutionary biologists, ecologists, anthropologists, psychologists, philosophers and historians of science, linguists, semioticians and researchers in artificial life, information theory and communication technology.

For further volumes:
http://www.springer.com/series/7710

Theresa Schilhab · Frederik Stjernfelt ·
Terrence Deacon
Editors

The Symbolic Species Evolved

Springer

Editors
Dr. Theresa Schilhab
Danish School of Education
University of Århus
Tuborgvej 164
DK-2400 København
Denmark
tsc@dpu.dk

Prof. Frederik Stjernfelt
University of Århus
Jens Chr. Skous Vej 7
8000 Aarhus C
Denmark
semfelt@hum.au.dk

Prof. Terrence Deacon
Department of Anthropology
University of Berkeley
Kroeber Hall 329
94720-3710 Berkeley California
USA
deacon@berkeley.edu

Additional material to this book can be downloaded from http://extras.springer.com

ISSN 1875-4651
e-ISSN 1875-466X
ISBN 978-94-007-9599-0
ISBN 978-94-007-2336-8 (eBook)
DOI 10.1007/978-94-007-2336-8
Springer Dordrecht Heidelberg London New York

Printed on acid-free paper

Springer is part of Springer Science+Business Media (www.springer.com)

Contents

Contributors

Niels Bonde Institute of Geography & Geology, University of Copenhagen, Øster Voldgade 10, DK 1350 Copenhagen, Denmark; Fur Museum, DK 7884 Fur, Denmark, nielsb@geo.ku.dk; niels.bonde@mail.tele.dk

Ingar Brinck Department of Philosophy, Lund University, Lund, Sweden, Ingar.Brinck@fil.lu.se

Richard W. Byrne School of Psychology, University of St Andrews, St Andrews, Fife KY16 9JP, Scotland, rwb@st-andrews.ac.uk

Harry Collins SOCSI, Cardiff School of Social Sciences, Cardiff University, Cardiff, Wales CF10 3WT, UK, CollinsHM@cf.ac.uk

Terrence W. Deacon Department of Anthropology, University of California, Berkeley, CA 94720-3710, USA, deacon@berkeley.edu

Peter Gärdenfors Department of Philosophy, Lund University, Lund, Sweden, Peter.Gardenfors@lucs.lu.se

Cathrine Hasse Department of Learning, Institute of Learning, University of Aarhus, Copenhagen, Denmark, caha@dpu.dk

Jesper Hoffmeyer Biological Institute, University of Copenhagen, Copenhagen DK 2200, Denmark, jhoffmeyer@me.com

Stefan Leijnen Institute for Computing and Information Sciences, Radboud University Nijmegen, 6525 AJ Nijmegen, Netherlands, stefan@leijnen.org

Mathias Osvath Department of Philosophy, Lund University, Lund, Sweden, Mathias.Osvath@lucs.lu.se

Irene M. Pepperberg Department of Psychology, Harvard University, Cambridge, MA 02138, USA, impepper@media.mit.edu

Ahti-Veikko J. Pietarinen Department of Philosophy, History, Culture and Art Studies, University of Helsinki, FIN-00014 Helsinki, Finland, ahti-veikko.pietarinen@helsinki.fi

Theresa Schilhab GNOSIS Research Centre, Danish School of Education, University of Aarhus, 2400 Copenhagen, Denmark, tsc@dpu.dk

Göran Sonesson Centre for Cognitive Semiotics, Lund University, Lund, Sweden, goran.sonesson@semiotik.lu.se

Frederik Stjernfelt Faculty of Arts, University of Aarhus, 8000 Aarhus, Denmark, semfelt@hum.au.dk

About the Authors

Niels Bonde is Professor emeritus at Department of Geography & Geology, University of Copenhagen. Trained as vertebrate-paleobiologist, born 1940, degree Mag. Scient. Paleobiol, Goldmedal Copenhagen University, lecturer and senior lecturer. Professional fields of interest are evolutionary history of fish, dinosaur and birds, humans, cladistics, species problems etc.

Editor and main contributor to a book on the history of evolutionary thoughts. Bonde, N., Hoffmeyer, J., & Stangerup, H. (eds.) 'Naturens historiefortaellere' [Nature's history-tellers], 2 vols (1984, 1986, Gad, Copenhagen – 2nd ed. 1996, 960 pp; in Danish). Bonde, N., et al. (2008) Danekrae – Danmarks bedste fossiler [Denmark's best fossils], 226pp. Gyldendal, Copenhagen (in Danish).

Ingar Brinck is Professor of Theoretical Philosophy at Lund University. She is the author of numerous articles and a book: The Indexical 'I'. The First Person in Thought and Language, Kluwer Academic Publ., 1997. Her main research interests are philosophy of language, self-consciousness, intentionality, attention and creativity. She is an Associate Member of Institut Jean Nicod, Paris.

Richard W. Byrne is Professor of Psychology and studies the evolution of cognitive and social behaviour, particularly the origins of distinctively human characteristics. Current projects focus on the gestural communication of great apes, and convergences in cognition between primates and distantly related species such as the African elephant, in collaboration with Dr. Lucy Bates. Previous work has included tactical deception in primates and its relationship to brain size and intelligence, welfare-related studies of cognition in the domestic pig, and the analysis of social learning and imitation. Postgraduates under his supervision have recently worked on primate gestural and vocal communication, comprehension of mechanical tasks, manual feeding techniques, including the effect of disability on chimpanzees' feeding techniques, and cognitive control of navigation within the home ranges of monkeys and apes. Professor Byrne was awarded the British Psychology Society Book Award 1997 for his O.U.P. monograph The Thinking Ape.

Harry Collins is Distinguished Research Professor of Sociology and Director of the Centre for the Study of Knowledge, Expertise and Science (KES) at Cardiff University. He has held visiting appointments in Brazil, China, California Institute of Technology, University of California at San Diego, Max Planck Institute for

History of Science in Berlin, and many more. He was awarded the 1997 Bernal prize for social studies of science. His sixteen books include The Golem: What you should know about science (1993/98, with Trevor Pinch) which won the American Sociological Association's Robert K. Merton book prize and was followed by a volume on technology (1998). Harry Collins is continuing his research on the sociology of gravitational wave detection, on the nature of expertise and on a new technique – the 'Imitation Game' – for exploring expertise and comparing the extent to which minority groups are integrated into societies. The Imitation Game research is supported by a European Research Council Advanced Grant.

Peter Gärdenfors is Professor of Cognitive Science at Lund University. His main research interests are concept formation, semantics and the evolution of cognition. He is the author of numerous articles and books, including Knowledge in Flux: Modeling the Dynamics of Epistemic States, Bradford Books, MIT Press, 1988. Conceptual Spaces, Bradford Books, MIT Press, 2000, and How Homo Became Sapiens: On the Evolution of Thinking, Oxford University Press, 2003. He is a member of the Royal Swedish Academy of Letters, History and Antiquities, The Royal Swedish Academy of Science, Academia Europaea and Leopoldina Deutsche Akademie für Naturforscher.

Cathrine Hasse has as a cultural psychologist and anthropologist for more than ten years studied cultural learning processes in university and other educational organisations. She did her PhD. On "cultural learning processes" in an institution where she followed a group of "newcomers", young male and female physicists' students, in their first year of study. This project developed into a longitudinal study where she has followed the same group of students more than 6 years. In her next project "The Cultural Dimensions of Science" she compared physics institutions in Denmark and Italy and she has co-ordinated an EU project, financed by EU 6th framework programme, UPGEM (Understanding Puzzles in the Gendered European Map) focusing on comparative aspects of universities as cultural workplaces. She is professor at the Danish School of Education (DPU), Aarhus University, working on "transdisciplinary technological literacy" and technological imaginaries in everyday life. In her work she takes a special interest in the relations between learning, gender, culture and academic institutions as work places. She is an active participant in the ISCAR – International Society for Cultural and Activity Research, and 4S network of Science and Technology Studies and has a long time relation with Women in Physics in Denmark.

Jesper Hoffmeyer is Cand. Scient. and Dr. Phil. and Professor emeritus at Biological Institute, University of Copenhagen. Named fourth Sebeok Fellow of the Semiotic Society of America 2000; President of the International Society for Biosemiotic studies. Recent publictions include Signs of Meaning in the Universe, Indiana University Press 1996; Edited: A Legacy of Living Systems: Gregory Bateson as Precursor to Biosemiotics, Springer 2008; Biosemiotics. An Examination into the Signs of Life and the Life of Signs, Scranton University Press 2008.

Stefan Leijnen is a Ph.D Student at Nijmegen University's Intelligent Systems department, interested in understanding and building creative systems. Before this, he worked with soccer-playing robot dogs, emotional multi-agent systems and symbol-using neural networks. His research interests include the process of creativity, mechanism and emergence in nature, and the ontology of computation. Currently, his research revolves around merging mechanisms of abiogenesis with connectionist models, in order to simulate learning through an artificial creative process.

Mathias Osvath Ph.D., is a researcher in Cognitive Science at Lund University. His main research interest is comparative cognition, and he is working on great apes and corvids. His publications on planning in primates has received high international recognition. He is also the scientific director of Lund University Primate Research Station.

Irene M. Pepperberg (PhD, Harvard, '76) is a Research Associate/Lecturer at Harvard and an Adjunct Associate Professor at Brandeis University. She has been a visiting Assistant Professor at Northwestern University, a tenured Associate Professor at the University of Arizona, and a visiting Associate Professor at the MIT Media Lab. She has received John Simon Guggenheim, Selby, and Radcliffe Fellowships, is a Fellow of AAAS, the American Psychological Association, the American Psychological Society, the Animal Behavior Society, and the American Ornithologists' Union. She is currently the associate editor of the Journal of Comparative Psychology.

Ahti-Veikko J. Pietarinen Ph.D, University of Helsinki, 2002 is Professor of Semiotics at the University of Helsinki and Director of the Helsinki Peirce Research Centre. His research interests include logic, philosophy of language, philosophy of science, history of philosophy, Peirce, and Pragmatism. Pietarinen's recent monograph is Signs of Logic: Peircean Themes on the Philosophy of Language, Games, and Communication (Springer Synthese Library, 2006) and a collection Game Theory and Linguistic Meaning (CRiSPI, Elsevier Science, 2007). His latest journal publications have appeared in the Journal of the History of Ideas, Semiotica, Perspectives on Science, Linguistics and the Human Sciences, Studia Logica, Cognitio, History of Philosophy.

Göran Sonesson is Professor of Semiotics at Lund University and the head of the Centre for cognitive semiotics at that same university. He has written extensively on visual semiotics, semiotics of gesture, epistemology of semiotics, semiotics of culture, and, more recently, cognitive semiotics with an experimental bent. His publications include *Pictorial concepts* (Lund: LUP 1989) as well as more than 200 papers, many of them is well-established journals such as *Semiotica, VISIO, Degrés, Sign System Studies, Protée*, etc., and in numerous anthologies and encyclopaedias.

About the Editors

Terrence W. Deacon holds a Ph.D. in Biological Anthropology and has been a Professor at Harvard University, Harvard Medical School, Boston University, and most recently the University of California at Berkeley, where he is currently Chair of the Department of Anthropology and a member of the Helen Wills Neuroscience Institute. His bench research has mostly focused on the evolution of human brains and their unique cognitive capacities, but has also included studies of comparative brain anatomy, neural development, and the transplantation of fetal neurons and stem cells. His Staley Prize winning 1997 book "The Symbolic Species: The Coevolution of Language and the Brain" synthesized neurological, evolutionary, linguistic, and semiotic approaches to understanding human brain and language evolution. His forthcoming book "Incomplete Nature: How Mind Emerged from Matter" (W.W. Norton & Co., Fall 2011) outlines a theory of what he describes as emergent dynamical processes and offers a radical reformulation of the nature of biological function, information, consciousness, and value.

Theresa Schilhab has a Ph.D in the philosophy of biological sciences from the Niels Bohr Institute, Copenhagen is Associate Professor at the GNOSIS Research Centre, University of Aarhus. She is author and co-editor of numerous books including the anthology Learning Bodies (2008) as well as scientific papers in cognitive ethology, embodied cognition and evolutionary neuroscience. Her research touches on evolutionary perspectives on tacit and explicit knowledge and their indeterminate interaction, evolutionary epistemology (i.e. the evolution of epistemological mechanisms), and specifically the impact of bodily experience on human learning.

Frederik Stjernfelt is Professor Ph.D. at the Center for Semiotics, Aarhus University, Denmark. Recent books: "Diagrammatology. An Investigation on the Borderlines of Phenomenology, Ontology, and Semiotics", Dordrecht: Springer Verlag 2007; "Semiotics. Critical Concepts" vols. I–IV (ed. with P. Bundgaard, London: Routledge 2010); "The Democratic Contradictions of Multiculturalism" (with J.-M. Eriksen, N.Y.: Telos Press, forthcoming). Editor of the journal KRITIK.

Chapter 1
Introduction – Searching the Missing Links

Frederik Stjernfelt, Theresa Schilhab, and Terrence W. Deacon

Terrence Deacon's "The Symbolic Species" came out in 1997 and became an important participant in the renewed focusing upon the issue of the origin of man. The basic Darwinian framework agreed upon by all serious research since early 20C had left the important problem of accounting for the evolution of man's special intellectual abilities, including human language, as compared to other higher animals in general and man's primate relatives specifically. The many competing theories of the origins of language along with the lack of empirical evidence to support either of them had, for many years, made speculations upon language origins obsolete – but with the increasing amount of knowledge about man's genetic evolution, historical linguistics, cognitive science, neuroscience, the archeology of early human migrations etc. created a new platform for taking up this old issue. Deacon's proposal was fourfold – based upon the combination of an evolutionary, a semiotic, a neurological, and an anthropological hypothesis. The evolutionary hypothesis was based upon so-called "Baldwinian" evolution – after the American psychologist James Mark Baldwin: the idea that in social species with individuals possessing a certain degree of ontogenetic learning abilities, new, acquired capabilities may assume a large degree of selective advantage for those individuals able to learn them. Thus, seemingly Lamarckian effects of inheritage of acquired characters may occur within a completely Darwinian framework: the acquired capabilities are not inherited, but the possession of them in some individuals provide a large selection advantage over those who have less ability to learn them. The example chosen in Deacon was, of course, human language: speakers will be strongly favoured at the expense of non-speakers, and thus the appearance of early, primitive language will speed up the process of evolution, eventually making the evolution of language and the evolution of the human brain two aspects of one basic process with intense feed-back between the two.

Baldwinian evolution was also rediscovered by other thinkers in the same period – such as Steven Pinker – but on top of this, Deacon added a neurological hypothesis based on brain scannings of preserved brains of a variety of species:

F. Stjernfelt (✉)
Faculty of Arts, University of Aarhus, 8000 Aarhus, Denmark
e-mail: semfelt@hum.au.dk

T. Schilhab et al. (eds.), *The Symbolic Species Evolved*, Biosemiotics 6,
DOI 10.1007/978-94-007-2336-8_1, © Springer Science+Business Media B.V. 2012

namely that a basic novum in the human brain is its comparatively expanded prefrontal cortex and its connections. This hypothesis, to Deacon, contradicted or at least relativized modularism, pointing to the idea that despite the importance of Wernicke's and Broca's areas for language production and understanding, lexical semantics and enunciation covered much larger amounts of brain capacities – facilitated by the human brain's connectedness. This idea, of course, placed Deacon against strong modularists with the central emphasis of linguistic grammar and its supposed core module – and made him argue that the novelty of human linguistic capabilities must have a broader semiotic character rather than a narrowly conceived grammatical character.

This, in turn, made him appropriate some central aspects of Charles Peirce's semiotics – an early attempt to integrate semiotic and linguistic capabilities with logic and cognitive ones. Thus, Deacon hypothesized that Peirce's distinction between Icons, Indices, and Symbols – signs referring to their object by means of similarity, actual connection, or habit – might be a key to the understanding of the specificity of human language. Making the hypothesis that although forms of iconic and indexical communication were present in many species but only humans built on these to communicate symbolically. Deacon proposed that the ability to produce and process Symbols in this special sense of the word is a key to the general, detached intelligence characteristic of human beings. Some intelligent species, like bonobos and gray parrots, might be able to process Symbols to some degree, but the systematic use of Symbols was taken to be the defining feature of human semiotic intelligence – hence the title of the book, The Symbolic Species.

What, then, was the specific selection pressure pushing early man over the threshold to Symbol processing? Here, Deacon – anthropologist by training and career – proposed an anthropological narrative to account for a specific set of selection pressures. The discovery of stone tool technology by an australopithecine ancestor some 2.5 million years ago made it possible to include a larger degree of meat in their diet. But this required male-male cooperation and the risk of predation made it increasingly difficult for women with children to participate in hunting, resulting in the classic Stone Age scenario of gathering women and hunting men. This, in turn, made the connection between the sexes fragile. How could the pregnant mother-to-be know that the father of her child would, in fact, return with parts of a corpse after having been away in many days with his gang of hunters? Conversely, how could the hunting man know that the mother of his child would not be unfaithful to him during his hunting absence, making it uncertain it was in fact his own genetic offspring he was busily catching protein for? Moreover, male cooperation is crucial, given the dangers of meat scavenges, and so sexual competition must be minimized. In short, all players in this anthropological dilemma have an interest in securing the link to one another. So to Deacon, establishing socially-mediated fidelity was what required symbols' capacity to represent possible future relationships and commitments the arch-example of speech acts, and which introduced selection pressure to evolve cognitive functions for aiding the acquisition and use of symbolic reference.

Thus, the overall argument of Deacon's book united evolutionary, neurological, semiotic, and social-anthropological arguments. Many competing accounts for the same issue, like Bruner-Tomasello's emphasis on joint and shared attention, Turner-Fauconnier's pointing to double scope blending or the Chomskyans' focus on linguistic grammar, would disagree on one or several points in this chain of arguments. Deacon's theory is unique, however, in its integration of these many critical threads of causality and also because of its focus on a semiotic cause. This unprecedented approach is what prompted his two co-editors of this book to organize a couple of conferences addressing the critical discussion of Deacon's chain of arguments. The Symbolic Species Conferences I and II took place in 2006 and 2007 and presented a wide variety of scholars each with a unique view on evolutionary cognition and the questions raised by Terrence Deacon. This book is not simply a conference proceeding; rather it is an attempt at concentrating and focusing the conference discussion around the issues highlighted by Deacon's bundle of arguments. This is why some of the papers thoroughly discuss aspects of Deacon's theories, why others address other, maybe competing approaches to the same issue. In order to focus these different contributions on Deacon's argument, we decided to give the overall structure of the book a Deaconian frame.

1.1 Presentation of the Chapters

The volume is divided into three sections, namely "The biosemiotic connection": "The prehistoric and comparative connection" and "The cognitive and anthropological connection".

The main focus of the first section is the biosemiotic view on human cognition with special emphasis on the analysis of the Deaconian perspective. In "The prehistoric and comparative connection", human descent, learning abilities and species-specific cognition is discussed in an evolutionary as well as comparative perspective. The third section; "The cognitive and anthropological connection", sheds light on various aspects of symbol use especially as this applies to natural language use such as linguistic immersion and embeddedness on the on the one hand and, on the other, the associated emergence of semantic freedom. The individual chapters will be introduced in the following.

1.1.1 The Biosemiotic Connection (Part I)

The introductory chapter of the book is a newly-written contribution by Deacon addressing the issue of the status of the Symbol – in some sense the central concept of the book tying together its biology, neuroscience, semiotics, and anthropology arguments. In this chapter The Symbolic Species hypothesis is truly revisited in three ways: 1. by more explicitly and precisely defining his conception of symbolic reference and its dependency on iconic and indexical processes, 2. by applying this analysis to a re-thinking of the concept of Universal Grammar as neither nature nor nurture, and 3. by demonstrating a role for relaxed selection in setting the stage for the synergistic reorganization of brain functions to support the demands of language. In the following chapter, Frederik Stjernfelt critically addresses Deacon's analysis of the symbol concept and its hierarchic dependence on iconic and indexical forms of reference hypothesis that pure Icons appear early in evolution, only later to combine to form Indices and eventually Symbols. Although for Deacon the icon-index-symbol sequence is not historical or evolutionary, it is a reflection of the increasing complexity of the cognitive demands of these modes of referring. Thus according to the Symbolic Species the symbolic threshold is only crossed when sufficient special interpretive capacities are in place that he identifies with his systemic conception of symbolic reference. Stjernfelt criticizes this analysis by challenging both the hierarchic dependency of these sign-forms and their supposed separability. His alternative conception does not view symbols as systemically mediated in the way Deacon describes, but instead locates an analogous semiotic-evolutionary threshold in a special form of mental abstraction. He thus opposes the view that Symbols per se are a key to the general, detached intelligence characteristic of human beings and argues instead that the true demarcation criterion seems to be what he describes (after Peirce) as our extensive ability to hypostasize.

In his chapter: "Peirce and Deacon on Meaning and the Evolution of Language", Ahti-Veikko J. Pietarinen investigates the influence from Peirce on the work of Deacon by focussing on the similarities and dissimilarities between Peirce's and Deacon's positions with particular reference to the notions of meaning and the evolution of language.

The last chapter: "Semiosis beyond Signs. On two or three missing links on the way to human beings" by Göran Sonesson is concerned with two (nearly) missing links in the progression from animal to man, that is the (principle of) relevance and the sign, as well as the act of imitation bridging them. Sonesson aims to distinguish stages in evolution and development, notably the relationship between imitation and sign.

1.1.2 The Prehistoric and Comparative Connection (Part II)

The first chapter of part two, "The natural history of intentionality: A biosemiotic approach-2" by Jesper Hoffmeyer, takes the rich occurrence of natural intentionality as its starting point to demonstrate the wealth of sign action and therefore semiotic realism pervasive to the living world.

The chapter "The evolution of learning to communicate: Avian model for the missing link" by Irene Pepperberg offers a comparative perspective on language that analyses to what extent language might be considered (or reconsidered) to be constructed from purely primate-specific qualities.

Similarly, but now based on studies of organized, purposeful actions in great apes, in the chapter "From parsing actions to understanding intentions" Richard Byrne aims to demystify the putative missing linguistic link between man and non human animals is presented. According to Byrne, behaviour parsing might be a necessary step on the road to seeing the world in an intentional-causal way.

In the following chapter by Niels Bonde, named "Hominid Diversity and 'Ancestor' Myths: *Homo, H. sapiens*, and Other Taxa from a Phylocladistic Viewpoint", we critisize up to date front-line consensus on our human descendence within contemporary palaeoanthropology and claims of 'direct fossil ancestors'.

Finally, part two is closed by the chapter "The tripod effect: Co-evolution of cooperation, cognition and communication" by Peter Gärdenfors, Ingar Brinck and Mathias Osvath. The chapter simultaneously addresses hominin cognition, cooperation, and communication to show how these interdependent factors mutually reinforce each other over the course of evolution.

1.1.3 The Cognitive and Anthropological Connection (Part III)

"Language as a repository of tacit knowledge" by Harry Collins highlights the linguistic exclusivity that allows almost unlimited knowledge exchange between competent language users. In this view language is not conceived of as merely a tool put to cognitive use but as a form of life.

Theresa Schilhab focuses on the situatedness of language in the chapter "Levels of immersion and embodiment" to expand on the relation between symbol use as it applies to the linguistic exchange in professional communities and the lack of first hand experiences of the concepts mastered to perfection.

In "Emerging symbols", Stefan Leijnen explores the difference between indexical and symbolic interpretation on the basis of a neural network simulation of a series of language training experiments with chimpanzees. Leijnen then discusses systemic requirements for crossing the symbolic threshold.

Finally, the closing chapter of the book "Gender in innovative techno fantasies" by Cathrine Hasse explores Deacon's idea of the evolution of language and the evolution of the human brain as two aspects of one basic process to argue that human agents have developed a particular capacity for creating their habitats according to their fantasies about how they would like to live in the future, especially in the case of technological tools.

Part I
The Biosemiotic Connection

Chapter 2
Beyond the Symbolic Species

Terrence W. Deacon

Abstract Confusions about the nature of symbolic reference are at the core of two major challenges to understanding human language. A failure to take into account the complex iconic and indexical infrastructure of symbolic interpretation processes has blocked progress in the study of language structure, language evolution, neural processing of language, and language acquisition. Simplistic notions of symbolic interpretation are critiqued, the semiotic infrastructure of symbolic interpretive processes is described, and some implication for understanding the universals of grammar and syntax are explored. Finally, the evolutionary problem of language origins is re-examined and an unexpected important role for relaxation of selection is demonstrated.

2.1 The Problem with 'Symbol'

In the years since the publication of *The Symbolic Species* (Deacon, 1997) one consistent source of confusion has persistently been used as a reason to take a critical view of the symbolic threshold as key to the human difference. This is in one sense merely a terminological problem with interpretations of the term 'symbol,' and yet it obscures a critical issue that if not resolved will be a roadblock to both the study of language and the further development of semiotic theory itself. The confusion superficially has to do with the concept of arbitrarily of reference, but more deeply it involves a tension between a structural and dynamic conception of the process of semiosis more generally.

I will first address the terminological dispute, which although a source of confusion in the literature, should be resolvable with a bit of care in defining terms and avoiding the attribution of one definition to uses where it does not apply. The conceptual dispute is much more subtle, and I think critical to sort out. Failure to do so will have two serious consequences. First, it will doom semiotic theories to the status of mere taxonomic exercises where different scholars are free to invent their

---

Parts of this chapter are reprinted from Deacon (2010a)

T.W. Deacon (✉)
Department of Anthropology, University of California, Berkeley, CA 94720-3710, USA
e-mail: deacon@berkeley.edu

T. Schilhab et al. (eds.), *The Symbolic Species Evolved*, Biosemiotics 6,
DOI 10.1007/978-94-007-2336-8_2, © Springer Science+Business Media B.V. 2012

own categorical principles without careful reflection on the underlying generative processes and constraints that determine the semiotic differences they hope to distinguish. This often ends up turning semiotic research into a renaming exercise, where commonly studied phenomena are redescribed in semiotic terms and it often devolves into battles over competing naming paradigms from the past. Second, and more serious, it will cut semiotic research off from the sciences of psychology, neurology, and biology due to a failure to come to grips with the *process* of semiosis; the dynamic of interpretive activity by which semiotic relationships emerge from other semiotic relationships and ultimately derive their grounding on the physical phenomena they thereby bring into consideration. The problem here is the tendency to imagine signs as things, or as synchronic relationships, whereas they are instead intrinsically dynamic phases in a generative process, and ultimately something apart from the artifacts being manipulated in this process.

The term 'symbol' has come to be used differently in different traditions, and so first we need to be clear what we are talking about. If all that is meant is a mark that need not share any specific quality with its object of reference, then the term has trivial consequences. This gloss of the concept makes it easy to dismiss its importance for evolution, and indeed this simplification has been the motivation for many language origins researchers to imagine that it is only syntax that demands explanation. This assumption about the concept of symbol is also reflected in many critics' claims that most species are capable of learning arbitrary associations (e.g. see Chapter 3, this volume) so claiming that the symbolic capacity divides humans from other species must be trivially false.

This focus on arbitrary correlation as the defining attribute of symbolic reference is a serious oversimplification that collapses critical distinctions between sign vehicle and referential properties. The common usage of a 'code' analogy in describing language reference also reflects this simplification, and for similar reasons leads to serious theoretical misunderstandings. A code does indeed involve an arbitrary mapping or correspondence relationship, but that is precisely why its reference is opaque and is the basis for encryption. A code is a mapping of a parallel set of sign tokens to a language, and typically a token-to-token mapping. So to describe language or any of its attributes, such as the basis for phonology, syntax, or semantics as a code, merely begs the question: what is the basis for this mapping relationship?

It is often argued, for example, that arbitrariness is a property of many animal calls. Consider the case of predator-specific alarm calls (which have been identified in species as diverse as vervet monkeys and chickens). The assumption that these calls 'mean' or 'name' a particular predator is as, the linguist Derek Bickerton (2010) has also argued, a 'back-projection of our own language-saturated view of the world.' Alarm calls are indexical, even though they don't sound like the predator they indicate and even though they are emitted to many similar types of predators. Their arbitrariness and generic reference are red herrings in this detective story. Their reference depends on and evolved from repeated correlations between the presence of a predator, the production of a call, and an appropriate escape behavior, and merely distinguished from other experiences, vocalizations, or behaviors.

A symbolic sign relationship is, in contrast to an iconic or indexical sign relationship, a doubly conventional form or reference. It involves a conventional sign type that is additionally conventionaly-mediated in the way it represents.

Arbitrariness is a negative way of defining symbols. It basically tells us that neither likeness nor correlation are necessary. But this is inadequate, even though it is a common shorthand way of characterizing symbolic reference. All sign relationships include some degree of arbitrarity, because those attributes that are taken as the ground for the sign-object linkage can be chosen from many dimensions. Thus, anything can be treated as iconic or indexical of almost anything else depending on the interpretive process.

For example, with a bit of imagination a face can be discerned on the full moon, or in a cloud formation, and it might even remind you of someone you know. But iconism can also be highly abstract, as in the complex way that a mathematical equation refers iconically, once you know how to discern its symbol-mediated isometry (e.g. between the structure of the equation and a corresponding geometric or dynamical relationship). An equation can be interpreted to be iconic (e.g. of a parabolic trajectory) only, however, if you know how to discern the way that differences in the values or operations directly correspond to differences in the geometric object of reference. So one first needs to be able to interpret the symbolic components before the diagrammatic iconism of the equation can be appreciated.

Indices refer by contiguity in space, time, or substrate. A simple correlation can therefore be the ground for indexical reference. A lipstick smear on a man's shirt collar can be a troublesome indication to his wife, a urine scent on a branch can be a sexual index to a female lemur, and the mobbing call of a small bird can indicate the present of a raptor. What gets correlated and how (accidental, cultural, evolutionary) can be arbitrary, only the fact of correlation is not. Thus, a rat in a Skinner box pressing a bar in response to a bell in order to get a water reward has learned that the bell is an arbitrary index of the state of the apparatus (an indexical legisign). These states are arbitrarily paired in the experimental design, but that doesn't make the one a symbol of the other.

So symbolic reference is not merely a function of arbitrariness, conventionality, and generality, though these features are properties that symbolic reference makes available. First of all, arbitrariness isn't required. For example, many symbols used to designate religious concepts employ obvious iconism and yet this doesn't undermine their potential to symbolize quite complex esoteric abstractions. This also demonstrates that the sign vehicles used for symbolic reference need not be widely understood as conventional. When first encountering an unfamiliar religious symbol it may only require a brief few comments to understand its symbolic import. And of course icons, such as the eye-spots on male peacock tail feathers or faces 'seen' in the clouds often bring to mind general types of objects, not just specific instances. These attributes are not sufficient determinants of symbolic function, either individually or collectively.

As Charles Peirce (1931) pointed out over a century ago, we must distinguish properties of the sign vehicle (which he terms a representamen), which can include being an arbitrarily defined (i.e. conventional) type of sign vehicle, from properties taken to link it to its object of reference. Thus although current vernacular has habitually termed alphanumeric characters "symbols" this usage ignores any referential relationship. If not used carefully, in recognition of this shorthand, it can lead to all manner of theoretical confusions.

Thus when your computer begins randomly spewing alphanumeric characters onto your screen they are indices of a malfunction, not symbols of anything. And likewise the typographical character combination ;-) does not refer symbolically, even though it is composed of conventional tokens designed for symbolic purposes. Peirce terms conventional sign vehicle types 'legisigns,' and argues that symbols must also employ legisigns. However he notes that legisigns can also serve iconic and indexical roles as well. Consider, for example, the conventionalized stick figure icons on restroom doors, or the use of red for traffic lights and road signs to indicate the requirement to stop (i.e. it indicates a convention—an injunction to act according to a rule—but it does not 'mean' "stop" in the way that this word does. Because legisigns are often created (or chosen) with a specific type of referential relationship in mind it is the arbitrary choice of the creator which properties are to be used referentially. This is why legisigns created for typographical use to symbolize the parsing and punctuation of written text can also be recruited for their iconic features (as in the case of the smiley face).

Of course communicative intention is also an interpretation, and this also does not fix the referential function of a sign vehicle. Whether something is interpreted iconically, indexically, or symbolically depends on what's going on in the mind of the beholder.

Recognizing that the same sign vehicle need not always be interpreted as intended, or as referring always in the same way is the first step toward reframing semiosis in diachronic, not synchronic, terms. A sign vehicle can be interpreted in multiple ways not because it is in some way a combination of sign types, a fractional mixture of iconic, indexical, and symbolic features, but because its semiotic significance is not vested in the sign vehicle at all. Although a given interpretation may depend on some feature intrinsic to that artifact for motivating its semiotic function, no semiotic attributes are invested in the sign vehicle itself. They are properties of it being interpreted (whether in its creation or its consideration). So given that the same sign vehicle can be interpreted differently by different individuals, or at different phases of considering it, worrying about whether *it* is a 'pure' sign of a given type or a 'mixed' sign commits the fallacy of misplaced concreteness.

As we will discuss below, a given sign relation is created by an interpretive process. It is a phase in this process in which the sign vehicle is incorporated in a particular way, but which may be transitory, leading to a different mode of considering that same sign vehicle. And at any given phase of this interpretive process there is no 'mixture' of semiotic characteristics. It is only when we attempt to analytically collapse this process into a single synchronic relation that we run the risk of confusing sign vehicle properties with semiotic properties and think of signs as simultaneously exhibiting iconic, indexical, and symbolic features.

Although it is far beyond the scope of this chapter to attempt a reframing of semiotic theory in process terms, carefully dissecting a few examples of interpretive processes can help to illustrate the difference between this and more synchronic forms of semiotic analysis and clear up confusions created by the 'compositional' account of symbolic reference presented in *The Symbolic Species* (Deacon, 1997). More importantly, exemplifying the process of hierarchic differentiation of referential form that constitutes an interpretive process allows us to see how semiotic analysis is directly relevant to understanding cognition, and by implication the evolution of symbolic cognition.

As a starting point for exhibiting the hierarchic dependency of the different modes of referential interpretation consider one of the classic examples of a symbolic form: the impression of a signet ring in wax used to seal a note and verify the sender's identity. Tracing the minute cognitive steps necessary to interpret this simple sign demonstrates that symbolic function depends on more than a simple arbitrary correspondence. First, the formal similarity between the impression and the ring is primary. This is iconic. But without the physical action of the ring-bearer pressing the ring into hot wax to produce this likeness, it would not indicate that this message, thus sealed, was produced by the bearer of that specific ring. The presumed connection between ring and bearer further indicates that a particular individual actually sealed the note. Finally, possession of such a ring is typically a mark of authority, royalty, etc. This status is a social convention. To interpret the wax impression as a symbol of social position, one must also understand these social conventions, because nothing intrinsic to the form or its physical creation supplies this information. The symbolic reference is dependent on already knowing something beyond any features embodied in this sign vehicle.

This dependency on an external system of relations within which the formal similarities and correlative aspects of the wax impression are embedded is a critical property of its symbolic reference. But without familiarity with this entire system of relationships, these non-symbolic components remain merely icons and indices. Indeed, if any link in this chain of referential inferences is broken, symbolic reference fails. So while the features comprising the sign vehicle are not necessarily similar in form or physically linked to what is symbolized, this superficial independence is supported by a less obvious network of other modes of reference, involving both iconism and indexicality.

Notice that the first step in this interpretive analysis involved recognition of an iconism. Only after this recognition was the implicit indexicality relevant and only after that was the social convention able to play a role in providing symbolic significance to the sign vehicle. This hierarchic dependency of symbols on indices on icons was the core semiotic argument of *The Symbolic Species*. But notice that it is not a simple compositional relationship. Indices are not *made of* icons and symbols are not *made of* indices. These are stages in developing and differentiating ever more complex forms of reference. Throughout the interpretive process described above there was only one sign vehicle: the wax impression. At first it is interpreted iconically, then indexically, and finally symbolically. The constructive nature of this interpretive process was what was critical. These semiotic relationships were not mixed in some fractional sense, they were distinct dependent phases in the process, and most of the relevant detail was supplied by the interpreting process not the wax impression.

This account leaves out many subphases of the interpretive process, but it captures the crucial architectonic that I believe is critical to understanding why there might be a cognitive threshold separating iconic and indexical forms of communicating, common to most mammals and birds, from symbolic communicating that is distinctive of humans. Interpreting something symbolically is simply more complex, and unlike iconic and indexical interpretation there is nothing inherent in the form or physical relationships of the sign vehicle to provide an interpretive clue. This must be supplied entirely by the interpretive process itself, and it is of the nature of a systemic relationship, not some singular object or event.

Before turning to language, it is worth exploring a few other simpler examples of this interpretive differentiation process in order to appreciate the generality of this hierarchic semiotic dependency.

Let me begin with a trivial index: a wind sock that indicates the strength and direction of the wind. What constitutes the interpretive competence to recognize this indexicality? Imagine that it is being seen for the first time through a window. It is iconic of cloth or clothing, and yet it is clearly not clothing or randomly fluttering cloth. Its distinctive shape and careful design, in contrast, indicate that it is likely designed for a purpose. Another iconic feature is its extended fluttering behavior, again iconic of clothing, but of clothing being blown by the wind fluttering on a clothes line. This iconism now brings to mind something that is not directly provided by the sign vehicle: wind. By virtue of developing these iconic interpretations then this sign vehicle is now embedded in a larger context in which something present points to something that it is not: the wind. And a further juxtaposition of iconisms that have involved other windblown experiences can eventually (quickly) lead to interpreting its behavior as an index of both the direction and intensity of the wind. The indexicality is not 'composed of icons' but rather emerges from the comparisons among iconic interpretations. Failing to recognize these iconisms, e.g. because of never having experienced the effects of wind, would make the indexical interpretation impossible to develop.

Next consider the interpretation of the chevron insignia on a military jacket. Initially, it appears just a colored shape, an iconic sinsign in Peircean terminology (a singular instance of something familiar). As similar shapes are seen on other shoulders, it develops from an iconic sinsign to an iconic legisign (shapes of the same type). As it is understood to distinguish the individual wearing it, it becomes interpreted as an indexical legisign (a type of sign vehicle pointing to something about this person). When its particular configuration is understood to designate that person's military rank it becomes interpreted as a symbolic legisign. The same sign vehicle thus is the locus for a sequence of interpretive phases in which both the relationship of the sign vehicle to other sign vehicles and the relationship of the sign vehicle to its reference are progressively developed.

Some of my favorite examples of this hierarchic interpretive dependence are captured in political cartoons and illustrations that make a general statement about

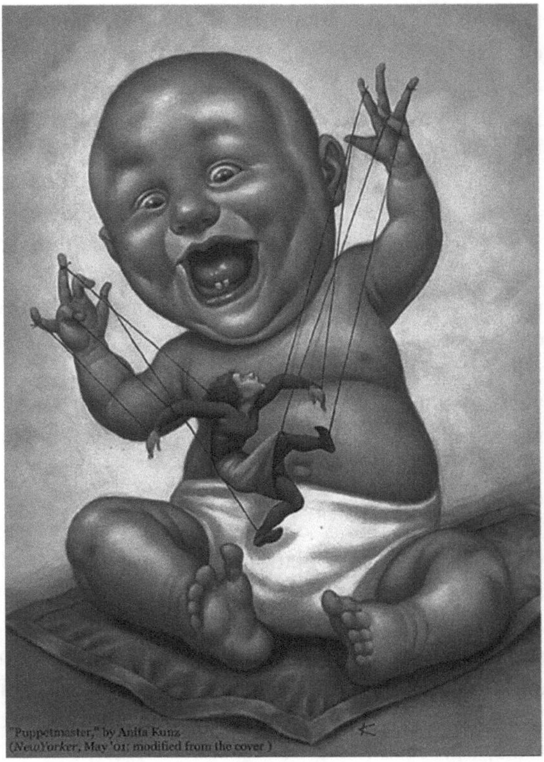

Fig. 2.1 Cartoon from the cover of New Yorker Magazine which exemplifies the progressive differentiation of iconic to indexical to symbolic interpretive phases (see text)

things by virtue of the atypical juxtapositions they employ. Consider the cartoon cover from the New Yorker Magazine in Fig. 2.1.

On first glance, as soon as the discordant features of the image are appreciated, one's mind jumps to an interpretation that is beyond anything depicted. It is commenting on a somewhat paradoxical aspect of motherhood. But how does it induce us to make this quite abstract interpretation? Seen in isolation an image of a mother and baby or an image of a child playing with a puppet do not 'say' anything, or even provide new information. But the violation of expectation created by the baby controlling the mother puppet is not merely interpreted iconically. Its inversion of expectation is interpreted indexically, pointing to its opposite: mothers control babies. This, in turn, reciprocally points back to the partial truth of the abstract relationship of baby controlling mother, and thereby to the paradox that both abstract relations are true, though the image is absurd. In this example, relationships between icons, one present another brought to mind by it, initiate an indexical interpretation of this relationship that ultimately leads the viewer to interpret this as being about something much more abstract and general. Although this interpretive process involves iconic, indexical, and possibly symbolic interpretive phases (the latter to the extent that it comments on the conventional cultural assumptions about motherhood), these are not vested in the sign vehicle and are not mixed or additive. They are distinct phases of interpretation in which the same complex sign vehicle is given progressively more differentiated and context embedded interpretations. Failure to initially interpret the iconisms would make it impossible to interpret any indexicality and failure to interpret the indexical relationships would make it impossible to ever assign any symbolic meanings to the image.

The import of these simple examples is this: to generate an indexical interpretation of any sign vehicle requires interpreting it iconically and interpreting this iconicity with respect to other iconic interpretations, and interpreting it symbolically requires interpreting it indexically and interpreting this indexicality in context with other indexical interpretations. A higher order interpretive process must in this way be supported by a lower order interpretive process, and so on down to the most minimal form. Although this analysis only focuses on this representational triad, it in fact captures an enigmatic aspect of Peirce's 9-part sign categorization system (shown in Fig. 2.2).

In this taxonomic scheme there are three levels of sign vehicle relationship three levels of sign-object relationship and three levels of relationship between a sign and its immediate interpretive semiotic effect (its *interpretant*). One of the strictures that Peirce imposes on the use of this taxonomic triad of triads is that the level of the sign vehicle must be at least as high as the level of the sign-to-object relationship and this must be at least as high as the relationship of the sign to its interpretant. But recognizing that an interpretant is itself, according to Peirce, another sign generation process (what I have above described as a phase of interpretation) we can now see that indeed a sign of a higher type depends on being interpreted (and thus its referential capacity generated) by the generation of lower order signs.

Language competence rests on a quite elaborate system of iconic and indexical relationships that necessarily come into play in the production and interpretation

	sign vehicle itself	sign to object	sign to interpretant
3	legisign	symbol	argument
2	sinsign	index	dicisign
1	qualisign	icon	rheme
	1	2	3

Fig. 2.2 Peirce's 9-part sign taxonomy. Each sign type is defined by the combination of one property from each column such that no property from a column to the right is at a higher level that that to its left. Thus there can be a rhematic indexical sinsign or a dicent symbolic legisign but not a rhematic symbolic sinsign or a dicent iconic legisign

of linguistic communication. What is remarkable about the semiotic infrastructure supporting the symbolic capacity of language is its incredible size and complexity. Its indexical character is made evident by the web of pointing relationships exemplified by a thesaurus, with its one-to-many reciprocal mapping relationships, or a dictionary in which each word or morpheme is mapped to a particular combinatorial relationship among other words. Indeed, a dictionary suggests that a language is a bit like an organism in which every molecule is created by combinations of other molecules interacting. It is this dependence on an underlying semiotic system of relationships that makes this threshold hard to cross for other species. But not only does this serve as the foundation for language reference, these underlying semiotic supports and requirements are unmasked, so to speak, when symbolic relationships are juxtaposed to form even higher order iconic, indexical, and symbolic complexes. Thus, like a circuit diagram that can only be seen as iconic of a type of electronic circuit when its component features are given correct symbolic interpretations, a sentence or narrative depends on first interpreting its symbolic components and then interpreting the higher order iconic and indexical relationships that their combinatorial relationships offer. These hierarchically embedded and emergent semiotic constraints turn out to be key to understanding the higher order logic of grammar and syntax.

2.2 The Semiosis of Grammar and Syntax

True symbolic communication and grammar are inextricably intertwined. They are hierarchically dependent. It is fundamentally impossible to have grammar without symbolic reference, though grammatical relationships don't automatically come to the fore with all forms of symbolic interpretation. Grammar and syntax are, however, intrinsic symbolic attributes that emerge into relevance as symbols are brought into various semiotic relationships with one another; e.g. in combinatorial referential processes. Once we overcome the tendency to treat symbolic reference as mere synchronic arbitrary correlation we can begin to discern the many contributions of the iconic and indexical supports of symbolic reference that have become incorporated into the constraints that define the grammar of language.

Because symbolic reference involves a complex higher-order interpretive development in order to emerge from more basic iconic and indexical relationships, there are implicit constraints that these supportive semiotic relationships impose on operations involving symbol combinations, such as phrases, sentences, arguments, and narratives. These constraints emerge from below, so to speak, from the semiotic infrastructure that constitutes symbolic representation rather than needing to be imposed from an extrinsic source of grammatical principles. Although this infrastructure is largely invisible, hidden in the details of an internalized system acquired in early experience, using symbol combinations in communicative contexts unmasks the iconic and indexical constraints that are implicit in this infrastructure. These semiotic constraints have the most ubiquitous effect on the regularization of language structure, but in addition there are sources of weaker less ubiquitous constraints also contributing to cross-linguistic regularities. These include processing constraints due to neurological limitations, requirements of communication, and cognitive biases specific to our primate/ hominid evolutionary heritage. Although none of these sources of constraint play a direct role in generating specific linguistic structures, their persistent influence over the course of thousands of years of language transmission tends to weed out language forms that are less effective at disambiguating reference, harder to acquire at an early age, demand significant cognitive effort and processing time, and are inconsistent with the distinctive ways that primate brains tend to interpret the world.

The list of sources of constraint on language structure can be broken down into four main categories as listed below. They each contribute a number of quasi-universal traits and highly probable language regularities, many of which are listed for a given category of constraint type. These categories and language consequences are listed below:

A. *Semiotic constraints*

1. Recursive structure (only symbols can provide non-destructive [opaque] recursion across logical types)
2. Predication structure (symbols must be bound to indices in order to refer)
3. Transitivity and embedding constraints (indexicality depends on immediate correlation and contiguity, and is transitive)
4. Quantification (symbolized indices need re-specification).
5. Constraints can be discovered pragmatically and 'guessed' prior to language feedback (because of analogies to non-linguistic iconic and indexical experiences).

B. *Processing constraints*

6. Chunking-branching architecture (mnemonic constraint)
7. Algorithmic regularization (procedural automatization)
8. Neural substrates will vary on the basis of processing logic, not linguistic categories

C. *Sensorimotor schemas & phylogenetic bias*

9. Standard schema/frame units (via cognitive borrowing)
10. Vocal takeover (an optimal medium for mimicry)

D. *Communication constraints*

11. Pragmatic constraints (communication roles and discourse functions)
12. Culture-specific expectations/prohibitions (e.g. distinctive conventions of indication, ways of marking discourse perspective, prohibitions against certain kinds of expressions, etc.)

2.2.1 Semiotic Constraints

The most important and ubiquitous source of constraints on language organization arise neither from nature nor from nurture. That is, they are not the result of biological evolution producing innate predispositions and they are not derived from the demands of discourse or the accidents of cultural history. Semiotic constraints are those that most directly reflect the grammatical categories, syntactic limitations, and phrasal organization of language. They are in a real sense *a priori* constraints, that precede all others. Consequently they are most often confused with innate influences.

In a recent and now well-known theoretical review of the language origins problem (Hauser, Chomsky, & Fitch, 2002) Noam Chomsky appeared to retreat from a number of earlier claims about the innate 'faculty' for language, but he repeated his long-term insistence that what makes the human mind unique is an innate capacity to handle recursive relationships. Like many related claims for an innate grammatical faculty, this one too derives from a reductionistic conception of symbolic reference. If we assume, in contrast, that non-human communication is exclusively mediated by iconic and indexical forms of reference and that only human communication is symbolic it becomes clear why recursively structured communication is only present in humans.

Symbolization enables substitutions that cross-logical-type (e.g. part for whole, member for class, word for phrase) levels in linguistic communications. Neither icons nor indices can refer across logical types because of the involvement of sign vehicle properties (e.g. similarity of form, correlation in space or time) in determining reference. But because of the independence of sign vehicle properties from the objects of reference, symbols can represent other symbolic relationships including even combinations of symbols forming higher logical type units (such as phrases, whole sentences, and even narratives). This is exemplified by pronominal reference and also includes recursively operating on iconic and indexical relationships.

In summary, recursion is not an operation that must be added to human cognition over and above symbolic capabilities, it is a combinatorial possibility that comes for free, so to speak, as soon as symbolic reference is available. But it is not possible when restricted to only iconic and indexical forms. So the absence of recursion in animal communication is no more of a mystery than its presence in human communication. The reason that it is not found in the communication of other species is simply due to their lack of symbolic abilities.

Though recursion is made available with symbolic communication, it need not be taken advantage of, and so its paucity in child language and pidgins and it absence in some languages (e.g. Everett, 2005) is not evidence that it is an unimportant feature of language. But it is an important means for optimizing communication. Recursion provides means for condensing symbol strings. By repeated recursive operations it becomes possible to refer to an extensive corpus of prior discourse. This not only optimizes communicative effort, it also reduces working memory load because a large corpus of material can be subsumed into the reference of single symbolic unit (such as a pronoun). However, recursion also creates new 'housekeeping' requirements that demand specialized forms of symbolized iconic and indexical operations (see below).

All languages require at least a dyadic sentential structure, i.e. something like a subject-predicate sentential form or a topic-comment structure. Although holophrastic utterances, commands, and expletives, are not uncommon, they typically are embedded in a pragmatic context in which what they refer to is made salient by non-linguistic means. Previous suggestions that this fundamental structure reflects an action-object, agent-patient, or what-where dichotomy have been easily refuted by demonstrating the ease with which these cognitive categories can be interchanged in their grammatical roles. In any case, this most general feature of language structure requires an additional explanation if language reference is treated as simple arbitrary correspondence.

Since Frege, it has been explicitly recognized that isolated terms express a sense but lack specific reference unless embedded in a combinatorial construction roughly corresponding to a proposition. The assignment of a specific reference to an expression or formula and thus to make an assertion about something is called predication. In logic a well-formed (i.e. referring) expression requires both a symbolic function and an argument (i.e. that to which the function is applied). In addition a complete 'predication' requires 'quantifying' the argument (unless it is a proper name). This latter requirement and exception is telling. In English, quantifiers include such terms as "a," "the," "some," "this", "these," and "all." Since a proper name refers to an individual thing or person, reference in this case is unambiguous as it is in such mass terms as "water" or abstract properties such as "truth" when speaking generally.

Why is this basic structure necessary and what are the linguistic consequences? Again, I believe that the answer is to be found in the complex structure of symbolic interpretation.

Consider propositional form and argument structure in logic. First order predicate logic is often considered the semantic skeleton for propositional structure in language, though its primary form is seldom explicitly exhibited in natural language. It is characterized by a "predicate(argument)" structure of the form F(x), where F is a function and x is a variable or "argument" operated on by that function. Such an expression is the basic atomic unit of predicate logic. Such an expression may refer to an event, state, or relationship, and there can be one-, two-, three- and no-place predicates determined by how many arguments they take. So for example the function "is green" typically is a one-place predicate, "is next to" is a two place predicate, and "gives" is a three-place predicate.

This suggests the following hypothesis: *Predicate (argument)* structure expresses the dependency of symbolic reference on indexical reference as in *Symbol (index).*

Once source of evidence for this semiotic dependency is implicit in the way that deictic procedures (e.g. pointing and other indicative gestures) are used to help fix the reference of an ambiguous term or description, and can even be substituted for the subjects and arguments of a sentence. Thus for example, uttering the word "smooth" in a random context only brings attention to an abstract property, but when uttered while running one's hand along a table top or pointing to the waveless surface of a lake, reference is thereby established. It can also refer even if uttered in isolation of any overt index in a social context where the speaker and listener have their joint attention focused on the same flawless action. In this case, as with holophrastic utterances in general, the symbolic reference is established by implicit indication presupposed in the pragmatics of the communicative interaction. Indeed, where explicit indexing is not provided, it is assumed that the most salient agreeing aspect to the immediate context is to be indicated. In general, then, any symbolic expression must be immediately linked to an indexical operation in order to refer. Without such a link there is sense but no reference.

This is a universal semiotic constraint (though not a universal rule) that is made explicit in logic and is implicit in the necessary diadic structure of sentences and propositions. It is a constraint that must be obeyed in order to achieve the establishment of joint reference, which is critical to communication. Where this immediate link is missing reference is ambiguous and where this constraint is violated (e.g. by combinations that scramble this contiguity between symbolic and indexical operations; so-called word-salad) reference typically fails.

This constraint derives from the unmasking of indexical constraints implicit in the interpretation of symbolic reference. Because symbolic reference is indirect and "virtual," by itself it can determine only ungrounded referential possibility. The subject, topic, or argument (= variable) performs a locative function by symbolizing an indexical relationship; a pointing to something else linked to it in some actual physical capacity (e.g. contiguous pragmatic or textual context). This reference determination cannot be left only in symbolic form because isolated symbols (e.g. words and morphemes) only refer reciprocally to their "position" in the system or network of other symbols.

The importance of immediate contiguity in this relationship reflects the principal defining constraint determining indexical reference. Indexical reference must be mediated by physical correlation, contiguity, containment, causality etc., with its object in some way. Indexicality fails without this immediacy. There are, of course, many ways that this immediacy can be achieved, but without it nothing is indicated. These constraints on indexicality are inherited by the grammatical categories and syntactic organization of sentences, propositions, and logical formulae.

To state this hypothesis in semiotic terms: A symbol must be contiguous with the index that grounds its reference (either to the world or to the immediate agreeing textual context, which is otherwise grounded), or else its reference fails. Contiguity thus has a doubly indexical role to play. Its contiguity (textually or pragmatically) with the symbolizing sign vehicle points to this symbol and their contiguity in turn point to something else. This is an expression of one further feature of indexicality: transitivity of reference.

Simply stated, a pointer pointing to another pointer pointing to some object effectively enables the first pointer to also point to that object. This property is commonly exploited outside of language. Thus the uneven wear on automobile tires indicates that the tires have not been oriented at a precise right angle to the pavement, which may indicate that they are misaligned, which may in turn indicate that the owner is not particularly attentive to the condition of the vehicle. Similarly the indexical grounding of content words in a sentence can also be indirect, but only so long as no new symbolically functioning word is introduced to break this linear contiguity.

Of course, every word or morpheme in a sentence functions symbolically and a word or phrase may take on a higher order symbolic or indexical role in its combinatorial relationships to other language units at the same level. This flexibility provides a diversity of symbolized indexical relations. So, for example, arguments can be replaced by pronouns, and pronouns can point to other predicates and arguments, or (via quantification) they can point outside the discourse, or if a language employs gender marking of nouns a gender-specified pronoun can refer to the next most contiguous noun with agreeing gender expressed in the prior interaction, even if separated by many non-agreeing nouns and noun phrases. A sentence that lacks inferrable indexical grounding of even one component symbolic element will be judged ungrammatical for this reason. However, the basis for this judgment by non-linguists is not determined with respect to either explicit rules or constraints. It is determined by the fact that the sentence doesn't have an unambiguous reference.

As mentioned above, both natural language and symbolic logic are constrained by the need to quantify nouns and arguments, respectively. This also exemplifies the need to ground symbolic reference via indices. Quantifiers are specifiers of virtual indexing. Words like "a" "the" "some" "many" "most" "all" etc., symbolize the virtual result of various forms of iterated indications or virtual ostentions (pointings). All quantifiers can be thought of as means for specifying the numerosity of potentially redundant forms of indexicality. They are effectively virtual pointings that take advantage of transitive correlation with other indexical relationships, such as proximity information ("this" "that") or possession information ("his" "your") to differentiate indexicality. One can even imagine a collection that is identified by a symbolized property, being pointed to en masse by a contiguous index, and then carrying out the quantificational operation by literally pointing to some, or few, or all members of this collection.

Analogous to the case of implicit presupposed indexicality in holophrastic utterances, there are also contextual conditions where explicit quantification in language may be unnecessary. This is most obvious in cases where the possibility of specifying individuals is inappropriate (as in some mass nouns; e.g. "a water," "all waters," "few waters"). Pronominal reference doesn't require quantification because it is supplied by the text that it indicates (transitivity of indication). But when general terms are substituted for pronouns or other words serving overt indexical functions (e.g. "this" or "that") they inevitably require the addition of quantification. There are also, of course, many other exceptions to the need for quantification. Proper names and numbers do not require quantification when they are used to refer to a type as a singular class because indicating would again be redundant.

The exception that proves the rule, so-to-speak, is exemplified by highly inflected and/or agglutinated languages where indexical marking is incorporated directly into word morphology. In comparison with English, which maintains the indexical grounding of most of its symbolic functions by strict word order constraints, these languages tend to have relatively free word order. This leads to a prediction: the more completely that indexical functions are incorporated into word morphology the less restrictive the syntax and vice versa.

So approaching this issue semiotically provides a functional account that can unify a wide range of grammatical and syntactic relationships. It also suggests that our naïve intuition about these linguistic regularities may be more accurate than the formal rule-governed approach would suggest. A naïve speaker seldom comments that an ungrammatical sentence breaks a rule, and is generally hard-pressed to articulate such a rule. Rather the usual comment is that it just sounds wrong or that it doesn't make sense said that way. Compare these examples of breaking the contiguity of symbol and index to knowledge of the rules invoked to explain them:

Implicit subject:

* " _ Roundly shining over flowing shimmering."

"_ Fire!"

"_ Hot!"

Island constraints:

"John found candy and gum in his shoe."

* "What did John find candy and _ in his shoe?"

Priority in argument structure:

* "John found surprisingly in his shoe some candy."

In these cases, and many others, naïve speakers know there is something wrong even if they can't articulate it, except to say that the ungrammatical sentences are awkward or difficult to interpret, and require some guesswork to make sense of them. Moreover, in everyday conversational speech, the so-called rules of grammar and syntax are only very loosely adhered to. This is usually because common interests and joint attention as well as culturally regularized interaction frames provide much of the indexical grounding, and so adherence to these strictures tends to be preferentially ignored. Not surprisingly, it was with the widespread increase in literacy that scholarly attention began to be focused on grammar and syntax, and with education in reading and writing these "rules" began to get formalized. With the written word shared immediate context, common pragmatic interests, and implicit presuppositions are minimally if at all available to provide indexical disambiguation and so language-internal maintenance of these constraints becomes more critical.

Finally, this semiotic functional analysis also provides an alternative understanding of the so-called poverty of the stimulus problem that is often invoked to argue that knowledge of grammar must be largely innate. Consistent with the fact that naïve speakers are generally unable to articulate the "rules" that describe their understanding of what is and is not a well-formed sentence, young children learning their first language are seldom corrected for grammatical errors (in contrast to regular correction of pronunciation). Moreover, children do not explore random combinatorial options in their speech testing to find the ones that are approved by others. They make remarkably prescient guesses. It has been assumed, therefore, that they must have some implicit understanding of these rules already available.

But in fact children do have an extensive and ubiquitous source of information for learning to produce and interpret these basic semiotic constraints on predication. First of all, discerning indexicality is a capacity that is basic to all cognition, animal and human. It requires no special training to become adept at the use of correlation, contiguity, etc., to make predictions and thus to understand indexical relationships. Second, although there is little if any correction of the grammar and syntax in children's early speech there is extensive pragmatic information about success or failure to refer or to interpret reference. This is in the form of pragmatic feedback concerning the communication of ambiguous reference. And this source of information attends almost every use of words. So I would argue that children do not "know" grammar innately, nor do they learn rules of grammar, and yet they nevertheless quickly "discover" the semiotic constraints from which grammars derive.

Although it is necessary to learn how a given language implements these constraints, the process is not inductive. It is not necessary for a child to derive general rules from many instances. Young children make good guesses about sentence structure—as though they already know "rules" of grammar—by tapping into more natural analogies to the nonlinguistic constraints and biases of iconicity and indexicality, and by getting pragmatic feedback about confused or ambiguous reference. Evolved predispositions to point or indicate desired objects or engage joint attention also make sense in this context. This universal human indexical predisposition provides the ideal scaffold to support what must be negotiated and must be progressively internalized to language structure. The early experience of communicating with the aid of pointing also provides additional background training in understanding the necessary relationship between symbols and indices.

Semiotic constraints should be agent-independent, species-independent, language-independent, and discourse-independent. They have been mistakenly assumed to be either innate structures or else derived from cognitive schemas or determined by sensorimotor biases and/or social communicative pragmatics. Though they are prior to language experience, and some are prerequisites to successful symbolic communication, they are neither innate nor socially derived. They are emergent from constraints that are implicit in the semiotic infrastructure of symbolic reference and interpretive processes. They are in this way analogous to mathematical universals (e.g. prime numbers) that are "discovered' (not invented) as mathematical representation systems become more powerful. Though each form of symbol manipulation in mathematics has been an invention and thus a convention of culture, we are not free to choose just any form if we want to maintain consistency of quantitative representation.

Assuming that symbolic reference lacks intrinsic structure has tricked linguists into assuming the need to postulate ad hoc rule systems and algorithms to explain the structural constraints of language. Failure to pay attention to the iconic and indexical underpinnings of symbolic reference has additionally exaggerated the complexity of the language acquisition problem. This myopic avoidance of semiotic analysis has led to the doctrine of an innate language faculty that includes some modicum of language-specific knowledge and this seeming logical necessity has supported an almost religious adherence to this assumption despite the biological implausibility of its evolution and the lack of neurological support for any corresponding brain structures or functions. Unfortunately semiotic theory has not been of much assistance, primarily because it has remained a predominantly structural theory tied to a static taxonomic understanding of semiotic relationships. But when semiosis is understood as a process of interpretive differentiation in which different modes of reference are understood as dynamically and hierarchically constituent of one another these many conundrums dissolve and these once apparently independent aspects of the language mystery turn out to have a common foundation.

These constraints are the most ubiquitous influences on language structure, and indeed they are even more universal than advocates of mentalese could have imagined, because they are not human specific. They are universal in the sense that the constraints of mathematics are universal. They would even be relevant to the evolution of symbolic communication elsewhere in the universe. But they are not exceptionless rules. Different languages, everyday spoken interactions, and artistic forms of expression can diverge from these constraints to varying extents, but at the cost of ambiguity and confusion of reference. In general, these constraints will probably be the most consistent regularities across the world's languages because means to minimize this divergence will be favored by the social evolution-like processes of language transmission from generation to generation.

However, the universality and non-innateness of these constraints does not mean that there aren't human-specific constraints that contribute to many of the nearly universal regularities that characterize the World's languages. These are the result of constraints of a different sort; some deriving from our biology and some from social processes. None determine language organization in a generative sense, but rather along with semiotic constraints they collectively constrain and bias the range of possible language variations.

2.2.2 Processing Constraints

Probably the most critical factor contributing to the structure of natural languages in addition to semiotic constraints is the need to communicate symbolically in real time. Brains are not computers. They are slow and limited in mnemonic and attentional capacity, and symbolic communication is extremely demanding in both of these domains of brain function. Basically, the ability to use language in real time demands the equivalent of computational optimization. The key to Noam Chomsky's original insight into the structure of language can probably be characterized as recognizing that natural language syntax can be modeled as a Turing machine. In abstract form a Turing machine can be understood as a set of rules for writing erasing and rewriting strings of characters, in which these rules are also encoded as character strings that can be treated the same way. Rendering an operation in these terms makes it possible to automate any finite determinate process (from robotic behavior to mathematical calculations), which is why it contributed to the design of modern computers. Because of the power of this methodology, this insight was not only valuable for developing a formalism for modeling syntax, it also became a driving force for the development of the cognitive sciences.

So in one sense it isn't surprising that natural language structure can be modeled by this formalism, however, by using this approach the remarkable systematicity of languages could also be clearly exemplified. Language structure could have been far more haphazard than it is, but what formal approaches have demonstrated is that languages are remarkably internally consistent despite their flexibility. Instead, the syntax of a highly grammaticalized natural language resembles a formal system or Turing machine architecture where all operations are systematically inter-defined and precisely complementary to one another, and where many operations are almost entirely structure-dependent and content-independent.

This requires an additional explanation, since language structures have evolved spontaneously without any attention to their design logic or optimality. As noted above, this regularization is almost certainly the result of a kind of cultural version of natural selection involving language "traits," in which the selection pressures that determine which forms get passed on and which forms go extinct are the various constraints of referential effectiveness and ease of use. Ease of use is determined by what can be described as processing constraints. So what are these constraints?

Because human brains have distinct limitations due to the nature of neural signal processing as well as distinctive cognitive biases that have been inherited from our primate ancestry, the way they must solve the challenge of online real-time symbolic communication has both biological and computational idiosyncracies.

Probably the most ubiquitous processing constraints have to do with the amount of attentional and mnemonic work that must be done to produce and interpret linguistic communications. Linguistic tricks that enable the various symbolic operations to be most efficiently and thoroughly automated will for this reason be highly favored over the history of a language's persistence. Indeed, the demands of making language functions nearly effortless may even be favored at the expense of easy interpretability. Thus there will likely be linguistic selection over time for what might be described as optimal computational design. This is essentially what skill learning is all about. And in many respects skills are predominantly associated with motor functions.

Automatization of behavior is acquired by extensive repetition. As a behavior is repeated again and again in slightly different contexts those features that are least variable from performance to performance become more streamlined. In this sense the behavior becomes increasingly algorithmic. The key to automatization is simplification and specifically a reduction of options.

The challenge to automatization of language is that symbolic relationships are dependent upon determining relative "position" in a vast web of associations. Taking the time to sample this vast search space with each new combinatorial relationship to interpret would result in an impossibly slow rate of communication, likely straining short-term working memory. This is the source of what might be described as costs of symbolic search. Because symbols are nodes in complex systems of multidimensional semantic relationships, selecting and interpreting symbol combinations may involve a very high dimensionality search for appropriate "blend" relationships between them. The semantic search space grows exponentially with the number of elements combined, the dimensions to be considered, and the ambiguity of the selection criteria. Contextual-pragmatic constraints (including extralinguistic indication of salient symbolic relations, and assessment of recipient knowledge/sender intention) may help to disambiguate the selection criteria and to reduce the search dimensions, but this cannot keep pace with the combinatorial explosion of the search space.

Partial automatization of language performance must therefore be achieved by strictly limiting the amount of symbolic search of memory that is required. So despite the power and flexibility of symbolic representation, the processes of selection at work during language transmission will tend to evolve means to reduce the density of symbolic operations per second in speech. Since the mnemonic and attentional demands of such a combinatorial search will depend on the numbers of dimensions of properties being represented, one way this can be reduced is simplification of certain common symbolic operations. Thus, improved automatization can be achieved by spontaneous linguistic evolution for what might be described as de-symbolization; a spontaneous degeneration of the semantic dimensions of selected symbolic elements to the point that their reference is reduced to virtual indexicality (i.e. pointing to a single simple symbolic relation). This aids the efficient formation, identification, and parsing of sentential subassemblies with very shallow symbolic search.

This trade-off between processing constraints and symbolic combinatorial analysis is the source of a curious paradox: often languages tend to change (evolve) away from communicative transparency. The historical process of grammaticalization often reduces semantic and functional transparency rather than increasing it. Lexical specificity is often degraded (sometimes described as "bleaching"). Highly grammaticalized language can include phrase fragmentation, multileveled embedding, and non-contiguous syntactic relationships. This structural complexity often results in reduced learnability by undermining interpretive iconicity, such as direct mapping of the temporal order of events and relations to word order.

If linguistic selection favored only clarity of communication this would not make sense. And this is one reason that some (including Noam Chomsky) have argued that natural language grammar did not evolve for communication but rather only for cognition. But this ignores the importance of reducing processing load. Structural relationships are effectively indexical operations and as such they can become automatized by virtue of having highly regularized unambiguously singular functions. To the extent that these operations play an indexical role in disambiguating which semantic dimensions are relevant in a given symbolic combination they also reduce the cost of symbolic analysis.

Increased automatization appears to lead to a minimal set of mutually exclusive, fully reflexive, indexical operations, whether embodied in morphology or syntax. Not surprisingly, the lexicon of most languages tends to be segregated into content words and function words, as well as primary morphemes and affixes, with different balances between these. Where this is mostly achieved by distinct word classes and syntactic relations the content words, like nouns, verbs, adjectives, and adverbs, comprise an open set that can be indefinitely added to, as need requires. They play the symbolic roles within a sentence. The function words, such as pronouns, determiners, prepositions, conjunctions, and the functional affixes like "-s" and "–ed" comprise a finite closed set and serve more-or-less inter-symbolic indexical functions, determining which semantic dimensions are relevant to consider when blending or differentiating symbolic relations.

In general we can distinguish between requirements for automatization and the various linguistic tricks to aid in meeting these requirements. Automatization requires a small, closed class of operators that are used in a stereotypic way, repeatedly (i.e. hundreds of thousands or millions of times each year). This invariant repetition is essential for developing a nearly unconsciously implemented skill. Moreover, any optional functions need to be reduced to no more than two or three. This may be aided by processes such as semantic bleaching, by agglutination or strict syntactic adjacency, by standardization of common thematic frames, by indexicalizing highly redundant and phylogenetically salient types (e.g. plurality, tense, possession, animacy, status, etc.), and so forth.

In semiotic terms, then, the index-symbol relationship also corresponds to a fundamental distinction between those aspects of language that can be automated and those that cannot, respectively. This has clear neurological implications.

In neuropsychological terms, automatization is characteristic of what is often called procedural memory. Procedural memories are mostly associated with highly regular activities or skills in which a sequence of component actions and assessments is made highly predictable and easily cued. These are effectively behavioral algorithms that have been acquired by constant repetition, to the point that they can be executed with a minimum of conscious monitoring.

Importantly, once a skilled behavior is well-ingrained it can be executed at a rate that is many times more rapid than if each component operation required monitoring. The result is that automated procedures tend to be automatically initiated by stereotypic cues, once initiated "run" autonomously to completion, are modular in the sense that dissecting them back into component actions is difficult if not impossible, and their structures tend to become inaccessible to introspection. These same characteristics have often been cited as evidence that grammatical functions must be innate, modular, and specific to language. Creating and executing procedural memory functions involves a distinct set of brain systems, typically associated with motor control: including particularly frontal cortex, striatal structures, and the cerebellum. A reciprocal connectivity and functional relationship between cerebral cortex and striatal structures is critical to both creation and implementation of such skilled autonomous operations.

In contrast to the procedural memory system that creates memories by constant repetition, the brain establishes memory traces of singular events or experiences using a very different set of interconnected systems. Remembering what you did immediately after breakfast two days ago, the structure of a narrative, or the meaning of a new technical term, cannot rely on extensive repetition to become ingrained. Recalling such one-off events or experiences or novel associations must therefore depend on a very different strategy for consolidation and recall. Instead of redundancy of performance or rehearsal consolidation of these memories must rely on redundancy of associations, i.e. linkage with many other related memories by innumerable commonalities and correlations. This is sometimes referred to as episodic or associative memory and is critically dependent on relationships between the cerebral cortex and the hippocampus. The associative memory system is thus ideally suited as a substrate for the storage of open-ended associative information and the procedural memory system is ideally suited as a substrate for the storage of a finite corpus of modular automated procedures. An interesting correlate of this segregation of automatized versus associative features of language processing is that brain damage that predominantly involves striatum and spares cerebral cortex has been found to preferentially impair the first language but not the second in some bilingual patients (Fabbro & Paradis, 1995). This is probably because the second language was not nearly so well automatized.

This functional segregation explains why indexical syntactic functions are performed with minimal effort, are largely unavailable to introspection, and have more-or-less modular organization, and why the analysis of more complex combinatorial symbolic relationships takes more mental effort and is generally the focus of attention. But it also suggests a way that language may provide a fundamental restructuring of cognition compared to other species. Not only are syntactic operations subject to automatization, but word-sound memory more generally is acquired in childhood by untold millions of repeated exposures and productions. So, like the relatively automatic processing of syntax, the production and recognition of words is also acquired like a deeply ingrained skill. Very little attention is paid to analyze and minimal effort is required to produce the familiar words of one's language. But although they may thus be generated from procedural memory, they cue associative loci in associative (episodic) memory. In this way language enables procedural memory traces to cue associative memory traces reciprocally, linking mnemonic strategies that in other animals are probably only minimally interdependent, and primarily with respect to external cuing. In contrast, in humans this acquired functional interdependence of memory systems provides an unprecedented internal reciprocal cuing mechanism for organizing experience. This ability to use a repertoire of acquired procedures to reliably access and organize life-episodes and abstract ideas is likely a major factor contributing to the human preoccupation with narrative.

One benefit of developing a functional account of the nature and origins of these language structures is that it leads to explicit predictions about how the brain processes language. One of the disappointments of the last four to five decades of formal linguistic theory is that while it provided unprecedented precision in describing language structures it has not been particularly useful in providing predictions about how language is processed in the brain. Instead, predictions about human-unique brain systems, dedicated language structures, modular isolation of language capacities from other cognitive functions, and the primacy of linguistic function over surface implementation of these functions, have not borne fruit. In contrast, reflecting on these semiotic and processing constraints a number of predictions spring immediately to mind.

- Hypothesis 1. In general, the way that language functions are neurologically distributed and localized does not respect "linguistic logic," but rather the processing logic determined by what constitutes a functional unit and how this can be manipulated.
- Hypothesis 2. The neural distribution of different classes of linguistic operations develops during language acquisition as different operations become automatized.
- Hypothesis 3. The linguistic evolution of more thoroughly grammaticalized forms aids the efficient distribution of language processing in the brain.
- Hypothesis 4. Pidgins are less able to be automated because they lack semiotic systematicity. Their functions will be more widely distributed in the brain, processing will be slower, and functions will be more transparently iconic or indexical in surface production.

This has implications for brain-language co-evolution. Whereas semiotic constraints do not evolve and have been ubiquitously present throughout hominid evolution, processing constraints have likely been subject to constant change and variations. Brains have been modified in evolution in response to both. But we need to keep them separate in our analysis, and indeed they are functionally and temporally (in evolutionary terms) asymmetrically related. Adaptations that aid processing of symbolic interpretive competence and linguistic communication are secondary to the presence of symbolic communication. The initial evolution of the symbolic capacity created the context in which both the socially evolved grammatical and biologically evolved neurological adaptations took place, though once this process was initiated adaptation to all of these constraints co-evolved as a complex to produce the modern symbolic species.

Finally, before turning to the issue of human evolution itself, we need to consider the last two sources of constraints affecting language structure: phylogenetic sensorimotor biases and the demands of communicative interaction. I will only very superficially outline these influences because they are somewhat more optional and contingent than semiotic and processing constraints, and therefore have less of a universal character and more of a context-dependent role to play in determining similarities shared by most languages.

An influential alternative approach to formal theories of grammar and syntax that takes a more functional perspective often travels under the name of cognitive grammar. Although this term is often used for a restricted programmatic approach to explaining grammar, I will here use it quite generically to describe all theories that explain grammatical operations as reflecting the structure of sensory, motor, behavioral, and social operations in their form, and thus arguing that grammatical and syntactic relationships have been motivated by these systems. This approach is also often used to explain for example the prevalence of visual metaphors and path-progress analogies built into vocabulary and syntax. It has even motivated a theory of the origins of subject and predicate functions of sentences based on the so-called "what" and "where" visual pathways in the brain. The basic idea is that the logic that organizes language structure is derived and abstracted from evolved embodied cognitive schemas. I think that it is without question that many aspects of language structure, lexical organization, and descriptive schemas have been shaped by these distinctively human cognitive biases, and also that there may even be culture-specific biases of an analogous sort that have served as linguistic selection biases causing parallel and convergent linguistic evolution in diverse historical contexts. While more contingent on human species-peculiarities than semiotic and processing constraints, these biases have inevitably also contributed to certain of the near universal regularities of human languages.

One nearly universal characteristic of language is its oral-vocal medium. The demonstration that the manual languages which evolved in deaf communities are indeed full-blown natural languages exhibiting features common to most spoken languages has undermined the universality of this feature of language. Nevertheless it is taken as an uncontroversial fact that language evolved as a vocal process, though it may have initially originated in a more gestural form.

One very telling piece of evidence supporting this scenario is the highly atypical human facility for skilled vocal behavior that is almost entirely absent in other land mammals and only modestly developed in cetaceans and certain bird groups. In *The Symbolic Species* (Deacon, 1997) I argue that this ability depends upon some quite unprecedented neurological relationships and that such a radical functional change must have been driven by significant selection advantages. But why this unlikely medium? I think that the answer is that it afforded an optimal medium for mimicry, and for a means of communication whose entire repertoire of sign vehicles must be acquired socially ease of mimicry is critical. It turns out that, despite what gets said in folk zoology, consciously learned mimicry is quite uncommon among animals. What monkeys see they seldom do. There is one general exception to this paucity of learned mimicry: singing in some songbird species, sound mimicry in parrots, mynahs, and mockingbirds, and song transmission in humpback whales, and there are probably other examples as well. Why this exception for oral-vocal communication? I think that the answer is that sounds heard can be behaviorly approximated without any need for any mental transformation. In contrast, gestural behaviors that are observed require a mental inversion before being reproduced. One needs to, in effect, imagine being the other producing this behavior. This shift of perspective is apparently not a trivial cognitive transformation.

Because of certain highly conserved phylogenetic limitations of nervous system organization, it seems reasonable to expect that our non-symbolic ancestors had as little control over vocal articulation, as is the case for other primates, and so the early stages of symbolic communication may indeed have involved more of a gestural embodiment. But once symbolic communication became a critical part of human social organization there would almost certainly have been a significant advantage to being able to shift manually signed symbolic communication to the oral-vocal domain. For this reason, I see this particular universal trait to be a relatively late emerging biological adaptation for symbolic communication, but one that set the stage for many processing adaptations, because of the immense advantages it created for rapidly expanding the sign vehicle repertoire and its combinatorial possibilities.

Finally, there are what I would describe as significant communication constraints that have also contributed to the convergences of language features worldwide. These are often not formally considered to be linguistic issues but rather associated with socio-linguistic and anthropological domains. Nevertheless they do play constraining roles that have shaped languages and provided a source of evolved parallelisms. Most significant of these are what I would lump into the category of pragmatic constraints. Language is used to convey information, to affect others' behaviors, to establish and restructure social relationships, to acquire information, and so forth. These functions and many more are universal simply by virtue of the fact of serving ubiquitous human social needs, and so the way they shape the various modes of organizing and interpreting symbolic communications will also exhibit shared attributes. And in addition there will be culture-specific expectations and prohibitions about how communication is to be used and information is to be shared. In these socio-cultural domains these pragmatic needs and customs are probably more variable and less tightly constraining than any of the other factors discussed, but the universe of these possibilities is probably quite limited and so we should expect these pragmatic constraints to contribute some further degree of constraint as well.

In summary, we should expect that many aspects of language come to exhibit near universal properties even despite its superficial arbitrarity of referential correlation, but not because of any innate set of rules or algorithms that generate these features. Language universals are a reflection of the many constraints that derive from the semiotic infrastructure of symbolizing and the processing demands this entails. The semiotic universals should be reflected in the symbolic communication of any species, should it evolve this competence, whether on Earth or elsewhere in the universe. However, the processing constraints that have influenced the structure of human language are less universal. So, for example, were we ever to find a way to engineer symboling minds in silicon, using electronic instead of chemical and ionic means of signal processing, we should expect some very different structures to emerge.

2.3 The Evolutionary Conundrum Posed by Language

In my work I use the phrase, symbolic species, quite literally, to argue that symbols have literally changed the kind of biological organism we are. I believe that we think and behave in many ways that are quite odd compared to other species because of the way that language has changed us. In many respects symbolic language has become a major part of the environment to which we have had to adapt in order to flourish. In the same way that our ancestors' bodies evolved in the context of the demands posed by bipedal foraging with stone tools and incorporating meat into the diet, their brains evolved in the context of a rich fabric of symbolic cultural communication. As it became increasingly important to be able to enter into the social web of protolinguistic and other early forms of symbolic social communication in order to survive and reproduce, the demands imposed by this artificial niche would have selectively favored mental capacities that guaranteed successful access to this essential resource. So rather than merely intelligent or wise (sapient) creatures, we are creatures whose social and mental capacities have been quite literally shaped by the special demands of communicating with symbols. And this doesn't just mean that we are adapted for language use, but also for all the many ancillary mental biases that support reliable access and use of this social resource.

But this claim depends on language-like communication being a long-time feature of hominid evolution. Theories suggesting that human language is a very recent and suddenly evolved phenomenon would not make this prediction. To them language is almost epiphenomenal. This is particularly true if the claim is that language appeared suddenly due to some marvelous accidental mutation that transformed dumb (but large brained) brutes into articulate speakers. This sort of scenario has become commonplace in recent years, though the evidence supporting it is mostly very indirect (e.g. archeological evidence of representational forms and objects for adornment, appearing in the Upper Paleolithic). I think that it is mostly a reflection of a caricatured view of the human/animal distinction and a sort of hero metaphor imposed upon the fossil evidence. The way that modern human brains accommodate language can be used as a clue to how old language is.

If language is a comparatively recent feature of human social interaction, that is if it is only, say, a hundred thousand years old or so, then we should expect that it had little effect on human brains. Any structural tweaks of brain architecture that evolved to support it would have had to be either minimal or else major but dependent on comparatively few genetic changes. A recent origin of language would give it little opportunity to impose selection pressure on human brains, so language function would not be supported by any widespread and well-integrated neurological changes. This would predict that language abilities are essentially an evolutionary after-thought, inserted unsystematically into an otherwise typical (if enlarged) ape brain. With little time for the genetic fixation of many supportive traits to occur, this adaptation would likely depend on only a few key genetic and neurological changes. As a consequence, language function should be poorly integrated with other cognitive functions, relatively fragile if faced with impoverished learning contexts, susceptible to catastrophic breakdown as a result of certain small but critical genetic defects, and severely affected by congenital mental impairment.

None of these seems to be the case.

On the other hand, if language has been around for a good deal of our evolutionary past, say a million years or so, that amount of time would have been adequate for the demands of language to have affected brain evolution more broadly. A large network of subtle gene changes and neurological adjustments would be involved, and as a result it should be a remarkably well-integrated and robust neurological function. Indeed, there is ample evidence to suggest that language is well-integrated into almost every aspect of our cognitive and social lives, that it utilizes a significant fraction of the forebrain, and is acquired robustly under even quite difficult social circumstances and neurological impairments. It is far from fragile.

The co-evolutionary interaction goes both ways. Languages also have to adapt to brains. Since the language one learns has to be passed from generation to generation, the more learnable its structures, and fitted to human limitations, the more effective its reproduction in each generation. Languages and brains will evolve in tandem, converging towards each other, though not symmetrically. Brain evolution is a ponderously slow and unyielding process in comparison to the more facile evolution of languages. So we should expect that languages are more modified for brains than brains are for language. Nevertheless, if we have been evolving in a symbolic niche for a million years or more, we should expect that human brains will have been tweaked in many different ways to aid life in this virtual world.

The world of symbols is an artificial niche. Its ecology is radically different than the biological niche we also find ourselves in (or at least our ancestors found themselves in). In the same way that beaver dam building has created an aquatic niche to which beaver bodies have adapted over their evolutionary history, our cognitive capacities have adapted to our self-constructed niche: a symbolic niche. This is not a new idea. Indeed the anthropologist Clifford Geertz (1973) suggested something like this many decades ago. I think that today we may be at a point in our evolutionary theorizing and our understanding of brains to begin to explore exactly what this might mean.

The most intense and unusual demands of this niche should be reflected in the ways that human cognition diverges from patterns more typical of other species. Although it has long been popular to think of the human difference in terms of general intelligence, I think this bias may have misled us into ignoring what may be a more important constellation of more subtle differences. These likely included differences in social cognition (e.g. joint attention, empathy, the ability to anticipate another's intended actions), differences in how we learn (e.g. superior transfer learning, a predisposition to assume that associations are bidirectional—known as stimulus equivalence, a comparative ease at mimicking) or even just unusual motor capacities (e.g. unprecedented articulatory and vocal control). These are members of a widely distributed and diverse set of adaptations that fractionally and collectively contribute to our language abilities.

With respect to the brain, we need to confront another mystery. How could these many diverse brain traits have become so functionally intertwined and interdependent as to provide such a novel means of communication? This is particularly challenging to explain because language is in effect an emergent function, not some prior function just requiring fine-tuning. Our various inherited vocalizations, such as laughter, shrieks of fright, and cries of anguish, are comparatively localized in their neurological control (mostly subcortical) as are other modes of communication in animals. In comparison, language depends on a widely dispersed constellation of cortical systems, each of which can be found in other primate brains, but evolved for very different functions. These brain systems have become collectively recruited for language only because their previously evolved functions overlapped significantly with some processing demand necessitated by language, though evolved for quite different functions altogether. Indeed, the neural structures and circuits involved in the production and comprehension of language are homologous to structures found ubiquitously in most monkey and ape brains: old structures performing unprecedented new tricks.

A related mystery concerns the extent to which this dominant form of communication depends on information maintained by social transmission. Even for theories postulating an innate universal grammar, the vast quantity and high fidelity of the information constituting even a typical vocabulary stands out as exceedingly anomalous from a biological point of view. How did such a large fraction of our communicative capacity wind up offloaded onto social transmission? And what explains the remarkable reliability of this process?

Perhaps the most difficult neurological feature to explain, however, is the evolution of the diversity of brain structures involved. The higher-order synergy of systems that contribute to language requires the cooperative functioning of many diverse brain regions. And it appears to paradoxically require that this synergy among diverse systems must already be in place in order for selection to have honed it for language.

The co-evolutionary niche construction scenario sketched above still does not account for the generation of the novel functional synergy between neural systems that language processing requires. The discontinuities between call control systems and speech and language control systems of the brain suggest that a co-evolutionary logic alone is insufficient to explain the shift in substrate. Recent investigation of a parallel shift in both complexity and neural substrate in birdsong may be able to shed some light on this (see also Pepperberg, Chapter 7, this volume).

In a comparative study of a long-domesticated bird, the Bengalese Finch, and its wild cousin, the White-Rump Munia, it was discovered that the domesticated lineage was a far more facile song-learner with a much more complex and flexible song than its wild cousin (discussed in detail in Deacon, 2010b). This was despite the fact that the Bengalese Finch was bred in captivity for coloration, not singing. The domestic/wild difference of song complexity and song learning in these close finch breeds parallels what is found in comparisons between species that are song-learners and non-learners. This difference also correlates with a much more extensive neural control of song in birds that learn a complex and variable song.

The fact that this behavioral and neural complexity can arise spontaneously without specific breeding for singing is a surprising finding since it is generally assumed that song complexity evolves under the influence of intense sexual selection. This was, however, blocked by domestication. One intriguing interpretation is that the relaxation of natural and sexual selection on singing paradoxically was responsible for its elaboration in this species. In brief, with song becoming irrelevant to species identification, territorial defense, mate attraction, predator avoidance, and so on, degrading mutations and existing deleterious alleles affecting the specification of the stereotypic song would not have been weeded out. The result appears to have been the reduction of innate biases controlling song production. The domestic song could thus be described as both less constrained and more variable because it is subject to more kinds of perturbations. But with the specification of song structure no longer strictly controlled by genetically inherited innate auditory and motor biases, other linked brain systems can begin to play a biasing role. With innate song biases weakened, auditory experience, social context, learning biases, and attentional factors could all begin to influence singing. The result is that the domestic song became more variable, more complicated, and more influenced by social experience. The usual consequence of relaxed selection is genetic drift—increasing the genetic and phenotypic variety of a population by allowing random reassortment of alleles—but neurologically, drift in the genetic control of neural functions should cause constraints to become less specific, generating increased behavioral flexibility and greater conditional sensitivity to other neurological and contextual factors.

This is relevant to the human case, because a number of features of the human language adaptation also appear to involve a relaxation of innate constraints allowing multiple other influences besides fixed links to emotion and immediate context to affect vocalization. Probably the clearest evidence for this is infant babbling. This unprecedented tendency to freely play with vocal sound production occurs with minimal innate constraint on what sound can follow what (except for physical constraints on vocal sound generation). Babbling occurs also in contexts of comparatively low arousal state, whereas laughter, crying, or shrieking are each produced in comparatively specific high arousal states and with specific contextual associations. This reduction of innate arousal and contextual constraint on sound production, opens the door for numerous other influences to begin to play a role. Like the domesticated bird, this allows many more brain systems to influence vocal behavior, including socially acquired auditory experience. In fact, this freedom from constraint is an essential precondition for being able to correlate learned vocal behaviors with the wide diversity of objects, events, properties, and relationships language is capable of referring to. It is also a plausible answer to the combinatorial synergy problem (above) because it demonstrates an evolutionary mechanism that would spontaneously result in the emergence of multi-system coordination of neural control over vocal behavior.

But although an evolutionary de-differentiation process may be a part of the story for human language adaptation, it is clearly not the whole story. This increased flexibility and conditionality likely exposed many previously irrelevant interrelationships between brain systems to selection for the new functional associations that have emerged. Most of these adaptations remain to be identified. However, if such a dedifferentiation effect has been involved in our evolution, then scenarios hypothesizing selection for increased innateness or extrapolation from innate referential calls to words become less plausible.

There is a much larger biological background behind my approach which of necessity has had to go unmentioned. It traces to my work on brain development and evolution, and more broadly it borrows from work that currently runs under the banner of "evodevo" and which has begun to illuminate once problematic issues in evolutionary genetics, molecular cellular biology, and epigenesis. My point is not to discount the contributions of natural selection, which I agree is the final arbiter of functional adaptation, but to bring attention to another unnoticed facet of the evolutionary process. Natural selection is explicitly NOT the generator of the biological phenomena that it prunes in the process that leads to increased adaptation. Not only are the variants of existing organismic subsystems generated irrespective of function (e.g. by genetic "damage") but the expression of these varieties of structure and dynamics depends on generative processes whose details we tend to hide in generic concepts like epigenesis and reproduction. New stuff, new structures, and new processes need to be generated so that there is raw material fed to the engine of natural selection. The second law of thermodynamics has to be locally tamed in order for this to be possible. And natural selection theory is so widely applicable precisely because it can be agnostic as to how any of this is achieved, so long as it is.

Surprisingly, despite our many disagreements about innateness, I find some resonance in Noam Chomsky's periodic suggestion that some of the complexity of grammar may have emerged from general laws of physics analogous to the way that the Fibonacci regularities exemplified in the spirals of sunflower seed and pine cone facets emerge. Natural selection has "found a way" to stabilize the conditions that support the generation of this marvelous regularity of growth because it has important functional advantages. But natural selection didn't generate it in the first place, geometric regularities that can become amplified due to center-out growth process are the ultimate source (as has now been demonstrated also in growth-like inorganic processes).

In closing, I would like to reflect on some of the more esoteric features of humanness that may be illuminated by the paired processes of symbolic niche construction effects and relaxed selection.

For example, I think it makes sense to think of ourselves as symbolic savants, unable to suppress the many predispositions evolved to aid in symbol acquisition, use, and transmission. In order to be so accomplished at this strange cognitive task, we almost certainly have evolved a predisposition to see things as symbols, whether they are or not. This is probably manifest in the make-believe of young children, the way we find meaning in coincidental events, see faces in clouds, are fascinated by art, charmed by music, and run our lives with respect to dictates presumed to originate from an invisible spirit world. Like the flight play of birds, the manipulation of objects by monkeys, the attraction of cats to small feathered toys, our special adaptation is the lens through which we see the world. With it comes an irrepressible predisposition to seek for a cryptic meaning hiding beneath the surface of appearances. Almost certainly many of our most distinctive social capacities and biases—e.g. tendencies to conformity and interest in copying the speech we hear as infants—are also reflections of this adaptation to an ecosystem of symbolic relationships. And of course there is literature and theater. How effortlessly we project ourselves into the experiences of someone else, feeling the joys and sorrows almost as intensely as our own.

Relaxation of selection, on the other hand, may have contributed to another suite of distinctively human traits. Widely distributed dedifferentiation at the genetic and epigenetic level would have increased flexibility of a variety of once phylogenetically constrained cognitive and motivational systems. Perhaps the most striking feature of humans is their flexibility and cultural variety. Consider the incredible diversity of marital and kinship organizations. Most species have fairly predictable patterns of sexual association, kin association, and offspring care, and although they are somewhat flexible, this variety is mediated almost entirely by individual motivational systems. In contrast, despite the evolutionary importance of reproduction, human mating and reproduction are largely controlled by symbolically mediated social negotiations. This offloading of one of the most fundamental biological functions onto social-symbolic mechanisms is perhaps the signature feature of being a symbolic species. Thus, because of symbols and with the aid of symbols, *Homo sapiens* has been self-domesticated and adapted to a niche unlike any other that has ever existed.

We have been made in the image of the word.

References

Bickerton, D. (2010). Commentary on Terrence Deacon (2010) *On the human: Rethinking the natural selection of human language*. Retrieved August 24th 2011, from http://onthehuman.org/2010/02/on-the-human-rethinking-the-natural-selection-of-human-language/

Deacon, T. (1997). *The symbolic species: The coevolution of language and the brain*. New York: W. W. Norton & Co.

Deacon (2010a). *On the human: Rethinking the natural selection of human language*. Retrieved August 24th 2011, from http://onthehuman.org/2010/02/on-the-human-rethinking-the-natural-selection-of-human-language/

Deacon (2010b). A role for relaxed selection in the evolution of the language capacity. *Proceedings of the National Academy of Sciences, 107*, 9000–9006.

Everett, D. (2005). Cultural constraints on grammar and cognition in Pirahã: Another look at the design features of human language. *Current Anthropology, 46*, 621–646.

Fabbro, F., & Paradis, M. (1995). Differential impairments in four multilingual patients with subcortical lesions. In M. Paradis (Ed.), *Aspects of bilingual aphasia* (pp. 139–176). Oxford: Pergamon Press.

Geertz, C. (1973). *The interpretation of cultures*. New York: Basic Books.

Hauser, M. D., Chomsky, N., & Fitch, W. T. (2002). The faculty of language: What is it, who has it, and how did it evolve? *Science, 29*, 1569–1579.

Peirce, C. S. (1931). *Collected papers of Charles Sander Pierce. Vol. II elements of logic* (C. Hartshorn & P. Weiss, Eds., CP 2.225). Cambridge, MA: Harvard University Press.

Chapter 3
The Evolution of Semiotic Self-Control

Sign Evolution as the Ongoing Refinement of the Basic Argument Structure of Biological Metabolism

Frederik Stjernfelt

Abstract This chapter argues that attempts to characterize semiotic evolution by phases corresponding to sign types from Peirce's triads do not hold – so as for instance the idea of an iconic, an indexical and a symbolic phase of evolution. This is because these sign types are not compositional so that indices are not composites of icons and symbols not of indices. Instead, the perspective should be turned 180 degrees. The "highest" Peircean sign types: propositions and their linking into arguments, are present from tbe beginning of biosemiotics, albeit in a rudimentary indistinct proto-form, corresponding to Peirce's idea that propositions are genuine signs, and the whole machinery of simpler signs are but degenerate signs which occur within propositions. Selection forces the survival of truth-bearing signs – propositions (Peirce: "Dicisigns"). Evolution then subdivides, sophisticates and articulates proto-propositions, gradually achieving growing autonomy of its parts. So, instead of an ongoing construction from building-blocks, semiotic evolution is the ongoing subdivision and autonomization of a reasoning process having its first proto-form in metabolism.

> *Given the oxygen, hydrogen, carbon, nitrogen, sulphur, phosphorus, etc., in sufficient quantities and under proper radiations, and living protoplasm will be produced, will develop, will gain power of self-control, and the scientific passion is sure to be generated. Such is my guess.*
> *(Peirce, CP 7.50, c. 1900)*

Terrence Deacon's achievement in the *Symbolic Species* (1997) and the inspiration stemming from that book is vast. Among its central and challenging ideas is the idea that the overall change in cognitive abilities from higher animals to man is a semiotic one and its description should include semiotic concepts. Another is the idea that hominid semiotic evolution forms an extreme case of Baldwinian evolution with feedback from the common semiotic capabilities of a hominid group onto the selection pressure on each group individual. The resulting hypothesis claims that

---

F. Stjernfelt (✉)
Faculty of Arts, University of Aarhus, Aarhus, Denmark
e-mail: semfelt@hum.au.dk

T. Schilhab et al. (eds.), *The Symbolic Species Evolved*, Biosemiotics 6,
DOI 10.1007/978-94-007-2336-8_3, © Springer Science+Business Media B.V. 2012

the co-evolution of human language, cognition, and brain is far more probable than the idea that language and human semiotic capabilities appeared as an accidental side product of a general growth in human cognitive abilities. And an ensuing idea is that human language capability does not form an isolated brain module but is rather involved in the overall evolution of the human brain, uniting formerly distinct capabilities. Finally, the important idea that central parts of linguistic structure are neither innate nor acquired – but rather constitutes a priori structures of any possible thought and communication – which evolution as well as development are forced to adapt to. This idea offers an important third possibility as compared to the ongoing nature-nurture cul-de-sac. Deacon's later development of concepts like autogen – the structure of primitive pre-cell proto-metabolism as an important step towards the earliest life and semiotic process – and his idea of alternating maskings and unmaskings of selection pressures in evolution add further conceptual tools to a unified explanation of the sophistication of semiotics within the framework of biological evolution.

Still, a conceptual problem remains in the heart of Deacon's hypothesis and the whole discussion revolving around it. Famously, he reinterpreted, in *The Symbolic Species*, Peirce's semiotics in order to find a conceptual skeleton scaffolding his overall evolutionary hypothesis – and picked upon Peirce's well-know triadic distinction between the sign reference types of Icons, Indices, and Symbols for that purpose. The icon-index-symbol distinction addresses the difference between signs which refer to their object by means of similarity (icons), by means of actual connection (indices) or by means of the habitual, conventional reference to a general idea (symbols), respectively. This distinction – like most of Peirce's semiotic distinctions – is not one of three separate classes of signs, like three different biological species. It is not the case that signs fall in three mutually exclusive classes named icons, indices and symbols. They rather form *aspects* of signs and may co-exist in empirical signs. The relation between these aspects is one of increasing complexity – so that most Indices contain Icons, and so that most Symbols contain both Indices and Icons, or, in any case, involves them in realizing its meaning in the ongoing chain of interpretation and reasoning. Certain scholars, Deacon among them, have attempted the obvious idea of mapping this increase in semiotic complexity onto the increase of biological complexity in evolution – so that purely iconic signs could be taken to precede indexical signs evolutionarily, and, in turn, indexical signs could be taken to precede the appearance of symbols in the course of evolution. The origin of symbols, finally, could then be taken to coincide roughly with the origin of man and the specific, plastic and fertile capabilities of human semiotics – so as to form a virtual semiotic missing link. Hence the very title of Deacon's "Symbolic Species" book.

Agreeing with the overall program of that book, this chapter addresses what I conceive as an important conceptual problem in the particular idea of mapping the icon-index-symbol scale onto the process of evolution – a problem whose solution, I shall argue, entails an important inversion in the conception of the place of semiotics in evolution. First, the symbol concept as defined in Peircean semiotics does not appear in human cognition and communication only. This might be taken to be a strife of words – maybe Deacon's reinterpretation of Peirce's symbol concept "gets it right", so to speak. But the ubiquity of Peircean symbols in pre-human biology points to a deeper problem in Deacon's seemingly small but significant reinterpretation of the relation between the three sign types: he takes them to be compositional so that indices are simply a special configuration of icons, and symbols, in turn, are but a special configuration of indices. But you can not compose the direct object reference of an index out of ever so many iconic likenesses. The summing up of predicates never results in an object reference. Existence is not a predicate, as Kant realized. And you can not compose the generality of a symbol out of ever so many singular indexical here-and-now references. The icon-index-symbol triad should rather be conceived of as "physiological" in the sense that icons and indices generally form *aspects* of symbols. This implies, conversely, that we should not expect to find "pure" icons and indices which will rather be limit categories. This "physiological" conception of Peirce's triad implies that the *most* complex parts of Peirce's basic triads should be expected to be realized even in the simplest biological processes – albeit with a lack of internal differentiation. This becomes clear from the wide applicability of the central sign in Peirce's semiotics, that of propositions or "dicisigns", in biology. Finally, this paves the way for conceiving of a more precise conceptualization of the semiotic "missing link" in terms of the ability for explicit control of semiotic processes in human beings. We shall attempt to flesh out this notion of "explicit control" in terms of its products, "hypostatic abstractions"; in terms of its structure, sign governing other signs; and in terms of its process in the shape of turn-taking dialogue, making possible the comparison and tentative integration of Deacon's hypothesis with Michael Tomasello's "joint attention" hypothesis.

3.1 Peirce's Symbol Concept Revisited

I have discussed the problems in Deacon's reinterpretation of Peirce's icon-index-symbol triad elsewhere (Stjernfelt, 2007, chapter 11) so I shall restrict myself to summing up that discussion. Peircean symbols are not restricted to human sign use. Symbols are described as signs which are general as to their object; they possess an *esse in futuro,* referring to a potential continuity of future objects; they refer to their object by means of a habit, natural or cultural; they comprise icons for their understanding and indices for their object reference; and they have propositions as an important subset. Thus, simple Pavlovian conditioning – dogs acquiring the habit of displaying eating behaviour by the ringing of a bell – will constitute a full-fledged Peircean symbol, not merely an indexical sign as Deacon would have it. It is not, like indexical signs, restricted to the here-and-now of actual connectedness between sign and object. The bell sound is a general type, referring, in turn, to another type, that of eating, a potential multitude of future eating situtations. The here-and-now of the particular bell sound token in a particular instant functions as an index incarnating that general meaning in the actual moment of the ringing. Even a case as simple as E. Coli swimming upstream in a sugar gradient as the result of its registration of molecules displaying a specific active site (Cf. Berg, 1988; Stjernfelt, 2007) must be described as symbolic in Peirce's sense: it is a habit (acquired phylogenetically, to be sure, in contrast to Pavlovian conditioning acquired ontogenetically), the habit connects a specific, typical aspect of molecular shape with a specific, typical action, that of oriented swimming and consumption. That the molecular configuration of the "active site" is general may be seen from the fact that it is not unique to a specific carbohydrate but covers a long range of different carbohydrates (Cf. Adler, Hazelbauer, & Dahl, 1973). This generality is facilitated by the chemoreceptors of the cell being geared to detect sugars by means of the presence of a specific active site on the surface of the molecule. And this generality, in turn, is what makes E. Coli sensitive to be fooled by the same artificial sweeteners as may human beings. So, Peircean symbols are not a human prerequisite only. Another related problem is that even if the icon-index-symbol triad is oriented from the simple towards the complex, it is not compositional. Deacon's reconstruction makes indices consist of a specific configuration of icons, and the symbol consist of a specific configuration of indices. But pure icons form a limit concept in Peirce – they will vaguely signify any possible object resembling them, because they are not connected to any actual object (that being the task of an index) – so an index could never result from any combination of such vague, dream-like signs. The pure index is also a limit category – like a push in the back or a pointing gesture directing attention to an object. Such signs are indeed possible, but they remain marginal, because neither the pure icon nor the pure index is able to *communicate* anything. In typical usage, indices are connected with icons in propositions, bearing information about the object which the index merely indicates without itself giving any information at all. Finally, a pure symbol without any iconical or indexical qualities are equally impossible – in order to be understood, a symbol must bear information in the shape of an icon and relate that information to an object by means of an index.

Thus, the collaboration of icons and indices within symbols is a way of describing the triad much closer to the actual functioning of signs than the focusing upon rare, detached specimens of the three aspects of sign use. This forms the basic reason why the tempting idea of mapping the icon-index-symbol triad onto the process of evolution is doomed to fail: pure icons, indices, symbols are marginal phenomena. So, there could never have been an evolutionary period where purely iconic signs prevailed – they are much too vague to communicate any information of value for biological processes, because their content is merely possible and does not relate to the actual world. And there could never have been a purely indexical period – indices being attention-directing and based on the here-and-now, they are unable to perform the central task of orienting and guiding biological activity into the future like the generality of the symbol.

I do agree, however, with Deacon in the conviction that a central part of the semiotic-missing-link issue is conceptual – and also in the contention that Peirce's semiotics may yield some of the conceptual tools needed for approaching that issue. The whole edifice of Peircean semiotics constructed in the years after the turn of the century rather forms a physiology of the semiotic reasoning process than forming a composition of complex signs out of simpler signs. Thus, in one of the important papers laying out the principles of his semiotics, the *Kaina Stoicheia* (New Elements) of 1904, Peirce describes symbols as "genuine" signs, by contrast taking simpler signs (like icons and indices) to be "degenerate" signs – indices degenerate to the first degree and icons degenerate to the second degree.[1] Degeneracy here means non-generic (and hence rare) at the same time as it means incomplete, in need of completion. As examples of symbols, single words or concepts are much too often selected; this is to some degree misleading because the central symbol subtype to Peirce is not the single word but the *proposition* (or Dicent, or Dicisign, or Pheme, in the ever-growing wilderness of terminology which constitutes part of the reason for the ignorance of his doctrine) – the sign that claims something about something.[2] So, the working of icons and indices within the confines of Dicisigns is the key to the relation between the sign types in pragmatic reality. In his essay on symbols in the present volume, Deacon actually approaches the issue of Dicisigns when he emphasizes the importance of the combination of indices with symbolic predicates in actual sign use (even if he leaves out of consideration the explicit discussion of Peirce's third triad – that of Rheme-Dicisign-Argument (or Term-Proposition-Argument)). By contrast I find this triad and its emphasis on reasoning is a key also to the icon-index-symbol issue.

3.2 The Ubiquity of Dicisigns

The wide applicability of the icon-index-symbol triad in semiotics often seems to have pushed other important aspects of Peirce's semiotics into the background – such as his third trichotomy, forming his version of the classic logical distinction between term, proposition, and argument. Peirce uses changing terminology about this triad, the most stable seeming to be Rheme, Dicisign, Argument.[3] Rhemes comprise predicate terms understood as unsatiated propositions – like "_ is blue" or "_ gives _ to _" where the filing-in of one or more of the empty slots (indicated by underscore) by a rhematic subject will give a proposition. Peirce's analysis of the proposition, hence, is a version of the traditional subject-predicate distinction – where each predicate may take several subjects (like 3 in the case of "give"), not unlike Frege's analysis of the proposition in terms of function and arguments.

[1] Peirce develops the notions of generic and degenerate in relation to his categories in "A Guess at the Riddle" (1888, CP 1.354), generalizing the terms from their use in geometry and the study of conic sections. In the "Minute Logic" (1902, only published in pieces, CP 1.203; EPII ch. 9, etc.), he applies them to signs.

[2] As to Peirce's notion of Dicisign, see especially the "Syllabus" (1903), partly published in EPII chs. 18-21 and in CP 1.180ff, 2.219ff, 2.233ff. 2.274ff. See also the discussion in Short (2007).

[3] I think Rheme-Dicisign-Argument is to be preferred to the traditional Term-Proposition-Argument in order to underline Peirce's special conception of the two former notions: Rhemes are predicate functions rather than isolated terms, and Dicisigns form a much broader notion of proposition than the received one, involving non-linguistic combinations of subject and predicate (e.g. a picture with a legend, the combination of a pointing gesture with a predicative gesture, etc.). I use proposition interchangeably with Dicisign to indicate that Peirce's notion comprise ordinary propositions as well. See Stjernfelt (forthcoming a) on Peirce's concept of Dicisign.

Dicisigns, now, are related to icons and indices by the fact that Dicisigns have two parts, subject and predicate, with indexical and iconical properties, respectively. The predicate part of the Dicisign must, directly or indirectly, convey some iconic content for the description of the propositional subject. It may work directly, by means of the presentation of a picture or a diagram, or indirectly, by means of a linguistic or other conventional predicate symbol, in turn referring to schematic iconic content. The subject part(s) of the Dicisign, on the other hand, must, directly or indirectly, bind this informational content of the rheme to some object(s) to which the Dicisign refers.[4] Again, this may take place directly, by pointing or by a rigid designator like a proper name, or more indirectly by means of a symbolic index like a pronoun or a common noun. The simple Dicisign "Alfred is happy" has a proper name as linguistic index which fills in the empty slot of the predicate "_is happy". A royal portrait is also a Dicisign, uniting the linguistic index of "Louis XIV" on the picture frame with the image predicate provided by the painted man on the canvas. This double architecture of the Dicisign, of course, is basically motivated by it *saying something* (predicate) *about something* (subject). These two aspects of the Dicisign now point, each in their way, description and reference, respectively, to the same object. So Peirce may describe the Dicisign as a special sign uniting two different signs both related to the same object, but in two different ways.[5] The possibility of the Dicisign to possess a truth value comes from the possibility of those two aspects actually, successfully, fitting the same object. It is this general truth-bearing capability of Dicisigns which makes them central to logic and reasoning – and which makes isolated icons or indices "degenerate" by comparison to full-blown symbols.[6]

These simple observations, however, make Dicisigns indispensable for biological sign use. Signs which may not convey truth are hardly efficient in biology: icons only indicating vague possibilities have little pragmatic efficiency in cognition and communication, just like isolated indices only able to indicate that something is happening but not what it is, may be of local use but not much more than that. This is why Dicisigns are ubiquitous in biology. This has been hard for both biologists

[4] Deacon addresses the same issue – without discussing Dicisigns or propositions – when he says that "every symbolic legisign must be immediately coupled with an indexical sign or else there is no specific symbolic reference" (Chapter 2, this volume) – by symbolic legisigns he here refers to general terms, like general predicates. See also Deacon (forthcoming).

[5] Again, Deacon makes the same point when he says that "The ability to replace noun phrases by pointing or other indicative gestures, or by indexical terms like this or that, demonstrates that noun phrases serve an indexical role, linking the predicate (as a symbolic core of the sentence) to some specific instance of reference." (Deacon, Chapter 2, this volume).

[6] This brief presentation glosses over some difficulties. Not all Dicisigns are symbols – there are certain Indices which may express propositions – but they are either single signs (unfit for regulating the recurring demands of simple biological behaviour) or indexical legisigns without general meaning. Correlatively, not all symbols are Dicisigns – simple rhemes not claiming anything about anything, e.g. But all these counterexamples must be deemed degenerate (Peirce mostly uses this about the lower signs in the icon-index-symbol triad, though), as compared to the full-blown symbolic Dicisign, able to put forward general propositions with truth claim possibilities. For further discussion of these details, see Stjernfelt (forthcoming b).

and biosemioticians to appreciate, probably because of the widespread idea that propositions require the judgment of a conscious "propositional stance" found only in human beings. Here, Peirce's purely semiotic definition of the proposition as a Dicisign combining two signs into one irreducible whole gives us a formal notion of Dicisign – neither presupposing consciousness nor explicit acts of judging. Thus, "Dicisign" in this definition is a much broader concept than explicit, linguistic "propositions" strictly; rather it corresponds to "proto-propositions" as defined by Hurford (2007) in a parallel argument for the existence of basic logical structure in animal cognition. Pragmatically, the existence of Dicisigns will be displayed by specific perception-action connections – in an organism's behavioral possibility of acting in a typical, categorized way prompted by the categorical perception of some biologically important, stable feature of its environment. Thus, when the E. Coli reads the perimeter of the carbohydrate molecule, its subsequent oriented swimming counts as the proof that a Dicisign combining the abstract shape of the active site with a here-and-now presence of such a site has been processed by the bacterium. Of course, the molecular surface configuration of this active site may appear in other inert macromolecules without any Dicisign to be realized – its "activity" is only granted by the bacterium. The decisive precondition is that the receptor molecules of the E.Coli make it change behaviour in a characteristic and typical way, oriented towards the continuation of its metabolism (and hence its survival).

This simple biological example gives us the important clue to what is semiotically basic in biology: the stable metabolism of an organism. The single phases of the metabolism, may, of course, be described by purely chemical means, but it is the fact that these phases form a circular, self-sustaining structure which provides the basic biological prerequisite for adaptations towards sustaining this process better – and to act so as to support metabolism (by swimming in the right direction for digesting carbohydrates, in this case). This formed the basic insight in von Uexküll's early biosemiotic notion of the functional circle of animals, binding together perception signs with action signs to form the fundamental cyclic interweaving of perceptions with guided action. Thus, the perceptual Dicisign of reading the active site on a carbohydrate molecule – a primitive version of the proposition "This is sugar" – is followed by the action Dicisign of swimming in that direction – to form an argument: "If sugar, swim in its direction. This is sugar. So, swim in its direction". That this forms a (very primitive) argument[7] – and not merely a cause-effect

[7] Calling it an argument is based on the fact that it displays the double structure of dicisigns. More precisely, as it may err, it must be an inductive argument which is probable only. But it is important to add that it almost completely lacks the quality which Peirce requires for an inference structure to count as real reasoning: namely that of self-control (see below). When I say "almost" it is because we could imagine the argument change due to the process of evolution over the long range of millions of generations. If the Umwelt of the bacterium were contaminated with a poisonous agent displaying the same "active site" as carbohydrates, this would form a grave challenge to E. Coli. If a mutation occurred, however, making some bacteria able to distinguish sugar from this toxic substance by means of other active sites on the periphery of molecules, of course, a mutated group of E. Coli might survive. Such adaptability could be interpreted as a sort of (very weak) self-control at the level of the lineage.

chain – is evident from the fact that the E. Coli may be fooled by artificial sweetener whose molecules possess the same molecular surface configuration as the active site in carbohydrates – but otherwise have a rather different chemistry without the easily releasable binding energy of carbohydrates.

This is not to say, of course, that this process is not underpinned by causal relations. The semiotic aspect of the process lies in the fact that the weak, local interaction makes a whole class of different stimuli give rise to the same, typical behaviour. Thus, it is the fact that the bacterium does *not* interact causally with the whole of the molecule (before consuming it, that is) but merely weakly interacts with a spot on its perimeter which is a precondition for its turning a semiotic and not merely causal process.[8] This argument structure binding together perception and action, of course, is close to being as primitive as it may get – and the explicit analysis of it into two distinct Dicisign phases is possible for the observer, but definitely not for the bacterium itself. It has no possibility to make any single aspect of the argument explicit nor autonomous – there are few chemical agents (besides carbohydrates, certain toxins) which the bacterium is able to categorize and react to. The automat-like character of the perception-action link testifies to its holist, yet undivided character. So the animal is not able to address the logical structure of its own perception-action chain as such, nor to substitute other perceptions or conclusions for those of sugar and toxin, or eating and fleeing, respectively. Still this basic argument structure is what makes it possible, during evolution, for higher animals to refine and spread perception-action cycles to much larger parts of their surroundings, therely enlarging their Umwelt, and, what is more, to isolate parts of the Argument as Dicisigns, and, in turn, parts of those Dicisigns as Subject Indices and Predicate Icons.

It is the fact than metabolism has an active perception-action phase – marginal in plants and funghi, central in animals – that introduces semiotics in the simple reasoning inherent in searching the environment for nutrients (and, in the E. Coli case, escaping toxins). The "reading" of carbohydrate and toxin gradients before a substantial concentration of either is present is what allows the animal the conclusion of going into the right direction for finding (resp. escaping) such concentrations.

3.3 Sign Action – A Process Differentiated Through Evolution

This discussion of E. Coli, then, serves to state the basic argument that biologically simples signs are not isolated icons or indices, only later to be composed into more complex signs.[9] Biologically simple signs, rather, are full-fledged perception-action

---

[8] Thus, the difference between the weak interaction of "reading" the active site on the one hand and the binding and breaking covalent bonds in chemical reactions is semiotically important. The former allows for categorizing the molecule without chemically interacting with it, due to the weak van der Waals bonds made possible by the variation of electric charge on the surface of the molecule.

[9] Please permit an analogy. To trace the origin of human architecture, you will have to turn to simple shelters and bivouacs and, before them, the nests of our biological cousins the great apes, and

arguments lacking explicit internal articulation – but bearing with them the possibility of later segmentation, articulation, autonomization, adaptation to further purposes, making it flexible, potentially loosening semiotic structure still more from its causal basis and eventually making the combination of separated Dicisigns, of predicates and subjects, possible. Parts of the metabolism may become relatively autonomous, forming organs – and parts of the perception part of metabolism may acquire their own parts, giving rise to cognitive plasticity, association learning, memory, recursivity and much more, just like the action part of metabolism may differentiate into motor limbs and tools able to support complicated action sequences, co-shaping the environment, depending in turn on this transformation.

The basic argument for this ubiquity of simple proto-propositions linking into arguments in biology is thus based on the observation that phylogenetically acquired habits – like bacteria swimming in the direction of sugar – must be both simple, stable, and true in order to support survival. If not simple, it would be beyond simple animals to process them. If not stable, they would not be able to address stable features of the environment (such as carbohydrate's combination of easily digestible binding energy and characteristic active sites). And if not more often true than false, they would lead to the perishing of the lineage rather than its survival. And Dicisigns are signs able to express truths. This points to the fact that semiotic evolution should not be seen as going from the simple to the complex in terms of beginning with atomic signs which later serve as building blocks for more complex signs. The process from simple to complex should be conceived of in a non-compositional way: the overall semiotic argument process structure is there from the metabolic beginning – and semiotic evolution rather takes the shape of the ongoing subdivision, articulation, and sophistication of primitive signs, an ongoing refinement of parts and aspects acquiring still more autonomy. Hence, on this view, semiotic compositionality rather forms an important *achievement* than it forms the starting principle: the ongoing autonomization of parts and aspects of Dicisigns and their combinations may make them more and more compositional – resulting in a growth of combination possibilities and hence increasing cognitive plasticity. Such segmentation of the argument process thus constitutes the overall shape of the increase in "semiotic freedom" during evolution, highlighted by Hoffmeyer (2010). The reason for taking, once again, the textbook example of E. Coli to illustrate basic sign use, is thus to insist on the fact that the kernel of semiotic cognition is the extremely simple piece of reasoning which connects perception and action. The fact that this process may err is what proves its character of (simple) reasoning. It also indicates

earlier, various biological hideouts and shells as protection devices, phylogenetically and ontogenetically built. Thus you begin with phenomena which perform the basic sheltering function of the whole building in a germlike form, rather than beginning with the development of bricks and planks which only much later assemble into full buildings. The full-building structure was there from the beginning, even if in a very primitive, unarticulated shape. Just like bricks, icons and indices primarily function within the wholes of dicisigns linked up into action arguments – and they only acquire semi-autonomous status much later, during the ongoing sophistication of argument structure through evolution.

that cognition begins long before organisms with central nervous systems arise, and even longer before the appearance of organisms with movable perception organs, binding of different sensory inputs into cross-modal perceptions and the construction of environment mappings. This should make us cautious with more or less automatic assumptions that nervous tissue such as found in the CNS of higher animals should be the privileged locus of cognitive processes. Such an idea is a sort of magic, ascribing special abilities of wonder to neurons – overlooking that cognition is a process connecting the whole of an organism to aspects of its surroundings. And such an idea is what Clark (2008), pointing to external elements of cognitive processes, calls "neurocentrism". Not only may "anthropocentrism" make us think that all accidental properties of human cognition are properties of cognition as such – but "neurocentrism" may repeat the error on a larger scale presuming that properties of nervous tissues are properties of cognition as such. Rather, cognition exists in the shape of perception-action cycles long before the evolution of multicellular organisms and the ensuing evolution of central nervous systems. Such evolution processes should rather be conceived of as adaptations to make the interface between perception and action more plastic, more versatile and add to the structure of dicisigns which the organism is able to process before turning to action. So, the specialization of certain cells to become neurons, interlinked in the CNS in multicellular organisms forms a way of adapting the organism not only to its specific surroundings – but to achieve still more complicated logical structures and reasoning capabilities. Neurons and CNS's are special adaptations to the requirement of complicated cognitive and logic processes – which is why they have to adapt to necessary structures of such processes. As Peirce says – bearing in mind his objective notion of "mind": "For we must remember that the organism has not made the mind, but is only adapted to it. It has become adapted to it by an evolutionary process so that it is not far from correct to say that it is the mind that has made the organism." ("Abstract of 8 lectures", undated, NEM IV, 141). The more varied the problems posed by the surroundings, including fellow conspecifics, become, the more complicated the intervening structure between perception and action must develop in order to adapt plastically – and the more that structure must conform to basic regularities of semiotics and logic: " Logic, for me, is the study of the essential conditions to which signs must conform in order to function as such." (Peirce, "New Elements", 1904; Peirce 1976 IV, 235ff) And, what is more, the more variation the environment presents within the small ontogenetic time-scale window of single organism life, the more of reasoning must be transported from the slow process of phylogenetic Darwinian adaption (teaching, e.g., bacteria, over millions of generations, the Argument habit of following sugar and avoiding certain toxins) to the comparatively extremely quick process of ontogenetic adaptive learning (teaching, e.g., apes, over a period of days, to acquire the habit of associating a specific location in the jungle with the presence of specific fruits). Here, biosemioticians must learn not to commit the time-scale error of automatically taking long-term habits for being non-semiotical while short-term habits are much easier seen as having a semiotical nature.[10] The actual sign-exchange, both in the bacterium and the ape example, always takes place in the vanishing moments of individual ontogenetic lifetime – whether the underlying habit is constituted in the faster timescale of phylogenetic adaptation or in the faster timescale of ontogenetic learning does not make any principal difference (albeit an enormous difference in behavioural plasticity, it is true).

Let us sum up then, the character of primitive, metabolic argument. It connects a perception Dicisign with an action Dicisign to an Argument which, again, forms part of the overall metabolism of the cell. The reason for calling it an Argument is its ability to attain truth – and to err, respectively.[11] As an Argument, of course, it lacks a series of important aspects characterizing explicit arguments made by human beings. The connection between its part has been established over the vast phylogentic timescale of evolution and could only be changed in the same way. There is no ontogenetic freedom to exchange the premises for other premises in an online trial-and-error process. No matter which consciousness definition you adhere to, there is no reason to assume any counscious access to the conclusion or to other parts or aspects of the argument structure. The argument appears as a behavioral gestalt, whose parts are only accessible as such to the external observer and analyst, not to the bacterium.

Finally, this overall argument implies that the distinction between man and animal must be sought elsewhere than in a distinction between icons/indices on the one hand and symbols on the other – namely in the growing degree of explicit control and metasemiotics, the ability for an organism to make explicit and control its own signs.

3.4 Hypostatic Abstraction

With the intensified research into human prehistory occupying many different disciplines, a Pandora's box of old questions has been reopened: the origin of language, the emergence of culture, the physical anthropology and evolution of human beings – and, conversely, the issue of communicative and cogntive abilities of other higher animals as compared to those of human beings. What is specific to human semiotic and cognitive abilities as compared to those of higher animals? A series of different answers to this issue of the semiotic or cognitive "missing link" between higher animals and human beings are already on the market. Symbol

[10] Hoffmeyer (1996) makes a similar argument. See also Hoffmeyer (2008).

[11] The notion of truth implied here, of course, is weaker than your average truth definition in terms of correspondence between an explicit proposition and an aspect of reality. Primitive biological truth might be described as adequacy of perception and correlated action – measured on the perception-action link's support of metabolism. If that link does not support metabolism, of course, it will be weeded out in the long run of natural selection. The overall argument here claims that such primitive adequacy truth forms the root of more developed truth types in higher animals and human beings.

use (Terrence Deacon), joint attention (Jerome Bruner, Michael Tomasello), language syntax (Chomsky), specific types of "blending" (Fauconnier & Turner, 2002), etc. The discussion is both electrified and muddled by the fact that these hypotheses range over different fields such as psychology, linguistics, semiotics, cognitive science, etc. This implies that the proposals mentioned are not even directly comparable – in order to be compared, they should so to speak be translated into each others' terminology. What would, e.g., the psychological notion of "joint attention" amount to if translated into the terminology of linguistics, semiotics, or neuroscience? The possibility exists that it might turn out to mean approximately or even exactly the same as one or several of the other proposals – this could only be decided after such a reconstruction process which is, by no means, a simple translation issue but a reconstruction which will only be possible after a process of conceptual and empirical development and which will, in itself, constitute a main part of a solution.

I myself have aired the idea that a good candidate for this semiotic-cognitive "missing link" might be Peirce's notion of "hypostatic abstraction" (Stjernfelt, 2007, chapter 11). Of course, no single semiotic feature may presumably be held responsible for all semiotic and cognitive differences between human beings and higher animals – still I find hypostatic abstraction to be one of the central candidate devices because permitting the making explicit and controling of various prehuman semiotic capacities and hence indispensable for the construction of human thought and language. I shall begin by presenting Peirce's discussions of the term[12] – even if the matter may of course not be solved by mere Peirce philology only.

Peirce never wrote a comprehensive treatise on the issue but returns to it over and over again in his mature work around the turn of the century.

Here, it refers a process as well as a product, to be found in a bundle of related semio-cognitive events:

Linguistically: the construction of an (abstract) noun from more concrete expressions, such as a adjective ("hard" → "hardness"), a verb ("give" → "giver", "gift", "given"), or a (more concrete) noun ("object" → "objecthood"), etc. From the sentence "The sky is blue", the sentence "The sky possesses blueness" is constructed.

Logically: the corresponding construction of a subject on the basis of a predicate, thus adding a new 2nd-order individual ("blueness") to the domain represented.

Mathematically: the application of a meta-level operation or object regulating other, more basic operations or objects (e.g., passing from the existence of different types of connections between entities to forming the concept of "relation" as a new abstract object. The properties of this object now become

[12] As with many of his interesting proposals, the discussions of "hypostatic abstraction" (or "hypostatis", "subjective abstraction", etc.) are scattered over his work, so a bit of reconstruction work is necessary.

open to investigation in higher-level hypostatic abstractions (the "symmetry", "transitivity", etc. of relations may now be investigated).[13]

Cognitively: the process of taking a thought for a thing, so that a new cognitive object is constructed on the basis of a thought – alternatively described as the "stiffening" of transient, fleeting cognitive content into a stable shape facilitating further reasoning pertaining to this new, abstract object.

Perceptually: the spatialization of a temporal perception process, such as forming the trajectory as an abstraction from the locations covered by the perceived movement of an object: going from "a point moves" to "the line traced by the moving point".[14]

---

[13] One of Peirce's own examples of hypostatic abstraction in mathematics concerns the successive abstractions of sets from elements, powers from sets, cardinal numbers from powers: "In order to get an inkling – though a very slight one – of the importance of this operation in mathematics, it will suffice to remember that a collection is an hypostatic abstraction, or ens rationis, that multitude is the hypostatic abstraction derived from a predicate of a collection, and that a cardinal number is an abstraction attached to a multitude. So an ordinal number is an abstraction attached to a place, which in its turn is a hypostatic abstraction from a relative character of a unit of a series, itself an abstraction again." (5.534; Peirce writes "collections" and "multitudes" for sets and powers).

[14] A Peirce quote giving many different examples of Hypostatic Abstraction is the following: "But hypostatic abstraction, the abstraction which transforms" it is light "into" there is light here, which is the sense which I shall commonly attach to the word abstraction (since prescission will do for precisive abstraction) is a very special mode of thought. It consists in taking a feature of a percept or percepts (after it has already been prescinded from the other elements of the percept), so as to take propositional form in a judgment (indeed, it may operate upon any judgment whatsoever), and in conceiving this fact to consist in the relation between the subject of that judgment and another subject, which has a mode of being that merely consists in the truth of propositions of which the corresponding concrete term is the predicate. Thus, we transform the proposition, "honey is sweet," into "honey possesses sweetness." "Sweetness" might be called a fictitious thing, in one sense. But since the mode of being attributed to it consists in no more than the fact that some things are sweet, and it is not pretended, or imagined, that it has any other mode of being, there is, after all, no fiction. The only profession made is that we consider the fact of honey being sweet under the form of a relation; and so we really can. I have selected sweetness as an instance of one of the least useful of abstractions. Yet even this is convenient. It facilitates such thoughts as that the sweetness of honey is particularly cloying; that the sweetness of honey is something like the sweetness of a honeymoon; etc. Abstractions are particularly congenial to mathematics. Everyday life first, for example, found the need of that class of abstractions which we call collections. Instead of saying that some human beings are males and all the rest females, it was found convenient to say that mankind consists of the male part and the female part. The same thought makes classes of collections, such as pairs, leashes, quatrains, hands, weeks, dozens, baker's dozens, sonnets, scores, quires, hundreds, long hundreds, gross, reams, thousands, myriads, lacs, millions, milliards, milliasses, etc. These have suggested a great branch of mathematics. Again, a point moves: it is by abstraction that the geometer says that it "describes a line." This line, though an abstraction, itself moves; and this is regarded as generating a surface; and so on. So likewise, when the analyst treats operations as themselves subjects of operations, a method whose utility will not be denied, this is another instance of abstraction. Maxwell's notion of a tension exercised upon lines of electrical force, transverse to them, is somewhat similar. These examples exhibit the great rolling billows of abstraction in the ocean of mathematical thought; but when we come to a minute examination of it, we shall find, in every department, incessant ripples of the same form of thought, of which the examples I have mentioned give no hint. (CP 4.235, "Minute Logic", 1902).

Peirce's ambitious ideas are that these rather different examples constitute different occurrences of the same basic cognitive-logical structure and process characterized by their result: the occurrence of a new, higher-level cognitive object. Most often, these aspects of hypostatic abstraction are merely mentioned as examples; they are not explicitly distinguished and interrelated as subtypes of the concept. The basic cognitive purposivity of hypostatic abstraction stands out as its raison-d'être: it facilitates the explicit reasoning and investigation pertainting to general issues which would otherwise remain implicit, transient or lost in concrete particulars (Of course, once hypostatic abstraction is possible it need not build on existing particulars and thus may refer to non-existing or fictive universals). The many different linguistic devices for hypostatic abstraction are tools which further develop, detail, and make explicit aspects of the ongoing cognitive process of reasoning. Thus, the adjective "red" basically refers to particular, concrete, here-and-now occurrences of that color and allows for their comparison, while the noun "redness" (or "the color red", "the red", etc.) constructs a new, stable, abstract object interconnecting these different occurrences and makes possible the further reasoning on this color as such, abstracted from its concrete occurrences, and on its relation to other colors, other properties etc.

Hypostatic abstraction may be described as a simple deduction from a premise "This object is red" to a conclusion "Redness exists (in this object)", so that it makes sense to say that the hypostatic abstraction is an entity whose being consists in the (purported) truth of a predicate expression:

> For by means of abstraction the transitory elements of thought, the {*epea pteroenta*}, are made substantive elements, as James terms them, {*epea apteroenta*}. It thus becomes possible to study their relations and to apply to these relations discoveries already made respecting analogous relations. In this way, for example, operations become themselves the subjects of operations.
>
> To take a most elementary example -- from the idea of a particle moving, we pass to the idea of a particle describing a line. This line is then thought as moving, and so as generating a surface; and so the relations of surfaces become the subject of thought. An abstraction is an ens rationis whose being consists in the truth of an ordinary predication. ("Relatives" in Baldwin's Dictionary, 1901; CP 3.642)

This should not be taken to imply that hypostatic abstraction expressions referring to non-existing objects may not exist. "Unicornicity" is a hypostatic abstraction from "unicorn" even if no unicorns exist – the implicaiton of Peirce's definition is that, in this case, the hypostatic abstraction does not have any "being", that is, it does not refer to any real possibility like in the cases of "redness" or "hardness" or "trajectory". The deductive character of hypostatic abstraction is not changed by this observation – like any deduction, its validity dependes on the soundness of the premiss invoked: "If and only if x exists, then x-ness has being". Thus, if no x exists, x-ness has no being, – but in many cases the validity of this claim may be investigated both by investigating x's and investigating x-ness. Thus, the hypostatic abstractive deduction forms no guarantee that the resulting abstraction has a *fundamentum in re* and refers to really existing kinds – the well-known examples of fallacious hypostatic abstractions in science such as "phlogiston" or "caloric" testify to that.

The fact that the hypostatic abstraction is a deduction has often been confused with the possibly abductive character of the reasoning process in which a hypostatic abstraction may take part. Making a hypostatic abstraction may, in many cases, be part of a trial-and-error reasoning process where the abstraction made is subject to further investigation so as to determine its degree of reality. Thus, the deductive step of hypostatic abstraction forms, in this broader perspecive, part of an abduction whose valitidy must be investigated by further de- or induction on the basis of the abstraction made. This is why hypostatic abstraction has sometimes been characterized as deduction, sometimes as abduction (cf. Pape and Short in Houser, Roberts, & van Evra, 1997; cf. Stjernfelt, 2007, p. 458).

It should immediately be added that Peirce takes great care to distinguish this process from what he calls "distinction", the attention ability which permits the focusing on a particular part or aspect of an object at the expense of other parts or aspects of that object – and which is often confounded with hypostatic abstraction. These focusing abilities come in three variants, nicknamed "dissociation", "prescission", and "discrimination", respectively. Dissociation is what permits the distinction between different independent qualities, such as "red" from "blue": prescission is what permits the distinction of a part which may be supposed to exist independently of another part, such as "space" from "color", while discrimination is what permits the distinction of a part which may be only imagined separately, such as "color" from "space". These two latter distinction types are important to the investigation of objects involving features dependent on each other in different patternings. The kind of attention they pertain to, however, involves imagining the object endowed with indeterminate parts:

> In general, prescission is always accomplished by imagining ourselves in situations in which certain elements of fact cannot be ascertained. This is a different and more complicated operation than merely attending to one element and neglecting the rest. (CP 2.428).

According to Peirce, it is of paramount importance to keep the distinctions apart from hypostatic abstraction – while the former pertain to the degree of particularity and generality and thus permit chains of increasing generality like "red" – "color" – "hue", the latter does not lead to higher generality but to the creation of new, abstract or ideal objects of thought or discourse.[15] The distinction between the two may be expressed as follows:

> But even in the very first passage in which abstraction occurs as a term of logic, two distinct meanings of it are given, the one the contemplation of a form apart from matter, as when we think of whiteness, and the other the thinking of a nature indifferenter, or without regard to the differences of its individuals, as when we think of a white thing, generally. The latter process is called, also, precision (or better, prescission): and it would greatly contribute to perspicuity of thought and expression if we were to return to the usage of the best scholastic doctors and designate it by that name exclusively, restricting abstraction to the former process by which we obtain notions corresponding to the "abstract nouns." (CP 2.427, "Terminology" 1893).

[15] As a realist, Peirce holds that some hypostatically abstract concepts refer to aspects of reality ("gravity") while others do not ("phlogiston"). See Haack (1992) and Stjernfelt (2007) ch. 2.

In most concrete cases, of course, the two procedures work closely in tandem: before the hypostatic abstraction of "redness", a distinction is required to isolate the property of "red" in the object (more generally, hypostatic abstraction is impossible without a preceding distinction) – but still, the working of the two must be kept analytically distinct. In contrast to many empiricist theories of abstraction, moreover, it should be added that Peirce does not identify any of the two with induction as the statistical investigation of properties in a sample of objects. Abstraction does not presuppose induction and it is perfectly possible to perform a hypostatic abstraction on the basis of one observed object only (even if it may be wiser to perform it after an induction summing up knowledge of a wider sample of objects).

An important aspect of hypostatic abstraction is that, in making a second-order object out of a thought, it gives it concrete form and thus facilitates cognitive and logic manipulation and investigation of it – as if it were a particular individual object:

> Intuition is the regarding of the abstract in a concrete form, by the realistic hypostatization of relations; that is the one sole method of valuable thought. Very shallow is the prevalent notion that this is something to be avoided. You might as well say at once that reasoning is to be avoided because it has led to so much error; quite in the same philistine line of thought would that be; and so well in accord with the spirit of nominalism that I wonder some one does not put it forward. The true precept is not to abstain from hypostatization, but to do it intelligently. . . . (CP 3.383, "A Guess at the Riddle", c. 1890).

This implies that the hypostatically abstract object may be seen as if it shared some of the characteristics of particular individuals: it has properties, it stands in various relations to other such objects, it may be subsumed by still higher genera – in that sense hypostatic abstraction is a simplifying device involving cognitive economy because it permits to use some of the same means for their investigation which we use interacting with particulars. Peirce also ascribes abstractions a seminal role in his famous distinction between corollarial and theorematic deductions,[16] the former only relying upon definition of concepts appearing in the premises, the latter requiring the introduction of additional elements in the shape of postulates to conduct the proof. Theorematical reasoning, of course, requires creativity and guessing, even if being deductive – and the most challenging theorematical deductions are taken to involve the introduction of abstractions:

> "Deductions are of two kinds, which I call corollarial and theorematic. The corollarial are those reasonings by which all corollaries and the majority of what are called theorems are deduced; the theorematic are those by which the major theorems are deduced. If you take the thesis of a corollary, – i.e. the proposition to be proved, and carefully analyze its meaning, by substituting for each term its definition, you will find that its truth follows, in a straightforward manner, from previous propositions similarly analyzed. But when it comes to proving a major theorem, you will very often find you have need of a lemma, which is a demonstrable proposition about something outside the subject of inquiry; and even if a lemma does not have to be demonstrated, it is necessary to introduce the definition of something which the thesis of the theorem does not contemplate. In the most remarkable cases, this is some abstraction; that is to say, a subject whose existence consists in some fact

[16] See Levy (1997); Stjernfelt (2007, pp. 107–08, 2011).

about other things. Such, for example, are operations considered as in themselves subject to operation; lines, which are nothing but descriptions of the motion of a particle, considered as being themselves movable; collections; numbers; and the like. When the reform of mathematical reasoning now going on is complete, it will be seen that every such supposition ought to be supported by a proper postulate. At any rate Kant himself ought to admit, and would admit if he were alive today, that the conclusion of reasoning of this kind, although it is strictly deductive, does not flow from definitions alone, but that postulates are requisite for it." (CP 7.204; "On the Logic of drawing History from Ancient Documents especially from Testimonies", 1901).

As to the discussion of the semiotic "missing link" it should be mentioned that many higher animals are able to make prescissions – the ability to isolate features in an object is the precondition for associative learning, linking up co-occuring such features – just like they are able to make deductions on the basis of phylogenetically inherited or ontogenetically acquired habits. Theorematic reasoning and Hypostatic abstractions, on the other hand, seem to be missing among animal proto-concepts.

3.5 Self-Control by Abstraction in Human Semiotics

In a central argument, Peirce links the special semiotic and cognitive abilities in human beings to a higher degree of self-control which is, in turn, connected to the ability to make hypostatic abstractions. Let us first scrutinize his notion of "self-control". An important idea here is that self-control is crucial for inferences to count as real reasonings, as he epigrammatically may say: "... reasoning is thought subjected to self-control ..." (CP 5.533 "Pragmaticism, Prag. [4]" c. 1905). This is why computers ("logical machines") are not taken to be able to reason – even if their actions may formally realize inference structures and they are able to produce outputs which are interpretable as truths – they do not possess any self-control. The potentiality of specific action is sufficient to count as a habit – but belief requires the self-control of habit: "[Readiness] to act in a certain way under given circumstances and when actuated by a given motive is a habit; and a deliberate, or self-controlled, habit is precisely a belief." (5.480).

Fully realized self-control, on the other hand, may have as its result the formation of mechanical-like thought habits: "The power of self-control is certainly not a power over what one is doing at the very instant the operation of self-control is commenced. It consists (to mention only the leading constituents) first, in comparing one's past deeds with standards, second, in rational deliberation concerning how one will act in the future, in itself a highly complicated operation, third, in the formation of a resolve, fourth, in the creation, on the basis of the resolve, of a strong determination, or modification of habit. This operation of self-control is a process in which logical sequence is converted into mechanical sequence or something of the sort. How this happens, we are in my opinion as yet entirely ignorant. There is a class of signs in which the logical sequence is at the same time a mechanical sequence and very likely this fact enters into the explanation." (8.320. letter to F.C.S.Schiller, undated).

This, however, is only possible as the result of a complex process involving standards, future acts, and a decision to modifiy thought habits. Such self-control is required for full mastering of reasoning – but is the result of a process with simpler biological antecedents. Importantly, Peirce sees a decisive aspect of self-control in the psychological ability to isolate a thought from other intrusions – a psychological equivalent to the logical notion of "distinction" discussed above:

> Contemplation consists in using our self-control to remove us from the forcible intrusion of other thoughts, and in considering the interesting bearings of what may lie hidden in the icon, so as to cause the subjective intensity of it to increase. (7.555)

The isolation of the iconic sign may count as a first step in human self-control – to be followed by the hypostatic taking of that sign to be a thing in itself. Hypostatization, now, is crucially connected to the particularity of human reasoning. Peirce himself only rarely discusses hypostatic abstraction in connection to the man-animal issue. The most important locus is the following quote:

> To return to self-control, which I can but slightly sketch, at this time, of course there are inhibitions and coördinations that entirely escape consciousness. There are, in the next place, modes of self-control which seem quite instinctive. Next, there is a kind of self-control which results from training. Next, a man can be his own training-master and thus control his self-control. When this point is reached much or all the training may be conducted in imagination. When a man trains himself, thus controlling control, he must have some moral rule in view, however special and irrational it may be. But next he may undertake to improve this rule; that is, to exercise a control over his control of control. To do this he must have in view something higher than an irrational rule. He must have some sort of moral principle. This, in turn, may be controlled by reference to an esthetic ideal of what is fine. There are certainly more grades than I have enumerated. Perhaps their number is indefinite. The brutes are certainly capable of more than one grade of control; but it seems to me that our superiority to them is more due to our greater number of grades of self-control than it is to our versatility.
>
> Doctor Y. Is it not due to our faculty of language?
>
> Pragmaticist. To my thinking that faculty is itself a phenomenon of self-control. For thinking is a kind of conduct, and is itself controllable, as everybody knows. Now the intellectual control of thinking takes place by thinking about thought. All thinking is by signs; and the brutes use signs. But they perhaps rarely think of them as signs. To do so is manifestly a second step in the use of language. Brutes use language, and seem to exercise some little control over it. But they certainly do not carry this control to anything like the same grade that we do. They do not criticize their thought logically. One extremely important grade of thinking about thought, which my logical analyses have shown to be one of chief, if not the chief, explanation of the power of mathematical reasoning, is a stock topic of ridicule among the wits. This operation is performed when something, that one has thought about any subject, is itself made a subject of thought. (...)[17] ("Pragmaticism, Prag. [4]" c. 1905, CP 5.533).

[17] In the left-out part of the quote given, Peirce goes into his recurring example of hypostatic abstraction, Molière's joke from *Le malade imaginaire* about the idle inference from "opium puts people to sleep" and to "opium possesses a virtus dormativa" where Peirce agues that this ridiculed inference does in fact represent a step (albeit very small) forward in reasoning, because it opens the issue of what this dormitive powers more exactly consist in, how strong it is as compared to that of other substances, etc., and thus facilitates further investigations. Without such further investigations, of course, the hypostatic abstraction remains idle.

The quote gives two important arguments. The first is that self-control comes in many grades and increase during evolution. Our hypothesis here will be that this increase corresponds to the increase of articulation and segmentation of the perception-action chain into detailed argument structures. Any autonomization of a part of that chain corresponds to an increase in self-control. On top of such grades of self-control which is already present in higher animals, Peirce presents an architecture of additional human self-control grades: (1) training (2) self-training, controlling one's own self-control, involving imagination (3) adoption of a rule guiding this meta-control, (4) improvement of that rule after some higher ethical standard, thus controlling the control over one's control (5) controlling, in turn, that rule after some aesthetic standard (Peirce's notion of aesthetics pertaining to all goals which are worth pursuing[18]). Every such step, of course, takes the former step as its object, thus creating a newhypostatic abstraction subject to variation and evaluation.[19] Many higher animals, it is well-known, may be subject to training, but the next, decisive step of self-training seems only rudimentarily accessible to higher animals.

The other crucial argument here is that such self-control is seminal to human thought and language – and that this self-control is facilitated by thinking of our signs as signs, by thinking about thought and thereby becoming able to criticize our own thought logically. Self-control involves the taking one's own thought as the object of a meta-level thought. But this is only possible by making the first thought an object – stiffening in the shape of a hypostatic abstraction. Such self-control even makes possible language. How should this be interpreted? – it is well known that natural language learning does not take place by the explicit memorizing of linguistic rules and that practicing knowledge of grammar does not entail any explicit insight in grammatical principles (much like Peirce's logical distinction between implicit *logica utens* and explicit *logica docens* which differ in that the former is interested in the result of reasoning, not the process, the latter vice versa). The work performed by self-control here is more basic – it is the ability to wonder and check whether a particular sign is suitably used, focusing upon the relation between sign,

---

[18] The special concept of aesthetics referred to here is discussed later in the quote: "And you, Doctor W., will see that since pragmaticism makes the purport to consist in a conditional proposition concerning conduct, a sufficiently deliberate consideration of that purport will reflect that the conditional conduct ought to be regulated by an ethical principle, which by further self-criticism may be made to accord with an esthetical ideal. For I cannot admit that any ideal can be too high for a duly transfigured esthetics. So, although I do not think that an esthetic valuation is essentially involved, actualiter (so to speak) in every intellectual purport, I do think that it is a virtual factor of a duly rationalized purport. That is to say, it really does belong to the purport, since conduct may depend upon its being appealed to. Yet in ordinary cases, it will not be needful that this should be done." (CP. 5.535) That "duly transfigured" aesthetics is the generalized doctrine of all ideals possible to pursue; the idea is that such ideals may play a role in thought even if not explicitly addressed at all.

[19] We remark in the passing that Peirce, in this nesting of control acts into higher-level control acts, seems to subscribe to an Enlightenment ideal of the moral autonomy of human reasoning.

object, and interpretant, upon its relation to other signs and their objects and interpretants. Such ability is taken to be the prerequisite to the establishment of grammar, fine-grained taxonomies, tuning of schematic content, expression-content couplings, etc. in the development of languages.[20]

It is important to Peirce's notion of self-control, now, that such self-control is a merely *restrictive* measure, selecting valuable inferences among less valuable inferences – thus, it presupposes the existence of inferences which it then, subsequently, turns into reasoning by controlling them:

> But self-control is the character which distinguishes reasonings from the processes by which perceptual judgments are formed, and self-control of any kind is purely **inhibitory.** It originates nothing. Therefore it cannot be in the act of adoption of an inference, in the pronouncing of it to be reasonable,
> that the formal conceptions in question can first emerge. It must be in the first perceiving that so one might conceivably reason. And what is the nature of that? I see that I have instinctively described the phenomenon as a "perceiving." I do not wish to argue from words; but a word may furnish a valuable suggestion. What can our first acquaintance with an inference, when it is not yet adopted, be but a perception of the world of ideas? (5.194 Lectures on Pragmatism 1903).

The question of the roots of inference is here answered phenomenologically – investigating the origin of inference structures as seen from the perspective of human mind. Before subjecting an inference to control and evaluating it, we must be able to perceive it in "the world of ideas". This idealist wording of course leaves out the fact that we access that world not by any direct, mystic, purely intuitive route, but only by the intermediary of diagrams, facilitated by imagination. The inhibitory work performed by the different levels of self-control, then, presupposes a wealth of possible inferences and abstract objects to chose between. The imaginative creation, variation and combination of such inferences and objects – at each of the control levels – is thus the prerequisite for inhibitory self-control to perform its function.

And posed as an evolutionary question, the basic pool of such inference structures is found in the perception-action habits refined through the evolution of animals – habits which have been subjected to increasing degrees of control already over the course of evolution, before they are made, in turn, the object of the vastly increasing human processes of self-control by means of hypostatic abstraction and diagram experimentation.[21]

[20] Reasoning as opposed to mechanical compuation is characterized by self-control. Given the tower of control of control discussed here, however, self-control appears as a matter of degree. Even if perfect self-control may be achievable on one level, this hardly involves all levels at the same time. Conversely, cases of intermediary control are possible, Peirce muses in a psychological argument: "If, however, as the English suppose, the feeling of rationality is the product of a sort of subconscious reasoning--by which I mean an operation which would be a reasoning if it were fully conscious and deliberate--the accompanying feeling of evidence may well be due to a dim recollection of the experimentation with diagrams." (2.172) Subconscious diagram experiment – controlled only to some degree – might lie behind non-substantiated evidence-feelings.

[21] It goes without saying that this overall evolutionary increase in self-control recruits further capabilities to create higher level, more efficient cognition and action – such as conscousness, emotions, episodic memory, human language etc.

It is an important corollary of self-control, as described here, that it always involves (at least) two levels, that of inhibitory controlling, and that of imaginary creativity (being controlled). This implies that the focus of control must alternate between the levels, evolving inferences on the lower level and pruning them on the higher level. This makes self-control a special case of Peirce's important idea of the dialogicity of logic. This, I think, makes it possible to compare the Peircean notion of self-control by hypostatic abstraction with Tomasello's well-known ideas of joint attention (Tomasello, 1999, 2008).

3.6 Hypostatic Abstraction and Joint Attention

For how does Peirce's idea of hypostatic abstraction fit the Tomasellian idea of joint attention? For a first glance, the two ideas may seem wide apart, but for a closer glance important similarities appear. Joint attention can not, of course, be reduced to two parties both of them intending the same object. It also involves the knowledge in each part of the other part's attention. But even that is not sufficient. As Kaplan and Hafner insist from the point of view of implementing joint attention in robot research, joint attention is not achieved even by robots tracking other robots' attention and coordinating that with their own attention (which is robotologically possible but does not entail joint attention). Rather, joint attention is a collaborative process, in which

> ... the agent must understand, monitor and direct the attentional behavior of the other agent. Joint attention can only be reached if both agents are aware of this coordination of "perspectives" towards the world. (Kaplan and Hafner 2006).

Joint attention thus requires for each agent to assume the famous "intentional stance" towards the other: the attention direction detected in the other agent is interpreted as a sign that an intention is directing that attention to some goal. But even that is not sufficient: each agent must be able to influence upon the other's attention, for instance by directing it by means of gesture, eye movements, linguistic cues, etc. And such influence is only possible based on a skill of social interaction: the agents must be able to master turn-taking, role-switching and ritualized games, as Kaplan underlines. If no turn-taking schema is active, the agents will not know who is directing whose attention at any given moment. Thus, the apparently simple phenomenon of "joint attention" entails a whole series of interrelated concepts – a molecule of social interaction. But something similar is the case with Peircean hypostatic abstraction. It forms, of course, one of the major techniques of letting "symbols grow", Peirce's brief version of the Enlightenment ideal of common, increasing knowledge construction. It does that by means of its ability to take other signs as its object, thereby making their content and role explicit, and hence the possible object of scrutinizing, comparing and controlling. And this whole process of thought, according to Peirce, has an irreducible dialogic structure:

> Accordingly, it is not merely a fact of human Psychology, but a necessity of Logic, that every logical evolution of thought should be dialogic. ("Prolegomena", 1906, CP 4.551).

This necessity lies in the articulation of logic in signs as the means for communicating them from a person in one moment to the same person in the next moment:

> All thinking is dialogic in form. Your self of one instant appeals to your deeper self for his assent. Consequently, all thinking is conducted in signs that are mainly of the same general structure as words; those which are not so, being of the nature of those signs of which we have need now and then in our converse with one another to eke out the defects of words, or symbols. These non-symbolic thought-signs are of two classes: first, pictures or diagrams or other images (I call them Icons) such as have to be used to explain the significations of words; and secondly, signs more or less analogous to symptoms (I call them Indices) of which the collateral observations, by which we know what a man is talking about, are examples. The Icons chiefly illustrate the significations of predicate-thoughts, the Indices the denotations of subject-thoughts. The substance of thoughts consists of these three species of ingredients. (6.338, Amazing Mazes 4, 1909).

And this gives rise to the possibility of performing logical arguments and proofs, the same person occupying alternately pro- and con-positions in an ongoing dialogic process. These important ideas have often been overlooked because Peirce's ideas on this are only scarcely represented in his published work and have not been much discussed in the Peirce literature, but it has been highlighted in the wake of the tradition of Hintikkan game-theoretical semantics (Jaakko Hintikka, Risto Hilpinen; most recently, Pietarinen 2006 has furthered the investigation of this issue).[22] This idea occurs in the context of Peirce's logic representation systems known as Existential Graphs whose Alpha and Beta parts are isomorphic to propositional logic and first order predicate logic with identity, respectively. Peirce's idea is that these representations reveal a dialogic structure inherent in logical arguments. One agent, the so-called Grapheus, is responsible for the construction of the discursive world, while the other, the so-called Graphist, is responsible for counter-arguing the single steps of its construction. The two agents thus collaborate in critically investigating a logical issue and take it to conclusion, and they may, of course, often be instantiated in one and the same mind during soliloquious thought processes. In Pietarinen's Hintikkan interpretation, they may be seen as playing a semantic game against each other, and the existence of a winning strategy on the part of one of them is the game-theoretical equivalent to the truth of that part's argument. We shall not here go deeply into the specific means the two agents use when interacting in Peirce's elaboration of the existential graphs, but in our context, Peirce has some important general developments of what is involved:

> Now nothing can be controlled that cannot be observed while it is in action. It is therefore requisite that both minds but especially the Graphist-mind should have a power of self-observation. Moreover, control supposes a capacity in that which is to be controlled of acting in accordance with definite general tendencies of a tolerable stable nature, which implies a reality in this governing principle. But these habits, so to call them, must be capable of being modified according to some ideal in the mind of the controlling agent; and this controlling agent is to be the very same as the agent controlled; the control extending even to the modes of control themselves, since we suppose that the interpreter-mind under

[22] See also the Hintikka and Hilpinen refs. in the References.

the guidance of the Graphist-mind discusses the rationale of logic itself. (MS 280: 30–32, quoted from Pietarinen 2006).

The dialogic structure facilitates control of the thought process, because one part's utterance in the game takes the other part's utterance as its object in a hypostatic abstraction.[23] What is visible, of course, is only the other part's manifest utterance, but that utterance is the response to the whole preceding game and, in that respect, indirectly refers to it – much like a move in a chess game implicitly refers to the whole preceding game and one player's interpretation of the other's intention as perceived from his move sequence. Of course, hypostatic abstractions are not possible within the representation systems offered in the Alpha and Beta parts of the Existential Graphs (hypostatic abstractions quantify over other signs such as precidates and thus belong to second-order logic). Peirce envisaged this second order part in his Gamma graphs which were to comprise a part aimed at the explicit representation of hypostatic abstractions. But naturally, this representation of them takes place in a hypostatic abstraction of second order.

If logical thinking necessarily possesses a dialogic structure, it forces the individual engaging in such thinking to divide so as to accomodate to it:

> There is no reason why "thought," in what has just been said, should be taken in that narrow sense in which silence and darkness are favorable to thought. It should rather be understood as covering all rational life, so that an experiment shall be an operation of thought. Of course, that ultimate state of habit to which the action of self-control ultimately tends, where no room is left for further self-control, is, in the case of thought, the state of fixed belief, or perfect knowledge.
>
> Two things here are all-important to assure oneself of and to remember. The first is that a person is not absolutely an individual. His thoughts are what he is "saying to himself," that is, is saying to that other self that is just coming into life in the flow of time. When one reasons, it is that critical self that one is trying to persuade; and all thought whatsoever is a sign, and is mostly of the nature of language. The second thing to remember is that the man's circle of society (however widely or narrowly this phrase may be understood), is a sort of loosely compacted person, in some respects of higher rank than the person of an individual organism. It is these two things alone that render it possible for you – but only in the abstract, and in a Pickwickian sense – to distinguish between absolute truth and what you do not doubt. (5.420–421, "What Pragmatism Is", 1905).

If this is the case, there seems to be a deep connection between the dialogic structure of reasoning and self-control on the level of semiotics and logic – and the central place enjoyed by joint attention on the level of human psychology according to Tomasello's hypothesis. The parent-child dyad interaction trains the child in the first human level of self-control with he parent as the teacher, of course, but with the continuous exchange of positions making it possible for the child to experience the

[23] Might such dialogue structures take place in evolutionary time? El-Hani, Queiroz, and Stjernfelt (2010) takes firefly signaling as an example of deception where one species may mimic that of another for predatory purposes – creating a virtual arms race between firefly species. In some sense, the changing of the signal of one species forms an inference stated in the slow progress of phylogenetic time, and the other species may be able to answering the inference by some countermove (if not going extinct in the meantime).

dialogue structure and internalize it for the benefit of its critical self-control abilities (and for later social interactions as well, of course).

Thus, a hypothesis can be stated that there is a connection between the human ability to use signs about signs and thereby exercize semiotic self-control, on the one hand, and the ability of human beings to engage in joint attentions with other subjects shaping a shared world informed by shared thoughts in the shape of shared diagram experiments.

The overall argument of this chapter, then, is that a the gradual appearance of logic and semiotic capabilities during evolution forms the backbone of the increase of cognitive competences from simple biology to higher animals and human beings. This appearance takes the shape of the ongoing articulation, subdivision and making explicit of a basic argument structure inherent in perception-action loops. The basic reason is that biological semiosis must be oriented toward adequacy truths for survival reasons, making biological cognition acutely dependent upon the ability to perceive and act in a way which is adequate to the environment, thus expressing the linking of proto-propositions. This is not to say that issues like the emergence of communication, awareness, consciousness, emotions, episodic memory, human language and much else are not important. Quite on the contrary, in this framework, such capabilities arise during evolution in order to enhance, speed up, widen, and control the basic, biological process of argumentative cognition.

Acknowledgement Thanks to Peter Harder, Jesper Hoffmeyer, and Theresa Schilhab for comments.

References

Adler, J., Hazelbauer, G. L., & Dahl, M. M. (1973). Chemotaxis toward sugars in escherichia coli. *Journal of Bacteriology, 115*(3), 824–847.

Berg, H. C. (1988). A physicist looks at bacterial chemotaxis. *Cold Spring Harbor Symposia on Quantitative Biology, 53*, 1–9.

Clark, A. (2008). *Supersizing the mind: Embodiment, action, and cognitive extension*. Oxford: Oxford University Press.

Deacon, T. (1997). *The symbolic species*. New York: W.W. Norton.

Deacon, T. (forthcoming) *Incomplete nature*. New York: W.W. Norton.

El-Hani, C., Queiroz, J., & Stjernfelt, F. (2010). Firefly femmes fatales. A Case Study in the Semiotics of Deception. *Journal of Biosemiotics, 3.1*, 33–55.

Fauconnier, G., & Turner, M. (2002). *The way we think. Conceptual blending and the mind's hidden complexities*. New York: Basic Books.

Haack, S. (1992). Extreme scholastic realism: Its relevance to philosophy of science today. *Transactions of the Charles S. Peirce Society, XXVIII*(1), 19–50.

Hilpinen, R. (1982). On C. S. Peirce's theory of the proposition: Peirce as a precursor of game-theoretical semantics. *The Monist 65*, 182–188.

Hintikka, J. (1983). C.S. Peirce's "first real discovery" and its contemporary relevance. In E. Freeman (Ed.), *The relevance of Charles Peirce* (pp. 107–118). La Salle, IL: The Hegeler Institute.

Hintikka, J. (1997a). The place of C.S. Peirce in the history of logical theory. In J. Brunning & P. Forster (Eds.), *The rule of reason* (pp. 13–33). Toronto: University of Toronto Press, (also in Hintikka 1997, 140–61).

Hintikka, J. (1997b). *Lingua Universalis vs. calculus ratiocinator. An ultimate presupposition of twentieth-century philosophy*. Dordrecht: Kluwer.
Hoffmeyer, J. (1996). Evolutionary intentionality. In E. Pessa, A. Montesanto, & M. P. Penna (Eds.), *Proceedings from the third European conference on systems science, Rome 1–4. Oct. 1996* (pp. 699–703). Rome: Edzioni Kappa.
Hoffmeyer, J. (2008). *Biosemiotics. An examination into the signs of life and the life of signs*. Scranton, PA: University of Scranton Press.
Hoffmeyer, J. (2010). Semiotic freedom: An emerging force. In P. Davies & N. H. Gregersen (Eds.), *Information and the nature of reality from physics to metaphysics* (pp. 185–204). Cambridge: Cambridge University Press.
Houser, N., Roberts, D. D., & van Evra J. (Eds.). (1997). *Studies in the logic of Charles Sanders Peirce*. Bloomington, IN: Indiana University Press
Hurford, J. (2007). *The origin of meaning*. Oxford: Oxford University Press.
Kaplan, F., & Hafner, V. V. (2006). The challenges of joint attention. *Interaction Studies, 7*(2), 135–169.
Levy, S. H. (1997). Peirce's theoremic/corollarial distinction and the interconnections between mathematics and logic. In N. Houser, D. D. Roberts, & J. van Evra (Eds.), *Studies in the logic of Charles Sanders Peirce* (pp. 85–110). Bloomington, IN: Indiana University Press.
Peirce, C. (1976). C. Eisele (Ed.), *New elements of mathematics* I–IV. The Hague: Mouton.
Peirce, C. (1998a) (1931–58) C. Hartshorne, P. Weiss, & A. W. Burks (Eds.), *Collected papers* [**CP**, references given by "CP" followed by volume number and paragraph], I–VIII. London: Thoemmes Press.
Peirce, C. (1998b). N. Houser & C. Kloesel (Eds.), *The essential Peirce*, Vol. II (1893–1913) [**EPII**, references given by "EPII" followed by page or chapter]. Bloomington, IN: Indiana University Press.
Short, T. L. (2007). *Peirce's theory of signs*. Cambridge: Cambridge University Press.
Stjernfelt, F. (2006). Two iconicity notions in Peirce's diagrammatology. In H. Scharfe, P. Hitzler, & P. Øhrstrøm (Eds.), *Conceptual structures: Inspiration and application. Lecture notes in artificial intelligence* 4068 (pp. 70–86), Berlin: Springer.
Stjernfelt, F. (2007). *Diagrammatology. An investigation on the borderlines of phenomenology, ontology, and semiotics*. Dordrecht: Springer.
Stjernfelt, F. (2011). Peirce's notion of diagram experiment. Corollarial and theorematical experiments with diagrams. In R. Heinrich, E. Nemeth, W. Pichler, & D. Wagner (Eds.), *Image and Imaging in Philosophy, Science, and the Arts* (Vol. 2, pp. 305–340). Frankfurt: Ontos Verlag.
Stjernfelt, F. (forthcoming a). Signs conveying information. On the range of Peirce's notion of propositions: Dicisigns. *Signs and Semiotic Systems*.
Stjernfelt, F. (forthcoming b). The generality of signs. Anti-psychologism in the foundations of semiotics – and its actual relevance.*Semiotica*.
Tomasello, M. (1999). *The cultural origins of human cognition*. Cambridge, MA: Harvard University Press.
Tomasello, M. (2008). *Origins of human communication*. Cambridge, MA: MIT Press.

Chapter 4
Peirce and Deacon on the Meaning and Evolution of Language

Ahti-Veikko J. Pietarinen

Abstract According to Charles Peirce's theory of meaning, known as pragmaticism, the meaning of signs is in the habitual practices and activities according to which we acquire information that connect signs with other signs and their objects. In *The Symbolic Species: The Co-Evolution of Language and the Brain* (henceforth SS, Deacon, 1997), Terrence Deacon takes meaning to be explicated by the uniquely human capacity for symbolic reference. The evolution of language is couched in adaptive co-evolution that overcomes the symbolic threshold by increased social selection pressures. Peirce, on the other hand, understood evolution "agapastically": it is not the selective mechanisms that direct the adaptation, say, of neural structures, but the growth of habits of action that are in continuous interaction with one another. I argue that Deacon's and Peirce's positions on the meaning of signs and the evolution of linguistic meaning share some similarities but also differ in a couple of fundamental respects.

This chapter charts the similarities as well as the dissimilarities between Peirce's and Deacon's positions. What do they take meaning to ultimately consist of? What are their views on the evolution of language? These questions stem from the suggestions in the earlier literature, including that of Deacon's book, that he is following the footpaths of Peirce on these issues.[1] My chapter points out in which senses this is and is not the case, with particular reference to the notions of meaning and the evolution of language.

---

[1] See Cowley (2002) and Lumsden (2002), among others. After the first version of the present paper was written in 2006, de Villiers (2007) published a rejoinder in which Peirce's and Deacon's uses of the term 'symbol' was greatly clarified and defended against the misinterpretations of Cowley (2002).

A.-V.J. Pietarinen (✉)
Department of Philosophy, History, Culture and Art Studies, University of Helsinki, FIN-00014 Helsinki, Finland
e-mail: ahti-veikko.pietarinen@helsinki.fi

T. Schilhab et al. (eds.), *The Symbolic Species Evolved*, Biosemiotics 6,
DOI 10.1007/978-94-007-2336-8_4, © Springer Science+Business Media B.V. 2012

4.1 Peirce on Meaning

Peirce and Deacon are close intellectual colleagues in a fundamental sense: both are practicing scientists who have learned their methods and ways of thinking in laboratories and in the fields rather than in an armchair. By training as well as by edification, both are thus in a good position to comment on such multidisciplinary topics as the meaning of signs and the evolution of language. Peirce's scientific interests ranged from astronomy to geodesy, from linguistics to experimental psychology, and from neuroscience to engineering. Virtually no nook of scientific inquiry of his era was foreign to him. Deacon's *Symbolic Species* follows suit and combines semiotics, linguistics, anthropology, neurosciences and cognitive sciences to formulate an intriguing novel account of meaning and the origins and the evolution of language.

No wonder, then, that not only is there a considerable interest from first-rate scientists in Peircean topics such as the logic of inquiry, scientific methodology and the theory of signs. Also the emerging philosophical positions are likely to converge. In Peirce's view, philosophy is science. Yet it is science in a uniquely methodological way that does not turn it a mere flat naturalistic endeavor. It ought to be pursued by analogous methods and analogous processes of reasoning to those by which hands-on scientists pursue their experiments in a laboratory, but without any claim to explain the philosophical concepts solely in specialised scientific terms.

Yet we may reasonably enquire into the extent to which such wide-ranging philosophical concepts such as meaning could converge with what scientific investigations have to tell about it. Peirce's answer is that yes, that extent is in fact very significant. The meaning of all expressions is grounded in the very experimental results they produce, as these expressions are used across situations of various kinds. This requires only that the proper investigative conduct and reasoning in setting up the experiments have been followed. This is what his Maxim of Pragmatism is, in effect, calculated to articulate. Here is a quotation from his late, 1905 formulation of that Maxim, which was renamed as the Maxim of Pragmaticism[2]:

> Consider what effects that *might conceivably* have practical bearings – especially in modifying habits or as implying capacities – you conceive the object of your conception to have. Then your (interpretational) conception of these effects is the whole (meaning of) your conception of the object. (MS 290: 33, 1905, *Issues of Pragmaticism*).

The only shift that takes place when we move from the realm of a laboratory to that of human thought is the shift from *observable objects* given by controlled experiments to *conceivably observable objects* given by those *human practices* and *activities* that seek answers to questions by making experiments in thought. Now in seeking answers to questions we put questions to some source of information. In the

---

[2] This account concerns Peirce's mature theory of meaning, which differs in some crucial respects from his early 1878 theory. On these points as well as on the general proposal for a reconstruction of Peirce's late proof of pragmatism, see Pietarinen and Snellman (2006).

laboratory science that source will be Nature and in the pragmatic theory of meaning it will be the utterer of the sign. To put a question to the source of information is a reasoning activity that uses abduction. Abduction is the capacity of generating scientific hypotheses as well as selecting among them. Abduction is closely linked to the reasoning processes of the inquirers, which essentially derive from their skills in setting up new prospects and perspectives as well as hitting on the right kinds of new goals to be pursued further.[3]

One essential feat when moving from laboratories to philosophy and from scientific experiments to thought experiments is that the shift is *continuous* rather than discrete. This follows from Peirce's ontological synechist view according to which matter is "effete mind" and is "hidebound with habits" (CP 6.158, 1892, *The Law of Mind*).[4]

In summary, then, according to Peirce meaning is what transpires in certain habitual practices and activities connected with experiments with signs, including linguistic signs. Just as evolution influences and modifies the habits of nature and of the physical world, it affects and impinges on the habits of human beings by which we act in certain ways in certain kinds of circumstances. Habits, undergoing constant changes, in turn affect and shape the contexts within which human beings, guided by them, must operate. Therefore, habits influence the experiences which we may have through signs. For Peirce, meanings evolve and grow contemporaneously with biological aspects of evolution. In the Sections 4.3 and 4.4 of this chapter, I will return to the important point of Peirce's evolutionary theory of meaning and its allied notions which I preliminarily investigated in Pietarinen (2005).[5]

4.2 Deacon on Meaning

Deacon takes meaning to be explicated by the *uniquely human mode for symbolic reference* (SS: 43). One of the key questions Deacon poses is the time-honoured one: How do words and thought manage to correspond to, or refer to, reality, the world? With reference to neuroscientific nomenclature, Deacon calls this the Grounding Problem (SS: 439).[6]

---

[3] Pietarinen (2007c) is concerned with aspects of abduction in Peirce's 1907 proof of pragmatism. Some steps in the reconstructed proof assume abductive reasoning and are closely related to question-answering structures in cognition (Pietarinen, 2010a).

[4] A very similar 'Law of Mind' was suggested before Peirce by John J. Murphy in his *Habit and Intelligence* (1869; see Pietarinen, 2009b), a work which Peirce applauded in unpublished draft versions of his 1887 article "Guess at the Riddle".

[5] Pietarinen (2005) is a study of Peirce's theory of meaning, which was presented in a conference on evolutionary epistemology. It is surprising how little of Peirce, let alone Baldwin, have been acknowledged in the works of those propounding that evolutionary approach to epistemological issues.

[6] I should mention that Peirce did keep an eye on the neurosciences of the day. At one point, he learned about Broca's groundbreaking discovery of 1860 – that Broca's and Wernice's areas are connected with the abilities to speak – and mentions it in relation to religious questions concerning

The same issue has been long-standing in the tradition of philosophy of language. I suggest that we may look at it anew through the Deaconian interpretation of Peirce's semiotics. Begin with Peirce's notion of symbolic signs. Symbols are the most highly evolved and developed kinds of signs, which typically incorporate indexical and iconic features. Deacon naturally observes these Peircean rudiments very well. Deacon asks: how come humans have acquired linguistic communication whereas animals have not? This is the puzzle from which Deacon begins to build his story in his book. My question is: what is the service we hope to get from Peirce's theory concerning Deacon's fundamental question? Does it answer what the uniquely human forms of communication are in the first place?

We must note that Peirce did not approve of any sharp distinction between human and other forms of communication. For him, communication of humans is continuous with the communication of any other sign user, including other species and even inanimate entities. A sign is a "species of medium of communication" (EP 2:390, 1906; see Pietarinen, 2006b). A sign user is anyone or anything that fulfils the function of the utterer or the interpreter of the sign. Language is one species of a medium among many others. According to Peirce's often quoted – and equally often misunderstood – statement, every thought is a sign, and thought is not necessarily connected with brains but "appears in the work of bees, of crystals, and throughout the purely physical world" just as it appears in the works of men (CP 4.551, 1906, *Prolegomena to an Apology for Pragmaticism*).

Nevertheless, Deacon's original insight is that symbolic signs – where the notion of a symbol is intended to be understood in the sense of Peirce's sign theory – pertain, predominantly if not exclusively, to the human realm of communication. Animals do not have symbols. Animals do not use symbols. Vervet monkey alarm calls are at best indexical, not symbolic, according to Deacon. This was Thomas Sebeok's conclusion in his life-long study of animal communication from the semiotic points of view as well. There is a generality in the nature of symbols that is not and cannot be found in the realm of signs that animals can use.

Peirce notices that symbols owe their "origin (on one side) to human conventions" (CP 5.532, c.1905, *Consequences of Critical Common-Sensism*). I believe what Peirce means is the common fact that the other side of their origin comes from indices and icons, which are forms of signs distinct from symbols, yet signs which symbols are capable of accommodating. That Peirce would have agreed with Deacon's point about symbols to be "uniquely human" is nevertheless contentious, for the two reasons implicit in what was said above: that communication continuously involves anyone or anything capable of uttering and interpreting signs, and

immortality in *Answers to Questions Concerning My Belief in God* (MS 845, c.1906; CP 6.520). He believed that the dependence of mental action upon the brain had been demonstrated by Broca's finding, and noted how lesions in Broca's area greatly affect the use of language: "when Broca's convolution is much diseased we always find the use of language is greatly affected" (MS 845: 5–6). In the same context, he remarked on the brain's plasticity: "When a part of the brain is extirpated we find the result is that certain faculties are lost. But after a time they are recovered. How can this be? The answer given is that other parts of the brain learn to perform those functions" (MS 845: 6).

that signs are continuous media of communication that on average involve iconic and indexical features.

Deacon adds to the nature of the symbolic sign the further requirement that the symbolic meaning is inherently and uniquely social. He takes language to be a social phenomenon, but he does not qualify whether this is a claim about the social character of grammar, semantic meaning, use, or understanding of language (SS: 115, 400). According to Peirce, however, symbols emerge as soon as habits of action are found that grow and evolve as signs are interpreted in the triadic framework of interpretation. Such signs need to be public or publicly available, to be sure, and this is certainly necessitated by their nature of being media of communication, but yet the public character of language does not imply that language could necessarily exist in social contexts only.[7] We do not get much support from Peirce for Deacon's additional qualification of the inherently social nature of language. Thinking and reasoning is for Peirce a form of dialogue, to be sure, and an interactive and collaborative kind of a dialogue as that, too, but to claim that these are inherently and necessarily social kinds of activities is to make some strong additional assumptions about the nature of language use and logical reasoning. It is true that Peirce at one point states that "logic is rooted in the social principle" (CP 2.654, 1893, *The Doctrine of Chances*), but we at the same time must notice that according to him, "to be social is one thing, to be gregarious is another: I decline to serve as bellwether" (CP 1.10, c.1897, *The Principles of Philosophy*). To claim socialisation was for Peirce just to oppose the notion of "selfish" logicians, and the idea of "the German logicians" of the day that the general psychological notions such as hopes, wishes and desired could have something to do with logical validity or fundamentally pragmatistic nature of linguistic meaning. In a late manuscript, Peirce even remarks that "one can establish conventions with oneself, which enable one to express the essence of what [one] has to communicate free from signs that are not essential" (MS 846: 5–6, 1910, *Notes for my Logical Criticism of Articles of the Christian Creed*).

My additional remark is that in SS Deacon does not address the *conventional* nature of symbols, let alone the origins of conventions, which is a bit surprising given the extent to which not only Peirce but so many later philosophers, linguists, and logicians have taken conventions to be constitutive of the meaning systems of human languages.[8] I have further comments on this in the last section.

[7] This point turns on the Wittgensteinian question of whether Robinson Crusoe can have a language. Peirce inevitably thinks he can. On this point, see the appended dialogue in Pietarinen 2006b, chapter 9: "Dialogue Foundations and Informal Logic".

[8] Lewis (1969) was an influential treatise on the origins of conventions, with its own origins in Erik Stenius's work on 'gardening games' (Stenius, 1968). Guldborg Hansen (2007)'s paper in the collection *Game Theory and Linguistic Meaning* (Pietarinen, 2007b) is a convenient survey of the status of Lewis-type signalling games throughout the contemporary scene. Lewis's theory is nevertheless nominalistic in ways interestingly similar to those accounts for symbols criticised by Peirce in his theory of signs for missing their general, habitual character.

Peirce took communication to be a phenomenon which is continuous from one media to another. What is predominantly human is the conventional mode of symbolic communication but not the use of symbols per se. Deacon, on the other hand, is prone to make the global symbolic realm the perimeter distinguishing human from non-human. It is doubtful whether Peirce, a synechist thinker as he was, would have accepted such a border phenomenon as a feature actually existing in the common and shared world.

My supplementary observation on human symbolic communication has to do with one more "major transition" that appears to await us in the future – namely *iconic* communication. Now that we have learned to think and express ourselves using the symbolic media, we are rapidly moving towards a visual and multimodal communication age. While such non-symbolic means of communication may seem harder for humans to use than linguistic symbols, is it not just because we have not yet mastered the economy of pictorial and other iconic means of communication? Before early hominids learned to speak, they used gestures. Sign languages are mostly symbolic, but there are other, essentially iconic means of communication and information sharing, such as diagrams, charts and maps of a massive variety of different kinds.[9]

Deacon takes icons to be situated on the lower level of "referential competence" than indices or symbols, evoked by the "interpretive process of re-cognition". By re-cognition, he means a duplication or reproduction of some observed relational structure in the relational structure of our cognition. However, icons are typically quite subtle and abstract, as logical diagrams (which I argue to be the "logics of our cognition" in Pietarinen, 2011) and creative uses of metaphors readily testify (Pietarinen, 2006a, 2010b). Therefore, iconicity should not simply be seen as a straightforward duplication in another representational media. Moreover, not all icons have image-like qualities attached to them (Pietarinen, 2007d).

Moving from the comparison between the notion of symbols in Peirce and Deacon to my second major point, Deacon's use of the notion of *reference* in the exposition of his "Grounding Problem" and his assertion that "meaning is uniquely human mode for symbolic reference" deserves a further comment. "Reference", as we have come to understand it in contemporary philosophy of language, is a very un-Peircean term. Especially in his later philosophy, reference is never used as an unarticulated simple term. Instead, there must always be something or

[9] Stjernfelt (2007) is a multi-disciplinary treatise accentuating the ubiquity of diagrammatic forms of representation in human culture, and comments on Deacon's theory in at length. Pietarinen (2009a) studies Otto Neurath's vision of the Isotype system as an anticipation of the 'visual revolution' which is currently changing the very foundations of human communication.Think of the use of emoticons in online chats, for example: they are certainly vaguer, milder, and much more roundabout forms of putting forth assertions than what is achieved with the symbolic, written or spoken, means of expression. Such kinds of pictures certainly do not communicate the exactly same propositional content and speaker's meaning as the linguistic expressions, such as "I love you!", "I am happy because of what you said", or "I am crying now...".

someone who mediates the reference.[10] Deacon seems to take reference in two senses: that which connects symbols to things and that which connects symbols with one another. Moreover, we find inclination to use the phrase "indirect" rather than "direct" reference. In philosophy of language, direct reference means purely denotative names, "labels attached to things". This is not what Deacon means, and indirect reference does not imply a belief in the existence of non-denotative names. Rather, in Deacon's parlance it means belief in the existence of some intra-linguistic referential relationships among symbolic signs.

In terms of Peirce's theory, this means that indirect reference pertains to Peirce's Speculative Grammar in his three-part division of the normative science of logic: Speculative Grammar, Logic Proper and Speculative Rhetoric. Speculative Grammar studies the relationships between signs not yet interpreted. It does not concern the study of the relationships between signs and objects as does Logic Proper, nor does it concern the study of the relationships between signs and interpretants as does Speculative Rhetoric.[11]

Next, Deacon discusses various kinds of interpretative processes and responses that account for the multiplicity of reference. He borrows from Peirce the glorious notion of "interpretant" to refer to such interpretive responses. Deacon writes that "an interpretant is whatever enables one to infer the reference from some sign or signs and their context" (SS: 63).[12] A considerable task that Deacon undertakes is to chart the mechanisms of one-to-many mappings from interpretants to their "reference".

It seems to me, however, that Peirce's puzzle was not so much to explain how it is possible that signs may correspond or refer to reality in such diversiform ways but to articulate what it is that mediates those relationships. His answer was that they are mediated by the very activities and practices that the utterers and the interpreters carry out when they encounter signs. Wittgenstein called these activities "language games" and "life forms". Peirce himself came pretty close to using the idea of games in the notion of our habits of acting in certain ways in certain kinds of situations and circumstances. Such habits are strategies for interpreting "intellectual signs", in other words signs that contribute to the validity of arguments (Pietarinen & Snellman, 2006).

Summing up, Peirce had little real use for the concept of reference or referent in order to explain symbolic meaning. While Deacon rightly embarks on his theory from largely Peircean terrains, he gradually slides closer to standard theories of reference in the application of his notions.

---

[10] And the idea of a Universal Grammar which Deacon attacks is, of course, equally un-Peircean.

[11] A great deal of overlap and continuity naturally exists between the three parts, and Peirce certainly did not embrace the trichotomy of syntax-semantics-pragmatics – unlike Morris and others later incorrectly supposed (see Pietarinen, 2006b).

[12] My quibble is that Deacon presumably here means to write "referent" instead of "reference".

Peirce and Deacon both emphasise the important role of *mental images* and *imagery* in language interpretation. This is true in the more general realm of interpreting non-linguistic signs, too, though Deacon is primarily interested in lexical meaning whereas Peirce wants to accommodate the meaning of all kinds of assertions. However, the two of them differ on the role which such images have in their overall theory. Deacon states that a mental image is that which is generated when an interpretation of a word makes it to "refer to" the image (SS: 63). I believe this is precisely what Peirce wanted to avoid in his pragmaticism, in other words the idea that the image would function as some stationary or inert "referent" or an "object" of something. Rather, in Peirce's theory, mental images are useful instructions or recipes for action. Their purpose is to show or guide those processes of how we should go about in interpreting our intellectual signs. They do this in terms of suggesting a variety of alternative ways of experimenting upon them. Imagery is needed to constitute the mental correlates to those possible sequences of actions or events that the mind entertains in interpreting signs.

The key role of mental imagery in the interpretation of signs does not, in itself, imply a lapse into psychologism or cognitive psychology. It would do so only if laws of psychology were argued to govern the formation of such images. In the Peircean spirit if not the letter, Deacon appears to agree that it is laws of logic rather than laws of psychology that are the leading ingredients in the formation of our ability to create and interpret language.

4.3 Peirce on the Evolution of Language

Peirce's speculative – or to some even outlandish – views on evolutionary metaphysics have been subject to some revived scrutiny (Hausman, 1993). However, we have yet to pay attention to whether he might have had something comparable to offer regarding the evolution of language. The following quote, which is from Peirce's manuscript *A Detailed Classification of the Sciences* (1902) is our lead:

> In linguistics, there is the question of the origin of language, which must be settled before linguistics takes its final form. The whole business of deriving ancient history from documents that are always insufficient and, even when not conflicting, frequently pretty obviously false, must be carried on under the supervision of logic, or else be badly done. (CP 1.250, c.1902, *A Detailed Classification of the Sciences*).

This is one of the only references Peirce makes to the puzzle of the origin of language, which he sees as one of the most difficult problems facing the "psychical sciences". But what about the evolution of language? Is it an offshoot of historical linguistics? Peirce's point seems to be that the proper methodology and logic needs to be discovered before the puzzle of the origin and, consequently, the puzzle of the evolution of language, may be tackled at all. "Deriving ancient history from documents" is certainly a concern of historical linguistics, but even in our times its methodology seems to be in considerable disarray (Pietarinen, 2007a).

So what is that methodology and logic? Peirce outlined the methodology of "psychical sciences" in terms of *purpose*. In the *Classification of the Sciences* paper quoted above he took evolution to be "nothing more nor less than the working out of a definite end" (CP 1.204). It is quite widespread nowadays to refer to the purposes that *language users* have as entailing the need for *teleological* explanations, and with respect to *language itself* as entailing the need for *teleonomic* explanations. Likewise, a commonplace teleological explanation of the human purpose of having and using language is the increased *communicative effectiveness* or *communicative fitness*. This is the "survival of the clearest" paradigm. The measure of fitness could be, for instance, the linguistic functionality within some specified contexts of use, such as oral, written, formal and informal contexts. Teleonomic explanations, on the other hand, refer to the language as a goal-directed and finalistic *system* that is independent of language users. Unlike agent-driven systems of language use, the functions of a system that is isolated from its users are given teleonomic explanations for the reason that such an inanimate system must be unconscious of the goals and purposes it may have.

It is nonetheless doubtful whether such a dichotomy of two kinds of finalistic explanations is ultimately defensible. Certainly, Peirce would have been very reluctant to subscribe to it. First, he did not think that semantics and pragmatics are separable units of language. Second, as far as sign meaning is concerned, signs cannot be separated from their utterers and interpreters. Third, as matter is "effete" mind and thought is not restricted to brains, what is conscious and what is not does not follow the borderlines between animate/inanimate. Final causes need not be conscious goals.

Peirce does not discuss this question with respect to language, but what he had in mind regarding the purpose of evolution at large is that it results in the increase of the *summum bonum*, the common idea-potential of humankind. It is part of his overall view of scientific methodology that contributions of all rational inquirers are destined to gravitate to some common idea-potential. If linguistics is to be a bona fide science, then the increase of that potential is the purpose of linguistics, too.

How ideas are made clear happens, according to Peirce, through the application of the Maxim of Pragmaticism. All ideas require experience, without which they could not be represented. There can be no ideas without representing them in signs. As soon as ideas are represented in symbols, they become intellectual and assertive. They acquire propositional content. And their meaning gets increasingly more clarified through changes in the self-controlled habits by which we go about interpreting the symbols.[13]

According to Peirce, for ideas to be represented according to our experiences, they will be "suspended in the medium of consciousness" (CP 7.554). Depending on a mind that is conscious of its ideas, these representations may reach the level of symbolic signs. Deacon, on the other hand, takes symbolic representation to have "produced an unprecedented *medium* for consciousness" (SS: 449) itself. The

[13] Deacon briefly mentions the important meta-theoretical aspect of self-control in the context of self-awareness or self-consciousness (SS: 451), but he does not link that with Peirce's self-controlled habits of action, which are needed for most interpretations of symbols.

emphasis is clearly different. Deacon presumes that all living neural systems to be conscious with respect to iconic and indexical representations, and that symbols require an additional threshold in the complexity. Like Peirce, Deacon advances the creditable view that there is no real problem of consciousness to be explained. Rather, the existence of consciousness is irreducibly linked to the different though continuous classes and levels of sign representation. On the other hand, Deacon takes symbolic representation to create an altogether new kind of medium of consciousness, whereas Peirce sees consciousness as the medium in which ideas to be represented are seized at different depths and voluminosities.

Others have argued that the *summum bonum* might merely be an unintentional by-product of what human communication has produced. According to that view, the course of the evolution of language is not destined towards such common idea-potential. This is the lesson of the "Invisible Hand" argument for the evolution of language proposed by Keller (1994). However, in Peirce's theory is there no need for such excess hypotheses. They are unconfirmed by experience and liable to flunk the test of the pragmatist maxim. Moreover, such explanations may work only in the behaviourist contexts that take into account only the singular intentions of human language users or communities of users. They also impose stringent conditions of causality on all steps of linguistic evolution. Therefore, the Invisible Hand argument would work with the assumption of "anancastic" evolution only (CP 6.302, 1893, *Evolutionary Love*). Anancastic evolution is a degenerate form of evolution that adopts new ideas "without foreseeing whither they tend"; a blind evolution "determined by causes either external to the mind, such as changed circumstances of life, or internal to the mind as logical developments of ideas already accepted, such as generalizations" (CP 6.307). Habits, in contrast, are real generals out there in the wild reality, according to Peirce. They wield counterfactual force and should be interpreted neither naturalistically nor anacastically. Peirce would have held invisible-hand types of explanations as indications of nominalist error.

Such extraneous hypotheses are not needed in Deacon's theory, either. The postulation of an Invisible Hand would be as mistaken as the postulation of some Grammar Module in the mind to account for complex linguistic competences of human primates.

One more point that Peirce makes in the opening quotation of this subsection warns us of another kind of fallacy that may tempt a historical philologist: "The whole business of deriving ancient history from documents that are always insufficient and, even when not conflicting, frequently pretty obviously false, must be carried on under the supervision of logic, or else be badly done". Peirce advices us not to assign implausible weights of credibility to evidence that may appear to be of great importance, and which we might wish to use as a basis of a confirmation of hypotheses upon which our theories are built, but which, being scarce, easily mislead the investigators. Historical linguistics and arguments concerning the origins of language are particularly vulnerable to that fallacy. The fallacy can be avoided, however, by the careful use of appropriate probabilistic methods concerning the ways in which we select among competing hypotheses in science.

4.4 Deacon on the Evolution of Language

Like Peirce, Deacon sees language evolution through ends and purposes. Like Peirce, Deacon sees a world that is meaningful and consequential. Now for Deacon "symbolic abilities do not necessarily represent more efficient communication" (SS: 379). He thus discounts "communicative fitness" as the ideal in language evolution. Instead, he remarks, these symbolic abilities "represent a radical *shift in communicative strategy*" (SS: 379; see also Deacon, 2003).

This is a very acute observation. However, Deacon's own story about the needs for the early humans to change their communicative strategies is anthropological. What started the evolution that catalysed the cycles of symbolic capacities of the brain had, according to Deacon, to do with social and communal relationships between the opposite sexes. Deacon explains co-evolution in terms of changes in those contexts and circumstances that embody *selection pressures* (social, ritual, institutional and reproductive pressures). These pressures either exhibit or inhibit the needed changes in the communicative strategies of early humans.

I would like to propose an alternative explanation. What "a radical *shift in communicative strategy*" means comes close to the Peircean idea of the habit-change potential. However, it is not *any* habit-change, but the self-controlled changes in the general ways of how to act. New signs acquire their meaning according to such actions and therefore new, symbolic communication emerges. And self-controlled habit changes are those that are conspicuously lacking in the organic habits of animals. Now the lack of self-control does not as such imply the complete lack of some anticipatory behaviour or planning, which some primates may well possess, though it is certainly contestable whether such behaviour requires genuinely counterfactual reasoning based on the idea of synchronic possibilities. The lack of self-control only means the lack of metasystematic or self-critical stance to one's actions. Consequently, communication with signs that are interpreted with the associated self-controlled habits of action would almost inevitably lead to reasoning which, in turn, would lead to the possibility of communicating with symbolic signs.

According to Peirce, the theory of self-controlled thought is logic. And if so, it is from logic, and logic alone, that we might hope to garner an answer to what caused the mind to exercise such self-controlled thought that gave rise to strategic thinking and action and, consequently, to the emergence of symbolic thought.

Let us also keep in mind that the entire machinery of evolutionary game theory has been developed to explain strategic behaviour which dispenses with the strong assumptions of rationality and common knowledge of rationality afflicting ordinary game theory. As soon as the logical underpinnings of evolutionary game theory become better understood, we may be equipped to answer Deacon's question concerning the origins of symbolic expressions of thought in an exact manner.

An alternative angle to the key question can be taken as follows. Peirce attributes changes not to pressures but to *love*. He borrows the Greek term *agape* from Plato and suggests an *agapastic* theory of evolution as the alternative to the prevailing Darwinian version. In a sense, it is a non-Darwinian theory, since it denies the "fortuitous variation" which in natural selection terminology means "the crowding out the weak" (CP 6.296). Instead, it boasts "evolution by the force of habit" (CP 6.300). The force of habit accepts, with certain notable qualifications, the "transmission of acquired characters", which Peirce takes to be "of the general nature of habit-taking" (CP 6.299). Habits coerce new forms and structures to take such "practical shapes" that are "compatible with the structures they affect" (CP 6.300).

At one juncture, Peirce associates "evolution by creative love" with Lamarckian evolution, but I wonder whether it may be close to Baldwinian evolution and what later became known as the Baldwin Effect (see e.g. Weber & Depew, 2003).[14] Peirce indeed notes that Lamarckian evolution is a misleading, nonsensical term. Habit-changes are generalising tendencies that enable individuals to modify the contexts in which they operate, and Peirce is not requiring these changes to be passed directly on to offspring but to be "energetically projaculated". By this, he means that new features first are established by habits and then these features are brought "into harmony with the general morphology and function of the animals and plants to which they belong" (CP 6.300). Learning and habit change are closely comparable notions.

Baldwinian selection would become Darwinian selection only if it is assumed that the general context and morphology in question is that of natural selection, which neither Peirce nor James Mark Baldwin to whom the idea is customarily credited of course do not assume. It is in fact plausible that Baldwin, who was a close and long-term colleague with Peirce, got the impetus to change his interpretation of his earlier ideas on the "organic selection" for his 1896 article "A New Factor in Evolution", in which what was later of termed the Baldwin Effect was suggested, from reading Peirce's 1893 *Monist* article *Evolutionary Love*.[15] According

---

[14] Similar ideas on how to accommodate learning and creative effects into Darwin's evolutionary theory were suggested those times by several others besides Baldwin. Conwy Lloyd Morgan (1896) applied a similar principle of "organic selection". They both regarded the Darwinian way of explaining the creative force of evolution in terms of natural selection as an unacceptably negative and mechanistic type of a force. A middle course that sought for a mode of evolution that would require neither natural selection nor inheritance of acquired characters was proposed by Henry Fairfield Osborn (1896) in his "coincidence selection" model. The term originates from John Thomas Gulick's work on ontogenetic selection processes (Hall, 2006) which Baldwin later acknowledges. Morgan's and Osborn's works built and commented on August Weismann's (1893) "germinal selection" model, which likewise calls for a new kind of positive and creative force as the one directing the adaptations. Much of these improvements and modifications on Darwin's theory of natural selection were at the same time rejoinders to Herbert Spencer's theory, which these men saw as an utterly mechanistic and Lamarckian stripe of evolution that no longer would fit in with the prevailing scientific conceptions. Peirce accompanies Osborn's public criticism of Spencerianism in an anonymous piece in the *New York Times* in 1890 entitled "'Outsider' Wants More Light" (W6: 402).

[15] More research is needed to establish the exact influences between Baldwin and Peirce which no one seems to have carried out so far; let us recall the multiple articles authored and co-authored, some of them with Baldwin himself, in Baldwin's *Dictionary of Philosophy and Psychology* by Peirce in 1901–1902; as well as a number of critical reviews of Baldwin's articles and books which Peirce published in the *Nation* in 1895–1908. Peirce did not review or comment much upon Baldwin's article "A New Factor in Evolution" but he wrote a commentary on his book *Mental Development in the Child and the Race* (1895). It is in that book what Baldwin in the

to Baldwin (1896, p. 451, emphasis omitted), "The ontogenetic adaptations are really new, not performed; and they are really reproduced in succeeding generations, although not physically inherited."

Deacon discusses Baldwinian evolution at length in the *Symbolic Species*, though not invoking Peirce in that context. He writes that "Baldwin proposed that by temporarily adjusting behaviours or psychological responses during its lifespan in response to novel conditions, an animal could produce irreversible changes in the adaptive context of future generations" (SS: 322–323). Deacon does not bring up Peirce's agapastic theory of evolution, but even so, my point is that "producing irreversible changes in contexts" is not very far at all from the continuous interaction of habits with their environment and the fact that such interaction changes the context in which the offspring lives on. If the epigenetic, inherited changes in phenotype or gene expression are really reproduced in future generations, then Peirce's agapastic evolution is capable of accommodating the Baldwinian notion of evolution.

It is of note here how the Baldwin Effect has recently spurred a mini-industry in the context of the evolution of language and communication (see e.g. Watanabe, Suzuki, & Arita, 2008). Equally interesting are the studies that attempt to explain the effect in such contexts by using the evolutionary theory of games (Zollman & Smead, 2010).

Let me remark in closing that a more neutral and fitting term to that of "love" to describe what is going on in Peirce's evolutionary agapasticism seems to me to be that of *cooperation*. Though I do not quite like John Dewey's overly naturalistic version that sees habits as acquired dispositions of a kind by which organisms and environments cooperate, cooperation is what David Lewis and many others in the philosophical tradition studying the origins of linguistic meaning and communication have argued to be the prime mover for the emergence of conventions. And ditto for Peircean symbols in simple situations of signal exchange, I might add. Moreover, cooperation need not rule out competition which, as has been well established by the many works in game theory can, and often is quite beneficial for it to, co-exist with cooperative action and behaviour.[16]

much better known article of Baldwin (1896: 451) called the "new factor" and the "influence of organic selection" first appeared, though they did not mean quite the same as in his follow-up work. As we can observe, all of Baldwin's writings on this matter nevertheless appeared right after the publication of Peirce's *Evolutionary Love*. On the other hand, these topics were discussed by several other authors around the same time as well, and some of the discussions took place even before 1893 (see the previous note).

[16] A brief justification for thinking of Agape as cooperation comes from the ancient history of ideas: In Homer's Odyssey some derivations of the word 'agape' mean that which creates contentment or affection with the speaker. We may take this meaning coming close to modern principles that aim at accounting for conversations as rational activities, such as Paul Grice's principle of cooperation in his theory of communication or Donald Davidson's principle of charity in his theory of interpretation.

4.5 Conclusions

To conclude, the conceptions of evolution advanced by Peirce, Baldwin, and Deacon provide rich sources of further study not only for a committed historian of intellectual ideas but also for evolutionary biologists, philosophers, developmental psychologists, cognitive scientists and anthropologists who wish to examine evolutionary phenomena from a wide enough perspective. Yet there is a great deal of continuity in these ideas, over and above some apparent gaps in the "great chain of meaning", inescapable as they are when history tells its tale. My aim has been to call attention to the interesting and manifold similarities between Peirce's and Deacon's views on the meaning and evolution of language, together with some noteworthy differences and some still-prevailing missing links. Deacon writes that "learned associations, arbitrarity, reference, and transmission of information from one individual to another – are not sufficient to define symbolic reference" (SS: 66). Peirce would agree wholeheartedly: for a symbol to emerge a three-way association between signs, objects and interpretants is required. But symbolic meaning depends on self-controlled habit-taking tendencies. Habits, both natural and conventional, are real and general strategic rules and regularities for action and learning which mediate the relationships between sign vehicles and their objects and interpretants. Those relationships occur in nature just as they occur in the human cortex. Deacon has likewise had much to say about those relationships in both realms – without invoking Peirce's or other 19th century pragmatists' and evolutionary biologists' concept of habits.[17]

Acknowledgments Supported by the University of Helsinki "Excellence in Research" Grant (2023031: *Peirce's Pragmatistic Philosophy and Its Applications*, 2006–2008, Principal Investigator A.-V. Pietarinen). I wish to thank the organisers and participants of the *First Symbolic Species Conference* in Copenhagen in 2006 as well as the reviewers of the earlier version of the present paper for comments.

References

Baldwin, M. J. (1895). *Mental development in the child and the race*. New York: Macmillan.
Baldwin, M. J. (1896). A new factor in evolution. *American Naturalist, 30*(441–451), 536–553.
Cowley, S. (2002). Why brains matter, An integrational perspective on "*The Symbolic Species*". *Language Sciences, 24*, 73–95.
Deacon, T. W. (1997). *The symbolic species: The co-evolution of language and the brain.* New York: W.W. Norton.

[17] The last point is equally true of those who have recently taken up the topic of Baldwinian evolution under revived scrutiny. One concept has disappeared from the discussion, the generalizing tendency to act according to habits, though we might see it as being reinstated through the backdoor of evolutionary theory of games in its use of stable strategies as the solution concept. For the connections between Peirce's theory and the stable strategies of EGT, see Pietarinen (2006c).

Deacon, T.W. (2003). Multilevel selection in a complex adaptive system: The problem of language origins. In B. W. Weber & D. J. Depew (Eds.), *Evolution and learning: The Baldwin Effect reconsidered* (pp. 81–106). Cambridge, MA: MIT Press.

de Villiers, T. (2007). Why Peirce matters: The symbol in Deacon's *symbolic species. Language Sciences, 29*, 88–108.

Hall, B. K. (2006). Evolutionist and missionary': The Reverend John Thomas Gulick (1832–1923). Part 1: Cumulative segregation-geographical isolation. *Journal of Experimental Zoology, 306B*, 407–418.

Hansen, P. G. (2007). Evolutionary games and social conventions. In A.-V. Pietarinen (Ed.), *Game theory and linguistic meaning* (Current Research in the Semantics/Pragmatics Interface 18, pp. 61–88). Oxford: Elsevier.

Hausman, C. (1993). *Charles Peirce's evolutionary metaphysics*. Cambridge: Cambridge University Press.

Keller, R. (1994). *On language change: the invisible hand in language*. London: Routledge.

Lewis, D. (1969). *Convention: A philosophical study*. Harvard: Harvard University Press.

Lumsden, D. (2002). Crossing the symbolic threshold: A critical review of Terrence Deacon's *The Symbolic Species. Philosophical Psychology, 15*, 155–171.

Morgan, C. L. (1896). *Habit and instinct*. London: Edward Arnold.

Murphy, J. J. (1869). *Habit and intelligence* (2nd ed. 1879), London: Macmillan.

Osborn, H. F. (1896). A mode of evolution requiring neither natural selection nor the inheritance of acquired characteristics. *Transactions of the New York Academy of Science, 15*, 141–148.

Peirce, C. S. (1931–1958). (CP) *Collected papers of Charles Sanders Peirce*, 8 volumes, vols. 1–6, eds. C Hartshorne & P. Weiss, vols. 7–8, ed. A. W. Burks. Cambridge, MA: Harvard University Press.

Peirce, C. S. (1967). (MS) *Manuscripts in the Houghton Library of Harvard University*, as identified by Richard Robin, *Annotated Catalogue of the Papers of Charles S. Peirce* (Amherst: University of Massachusetts Press, 1967), and in "The Peirce papers: A supplementary catalogue". *Transactions of the C. S. Peirce Society, 7*(1971), 37–57.

Peirce, C. S. (1998). (EP) *The essential Peirce: Selected philosophical writings*, Volume 2, The Peirce Edition Project. Bloomington, IN: Indiana University Press.

Peirce, C. S. (2000). (W) *The writings of Charles S. Peirce*. Volume 6, The Peirce Edition Project, Bloomington, IN: Indiana University Press.

Pietarinen, A.-V. (2005). Evolutionary game-theoretic semantics and its foundational status. In N. Gontier, J. P. Van Bendegem, & D. Aerts (Eds.), *Evolutionary epistemology, language and culture: A nonadaptationist systems-theoretical approach* (Theory and Decision Library B, pp. 429–452), Dordrecht: Springer.

Pietarinen, A.-V. (2006a). Peirce and the logic of image. *Semiotica*. In press.

Pietarinen, A.-V. (2006b). *Signs of logic: Peircean themes on the philosophy of language, games, and communication* (Synthese Library 329). Dordrecht: Springer.

Pietarinen, A.-V. (2006c). The evolution of semantics and language games for meaning. *Interaction Studies: Social Behaviour and Communication in Biological and Artificial Systems, 7*, 79–104.

Pietarinen, A.-V. (2007a). On historical pragmatics and Peircean pragmatism. *Linguistics and the Human Sciences, 2*(1), 123–143.

Pietarinen, A.–V. (Ed.). (2007b). *Game theory and linguistic meaning* (Current Research in the Semantics/Pragmatics Interface 18). Oxford: Elsevier.

Pietarinen, A.-V. (2007c). Abductive issues in Peirce's proof of pragmaticism. In O. Pombo & A. Gerner (Eds.), *Abduction and the process of scientific discovery* (pp. 303–320). Lisboa: Centro de Filosofia das Ciências da Universidade de Lisboa.

Pietarinen, A.-V. (2007d). "Getting closer to iconic logic". In G. Dodig-Crnkovic & S. Stuart (Eds.), *Computing, information and cognition: the nexus and the liminal* (pp. 53–74). Cambridge: Cambridge Scholars Press.

Pietarinen, A.-V. (2009a). Principles and practices of neurath's picture language. In O. Pombo, S. Rahman, & J. M. Torres (Eds.), *Essays in honour of Otto Neurath* (Logic, Epistemology, and the Unity of Science) (pp. 71–82). Dordrecht: Springer, in press.
Pietarinen, A.-V. (2009b). Murphy's law of mind. To appear.
Pietarinen, A.-V. (2010a). Question-answer structures in cognition. *Proceedings of the 10th World Congress of Semiotics*. La Coruña. To appear.
Pietarinen, A.-V. (2010b). An iconic logic of metaphors. *Journal of Cognitive Science*, To appear.
Pietarinen, A.-V. (2011). Existential graphs: What the diagrammatic logic of cognition might look like? *History and Philosophy of Logic, 32*, 265–281.
Pietarinen, A.-V., & Snellman, L. (2006). On Peirce's late proof of pragmaticism. In T. Aho & A.-V. Pietarinen (Eds.), *Truth and games* (pp. 275–288). Acta Philosophica Fennica, Helsinki: Societas Philosophica Fennica.
Stenius, E. (1968). Mood and language game. *Synthese, 17*, 254–274.
Stjernfelt, F. (2007). *Diagrammatology: An investigation on the borderlines of phenomenology, ontology, and semiotics* (Synthese Library 336). Dordrecht: Springer.
Watanabe, Y., Suzuki, R., & Arita, T. (2008). Language evolution and the Baldwin effect. *Artificial Life and Robotics, 12*, 65–69.
Weber, B. H., & Depew, D. J. (2003). *Evolution and learning: The Baldwin effect reconsidered*. Cambridge, MA: MIT Press.
Weismann, A. (1893). *The germ-plasm: A theory of heredity* (W. Newton Parker and H. R. Ronnfeldt, Trans.). London: Walter Scott and Scribner's.
Zollman, K. J. S., & Smeads, R. (2010). Plasticity and language: An example of the Baldwin effect? *Philosophical Studies, 147*, 7–21.

Chapter 5
Semiosis Beyond Signs. On Two or Three Missing Links on the Way to Human Beings

Göran Sonesson

Abstract Human beings are special in mastering, apart from signs, a number of semiotic resources embedded already in perception, which is not differentiated, but which may still be iconic, indexical, or symbolic. The sign is no doubt one of the missing links between human beings and other animals. An even earlier breaking point between (some) animals and human beings may be the ability to distinguish type and token, that is, to have access to a principle of relevance. Somewhere on the border between relevance and the sign is found the act of imitation. The Peircean sign, which is so much more (and less) than a sign, may be able to account for the emergence of imitation and its accomplishment in the sign function, in the restricted sense.

Contemporary studies of evolution suggest that not only human language, but also the capacity for using pictures, as well as many kinds of mimetic acts and indices, are (at least in their full, spontaneously developed form) uniquely human. It is clear that semiosis itself must be manifold and hierarchically structured, in ways not yet dreamt of in our philosophy. In order to grasp some of the discontinuities between human beings and other animals, it is useful to start out from the conception of phylogeny suggested by Merlin Donald (1991, 2001), which may be supposed to have a least some rough parallels in ontogeny.

In Donald's evolutionary scale, stages of episodic, mimetic, mythic and theoretic culture correspond to types of memory (Fig. 5.1). According to this conception, many mammals, which otherwise live in the immediate present, are already capable of episodic memory, which amounts to the representation of events in terms of their moment and place of occurrence. The first transition, which antedates language and remains intact in language impairment (and which Donald identifies with *Homo erectus* and wants to reserve for human beings alone) brings about mimetic memory, which corresponds to such abilities as tool use, miming, imitation, co-ordinated hunting, a complex social structure and simple rituals. Without even taking into account intricate phenomena such as social structure, ritual, and hunting, one cannot avoid observing the heterogeneity of this list: in some cases, such as most clearly tool use and some instances of imitation, no sign structure, with a clear distinction

G. Sonesson (✉)
Centre for Cognitive Semiotics, Lund University, Lund, Sweden
e-mail: goran.sonesson@semiotik.lu.se

T. Schilhab et al. (eds.), *The Symbolic Species Evolved*, Biosemiotics 6,
DOI 10.1007/978-94-007-2336-8_5, © Springer Science+Business Media B.V. 2012

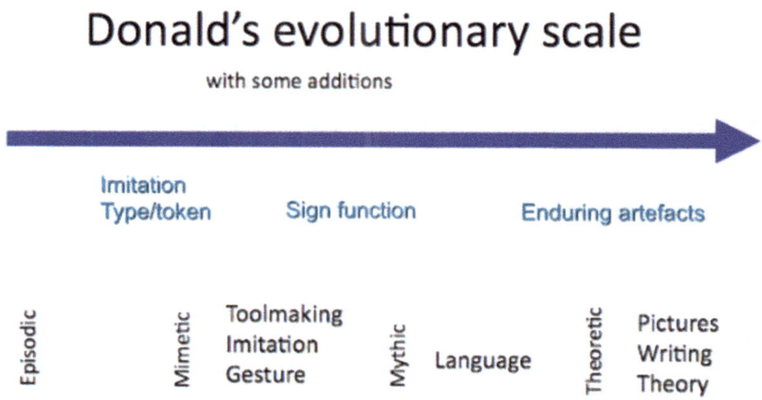

Fig. 5.1 Donald's model of evolution related to some further discontinuities: type/token, the sign, system character, and organism- independent artifacts

of expression and content, is required, but simply the conformity of tokens to a perceived or remembered type, but in other cases, exemplified by other instances of imitation, and by miming and other gestures, the sign function would seem an absolute prerequisite. If early mimesis may give rise to the organization of tokens into types, the sign would seem to emerge at the later mimetic stage.

Only the second transition brings about language (which, Donald muses, may at first have been gestural) with its *semantic* memory, that is, a repertory of units which may be combined. This kind of memory permits the creation of narratives, that is, mythologies, and thus a completely new way of representing reality. Although Donald is not very clear about it, his description of semantic memory could be taken to imply the presence of system character, that is, an organization in which signs mutually define each other. It is quite conceivable for language (but perhaps in an earlier gestural form) to be the first extant sign system.

Interestingly, Donald does not think development stops there, even though there are no further biological differences between human beings and other animals to take account of (however, the third transition obviously would not have been possible without the attainment of the three earlier stages). What Donald calls *theoretical* culture supposes the existence of external memory, that is, devices permitting the conservation and communication of knowledge independently of human beings. The first apparition of theoretical culture coincides with the invention of drawing. For the first time, knowledge may be stored externally to the organism. The bias having been shifted to visual perception, language is next transferred to writing. It is this possibility of conserving information externally to the organism that later gives rise to science. This, again, would seem to be a breaking point on the way to human beings: the possibility of memory as an external record, which perdures independently of the human organism.

Elsewhere, I have used Donald's conception of evolution, as rendered in the model above (Fig. 5.1), to discuss the curious fact that iconicity (and indexicality) are present already at the second stage, as mimetic gesture, but then makes an renewed appearance at the fourth stage, in the shape of pictures (Sonesson, 2006, 2007a, in press). I have also discussed, within the same framework, the final "missing link" in the progression from animal to man, the emergence of organism-independent artefacts (Sonesson, 2007a, 2007b, 2010a, 2010b, in press). In the following, however, I will be concerned with two other, (nearly) missing links, the (principle of) relevance and the sign, as well as the act of imitation bridging them.

5.1 A Sign Concept for Integral Semiotics

The most serious problem of semiotics is that both the Saussurean and the Peircean brands of received semiotic theory do not explain *what* a sign is; they simply take it for granted. It is not enough to say there are signifiers and signifieds, or representamen, object, and interpretants, without specifying the requirement for something to fall under one of these categories. A useful concept of *sign* designates a kind of meaning, but it does not cover all meanings. Perception is clearly meaningful to animals and infants alike, but it seems reasonable to suppose that the capacity for *sign use* is a much more exclusive property. Conceptualizing the capacity of sign use in this way may help us to distinguish stages in evolution and development, notably the relationship between imitation and sign.

We will say that the sign is a meaning which is made up of two parts, traditionally known as *expression* and *content*. That the sign consists of two parts implies that the parts are separated. In Piaget's (1945, 1967, 1970) terms, they are "differentiated from the point of view of the subject". This it not to say that the differentiation is "subjective", in the ordinary language sense – in most cases, the differentiation is part of what is learnt by the child growing into his particular culture. However, what is differentiated within the sign may or may not consist of several objects in the "objective" common sense world (where "objective" is that which is taken for granted in the dealings of ordinary life). Contrary to what Piaget suggests, we will therefore conclude that a thing which is immediately continuous to another or which is a part of another in the common sense world may very well be differentiated within the sign (cf. Sonesson, 1989, 1992a, 1992b, 2010b, in press). We can imagine the same child that in Piaget's example uses a pebble to stand for a piece of candy having recourse instead to a feather in order to represent a bird, or employ a pebble to stand for a rock, without therefore confusing the part and the whole: then the child would be employing a feature, which is *objectively* a part of the bird, or the rock, while differentiating the former from the latter *from his point of view*. Only then would he be using a true sign. In terms of socially better-established signs, a similar example would be the bull's head used to indicate, above a market stand, that beef is sold there. Although in France, for example, cast heads of bulls or horses are employed outside the relevant shops, it is still possible to find real heads used in traditional markets in some countries. In a parallel fashion, things that are similar to each other can be differentiated within the sign. Thus, there can be indexical (contiguity-based) and iconic (similarity-based), as well as symbolic (rule-based) signs. If I see a branch sticking up over the house and conclude that there is a tree behind the house, this is a mere indexicality; but the marks on the ground left by the animal are indexical signs, clearly separated from the (part of) the animals having produced them. And the photographic print of a person I know is clearly differentiated from the person seen in the picture.

Indeed, a further differentiation may have to be made for certain purposes. The marks on the ground tell me "an elk was here before", and this is something distinct from the marks, as well from the elk, which is now somewhere else. Similarly, the colour configuration on the photograph is distinct from the perceptual impression of my wife, and the photograph is here with me now, while my wife is at her working-place. This is why we really have to separate three parts of the sign, *expression*, *content*, and *referent*, where content is the standpoint taken on the referent by the sign user, as codified in some semiotic resource.[1] To the hunter, it is important to identify the marks on the ground (expression) as being those of an elk (indexical content), but, being a hunter, he cannot be satisfied with this; he will follow the traces left by the animal until he finds the real elk (referent). Looking at the photography, I have no trouble (unlike small children and animals) to distinguish the colour spots on the paper (the expression) from the vicarious perception it suggests, e.g. of my wife fifteen years ago dancing Jalisco in a ample, pink skirt (content), nor from the real person I have known for twenty-six years and with whom I share so many memories (the referent, the real, continuous person in my personal Lifeworld).

But differentiation is not a sufficient criterion. Each time we actively and consciously put together a set of items that we have perceived, we must first differentiate the items to be joined – as opposed to the obliteration of their difference in categorical perception. But categorization is not as such a kind of sign use. Contiguity and factoriality are present everywhere in the perceptual world without as yet forming signs: we will say, in that case, that they are mere *indexicalities*. An index, then, must be understood as indexicality (an indexical relation or ground) plus the sign function. Analogously, the perception of similarities (which is an iconic ground) will give rise to an icon only when it is combined with the sign function. As always, there are passages in Peirce's work, which may be taken in different ways, but it makes more systematic and evolutionary sense to look upon iconicity and indexicality as being only potentials for something being a sign.[2] Iconicity, indexicality, and symbolicity only describe *that which connects two objects*; they do not tell us whether the result is a sign or not (Fig. 5.2). These considerations allow us to separate the study of the phylogenetic and ontogenetic emergence of iconicity, indexicality and symbolicity from that of the corresponding signs (cf. Sonesson, 1998, 2001, in press).

[1] This is of course not the Peircean triad, but rather corresponds to the representamen, and to the immediate and dynamical objects, respectively (as well as to the corresponding interpretants).

[2] In relation to the standpoint of many other semioticians, I have to spell out here (as in many of my earlier publications, but perhaps most explictly in Sonesson, 2009), that I am not interested in finding out what Perice "really said". To give an all to simple expression to a complicated issue, I will just say that I use Peirce as a source of inspiration, just as I do with many other writers on the theme.

	Firstness	Secondness	Thirdness
Principle (Firstness)	Iconicity	—	—
Ground (Secondness)	Iconic ground	Indexicality = indexical ground	—
Sign (Thirdness)	Iconic sign (icon)	Indexical sign (index)	Symbolicity = symbolic ground = symbolic sign (symbol)

Fig. 5.2 The relationship between principles, grounds, and signs, from a point of view inspired by Peirce as spelled-out in Sonesson (1996, 2007a, 2007b)

The sign as such is thus a whole made up of two parts, expression and content, and there is a *double asymmetric relationship* between them. First, from the point of view of immediacy, expression is more accessible to consciousness than content. In the second place, content is more in focus (more prominent, more important) than expression. When I look at the photograph, I am normally interested in the person depicted (my wife, either at the exact moment she was dancing Jalisco, or as an enduring person of my personal Lifeworld). My wife does not represent the photograph.[3] The phenomenologist Edmund Husserl (1939) formulated the definition of the sign (more precisely, "appresentation") more or less in these terms, but a similar view is implicit already in Augustine's conception of the sign (in our terms, the expression) as something which, by becoming conscious, makes us aware of something else (the content; cf. Deely, 2001).[4]

However, Bates (1979, p. 43) has hinted at the idea that the sign (our expression) and its referent (which would seem to correspond to both what I have called content and referent) must be conceived as being both similar and separate for a sign relationship to obtain. Bates' somewhat convoluted definition is later unpacked by Daddesio (1995, p. 117):

> Given a physical mark (sound, movement, shape, etc.), *a*, and a particular class of things, *b*, that *a* is thought to stand for, let us consider three possible ways which an organism can relate *a* and *b*. In the first instance, the organism fails to grasp any relation whatsoever between the two. /---/ In the first case, semiosis is thus absent. In the second case, the organism would be capable of relating the two, but instead of apprehending a relation between two distinct entities, it would simply react in the same fashion if presented *a* and if presented *b*. /---/ In the third case, the organism would recognize *a* and *b* as distinct but related.

[3] Seeing her now, I may of course be reminded of when I took that photograph, or when she made that dance, but this does not change the asymmetric structure of the sign, only my mental use of it.

[4] This does not preclude other relations between expression and content being symmetric. It is common to suppose a substitutive relationship, which is a symmetric relation, between expression and content, but this may be misleading, since expressions are rarely used for the same purpose and in the same context as their contents.

Nevertheless, it is in fact impossible to conclude from an individual treating *a* and *b* as being distinct, that the particular relationship between *a* and *b* is necessarily one of appresentation (sign function). Daddesio's second case is that of categorization, which is important to perception. Given a prototype conception of categories, *a* and *b* may be treated as different just because they are differently central to the category of which they are perceived to be a part. Or they may be attended to differently, merely because one contains more, and more interesting, perceptual properties than the other (and, indeed, sign vehicles would tend to be "degraded stimuli", when compared to what they are signs of). The problem of separating the expression and the content of a sign becomes particularly acute in the case of an iconical sign, in which, by definition, expression and content must share at least some properties (Cf. Sonesson & Zlatev, in press).

The sign, then, consists of two intrinsic parts, expression and content, which are related to a third, the referent. The relation between these parts may be iconic, indexical, or symbolic, but it always supposes a differentiation of the parts, from the point of view of the sign user. The sign relation is asymmetric in a double sense: what we call expression is always more directly perceived than the content, and the content is more accessible than the referent. On the other hand, it is the referent and/or the content that is in focus, at least more so than the expression.

5.2 Imitation as Token and as Sign

The characterization of the sign above is partly inspired in Piaget's notion of the "symbolic" (later the "semiotic") function, which is a capacity acquired by the child at an age of around eighteen to twenty-four months, which enables him or her to imitate something or somebody outside the direct presence of the model, to use language, make drawings, play "symbolically", and have access to mental imagery and memory. The common factor underlying all these phenomena, according to Piaget, is the ability to represent reality by means of a signifier, which is distinct from the signified. The sign function thus characterizes a stage of child development, though Piaget himself chooses to describe this stage only negatively, that is, a being pre-operational. Imitation, or, more exactly, "representative imitation", is claimed by Piaget (1945) to be at the origin of the semiotic function. When more closely scrutinized, some instances of imitation actually turn out to be signs already, while others clearly are not.

Donald places imitation within the second stage of human development, mimesis. In his view, mimetic culture starts out with the emergence of "conscious, self-initiated, representational acts, which are intentional [i.e. voluntary] but not linguistic" (1991, p. 168). The examples given by Donald are such things as gesture, dance, ritual, mime, play-acting, and (precise) imitation, but also tool use (or perhaps rather the social generalization of tool use) and skill. Somewhere in between mimesis and language the semiotic function arises, though Donald addresses this only obliquely, mentioning the use of intentional systems of communication and the distinction of the referent. In fact, this certainly happens between animal camouflage and pictures. According to Deacon (1997, pp. 74ff), however, iconicity as found in "a portrait" is "not basically different" from the fact of there being no distinction at all, that is, it would seem, from mere identity. On the following pages, Deacon then goes on to maintain that a number of phenomena which could otherwise appear to be completely different are in fact equivalent: the perception of the same "stuff" over and over again (seeing something that does not change into something else), camouflage as exemplified by the case of the moth's wings being seen by the bird as "just more tree", "stimulus generalisation", and even recognition, that is, the identification of something as pertaining to the same category. Although all or most abilities subsumed under the mimetic stage depend on iconic relations, only some of them are signs, because they do not all involve some asymmetric relation between an expression and the content for which it stands.

In fact, in his early book, Donald (1991, pp. 168f) opposes mimesis to mimicry and imitation, both of which are said to be quite common in animals but lacking "a representational dimension". Though the import of this claim is not clear, it could be taken to mean that mimicry and imitation, in this sense, lack differentiation. In Donald's (2001, pp. 260f) later book, however, "(precise) imitation" is an instance of mimesis. This would no doubt exclude the kind of automatic imitation in the infant ("neonatal mirroring"), discovered by Meltzoff, such as sticking out the tongue to one who does just that (Cf. Gallagher, 2005; also see Donald, 2001, pp. 264ff). It is less clear whether Donald would follow Tomasello (1999) in making a distinction between the imitation of goals (called "emulation"), of which he believes apes to be capable, and the imitation of means, which is a capacity Tomasello would like to restrict to human beings, although he later on (in Tomasello, 2008) recognizes its presence in at least some apes.[5] At first it may seem strange that imitating the goal is presented as being easier than imitating the means by which the goal is achieved. But no doubt it is less demanding to recognize the interest of the aim (getting the banana) than the interest of the requisite steps for realising the goal. At another level, it is like attending to the content, not the expression, of a sign. Indeed, it is an instance of quite ordinary Lifeworld behaviour.

One may wonder why tool use and skill are thought to be part of mimetic culture and not just "routine locomotor acts" or "procedural memory" which Donald (1991, p. 168) elsewhere takes pains to separate from mimesis. No doubt Donald (1991, pp. 171ff) would answer that they are different because they comply with his criteria for mimetic acts: they are "intentional" (that is, voluntary), "generative" (that is, analysable into components which may be recombined into new wholes), and "communicative" (or at least, as we shall see "public"). Moreover, they have reference ("in mimesis the referential act must be distinguished from its referent", that is, in our terms, there must be differentiation), stand for an unlimited number of objects, and are auto-cued (produced without an external stimulus). Generativity

[5] A study of imitation of actions from static pictures, reported in Hribar, Call, and Sonesson (in press) would certainly seem to suggest that apes may be capable of imitating means as well as goals, at least in one sense of these terms. In his most recent book, however, Tomasello (2008) seems to downplay even more the capacity for imitation in apes.

is a property of many kinds of meaning, which are not signs. However, it is not clear in what sense tool use and many other kinds of skill are "communicative", and therefore, in which way they have reference and stand for an unlimited number of objects.

After introducing "communicativity" as a criterion of mimesis, Donald (1991, p. 172) goes on to say that "although mimesis may not have originated as a means of communication, and might have originated in a different means of reproductive memory, such as tool-making, mimetic acts are, by their nature, usually public and inherently possess the potential to communicate." This, though, is very different from imitation as a sign, which is what is realised by the actor, who presents his acts to a specific public; it is even different from the child's symbolic play, which must be available to and shared with other children. What we have here is, first, the extraction of a token from a type, which supposes treating the other as a spectacle; and second, the realisation of the tool act, which is not public-directed, but can be made available to the public (Fig. 5.3). The use of the tool does require the separation of the typical properties from the single act occurring in the here and now, i.e. relevance. In order to learn the use of a tool, you must at least be able to isolate the properties that should be imitated from those which are of no avail. However, even though this act of imitation may be observed, it is not part of its purpose to be observed. When the actor who has the part of Hamlet lifts up the skull of "Poor Yorick", then his act does not only consist in imitating what a man having that name supposedly did in Renaissance Denmark, but also in presenting this act as something to be seen, as a spectacular act (cf. Sonesson, 2000a). The symbolic play of children may perhaps be considered to be some kind of intermediary case, because its spectacular character is not its ultimate goal, but is only instrumental in making the play function as play; indeed, it is not intentionally offered as a spectacle for individuals not participating in the play.

	Imitation (Token/Type)	**Imitation as Learning (Extracting Type from Token)**	**Symbolic play (Expression/ Content)**	**Play-acting (Expression/ Content)**
	Instantiates a type of act	*Extracts a type from one or several (novel) token acts*	*Represents a type of (habitual) act - or perhaps token outside of time and space*	*Represents an individual act in time and space*
vehicle	Using the typical means for realising the type hammering the nail	Observing the hammering (first token) extracting the type for doing hammering (second token)	Realising the typical acts of the mother part	Creating the appearance here and now of being Hamlet doing Hamlet things
tenor	Doing the type of act having as goal to hammer a nail	Extracting the type of hammering a nail	Doing what mothers usually do to their babies	Doing as Hamlet did in Helsinoer during the Renaissance

Fig. 5.3 From imitation as token for a type to imitation a sign

Thus, tool use and other kinds of skill as such are not mimesis, because they are not communicative, but they are "public", and they lend themselves to imitation – which leads to generalization of tool use and skill in society. This is where they become different from routine acts and procedural memory. They are socially shared. But this is only possible if the act can be separated from the unique tool user and transferred to another user. That is, the act as token must be abstracted to a type in order to be realised in another token. What is shared is the type, in other words the scheme of interpretation, which defines the principle of relevance (in the sense of a rule that picks out the properties of one object being mapped onto another). In this sense (not in the sense of reference), a single mimetic act may correspond to various events.

It is therefore by means of imitation that the "extension of conscious control into the domain of action" (Donald, 2001, p. 261) may be obtained. But the act of imitation, in this instance, is in no way a sign. If I see somebody use a stone as a tool to crack open the shell of a nut, I may do the same thing, not to bring into mind the act of the other person I have observed, but to obtain the same effect. I attempt to realise the same act as he did, that is, to open the shell up, so that I can take out the nut and eat it. Instead of producing an expression that is non-thematic but directly given which refers to a content that is thematic but indirectly given, I am realising a new instance of the category of acts consisting in cracking open a nutshell. Like Tomasello's apes, I may of course try to obtain the same effect without attending to the adequate means, which would produce a failed act of imitation. Or, I may merely simulate the outer actions of cracking the shell open, without letting them have a sufficient impact on the physical environment, in which case I may either be engaged in symbolic play, play-acting, or simply practicing the movements.

Imitation, in this sense, may thus be said to be differentiated, in the sense of separating the mediator and that which is mediated, but it is not asymmetric, neither in the sense of focus, nor in that of directness. Indeed, it is really the type that is mediated by the token. This also means that the purpose of the act of imitation is not to present the original act to another subject (or even to oneself). Bentele (1984) in fact argued against Piaget that imitation does not manifest the semiotic function, but is a prerequisite for it: indeed, it will function as a sign only to the extent that it is taken to refer back to the imitated act, instead of just being another instance of the same kind.

Acts of imitation in this sense have two interesting properties: they are "public", in the very broad sense characterized by Donald, i.e. they may be perceptually, often visually, inspected; and they can be copied by means of the observer's own body, with or without some additional implement such as a stone. In both these ways, imitation is different from episodic memory; and it is different from procedural memory in being a public record. Like in procedural memory, the record is located in one's own body, but it can only function as memory to the extent that it is somehow separable from the body as such. In fact, this can only be so, to the extent that memory traces are instantiated in other bodies as well as in one's own body. This supposes a distinction between token and type (that is, relevance) preceding that of the semiotic function.

5.3 The Peircean Sign or the Observer Observed

The Peircean sign is a sign only in a very Pickwickean sense of the term. It is one of three specifications of Firstness, Secondness, and Thirdness. It might be said to be concerned with interpretation in a more generic sense than the sign: "semiosis". Perhaps this is what Peirce was thinking about when, at a later stage, he complained that his notions were too narrow, and that, instead of referring to signs, he should be talking about mediation or "branching" (CP 4.3. and MS 339 quoted in Mertz & Parmentier, 1985).

Conceived in this way, Peirce's theory appears to be about the situation of communication, but much closer to what we now would describe as a hermeneutical model than to the model known from the theory of information. In this sense, "a sign [or rather semiosis] is whatever there may be whose intent is to mediate between an utterer of it and interpreter of it, both being repositories of thought, or quasi-minds, by conveying a meaning from the former to the latter" (MS 318, quoted by Jappy, 2000). In many passages, the object is not described as that which the sign is about, that is, to which it refers, in the sense in which this term is used in linguistic philosophy: instead, it is that which incites somebody to produce a sign (which may or may not coincide with the referent). It is in this sense that the object is Secondness: it concerns the relation between the reality perceived and the expression produced. Similarly, the interpretant must be seen as the result of the receiver taking in the whole event of the utterer creating an expression starting out from some feature of his experience. Because it refers to the relation between the utterer and that which he reacts to, it is not only an elementary relation, it is Thirdness. Indeed, this idea is very well illustrated by the notion of "branching", which Peirce used to characterise his later concept of mediation.

Even describing that which Peirce is concerned about as an act of communication may amount to being too specific. Instead, it could be characterized as an observation being observed. Summarizing all of Peirce's different attempts at pinning down the nature of Firstness, we could probably say that it is something that appears or may appear (without connection to anything else). It is thus prior to all relationship. Secondness is not only the second term that comes into play, but also it is made up of two parts, one of which is a property, and the other a relation. It is something the function of which it is to hook up with something already given as a possibility. In this sense, it is a reaction, in the most general sense, to Firstness, where the first part is the connection to the property independently appearing and the second part describes the nature of this relationship. Thirdness is not only the third term which is ushered in, but it consists of three parts, two of which are relational: one which is hooked up to the term of Firstness and another which is connected to the relation of Secondness, together with which we find a third term describing the relationship between these two terms. It is thus an observation of the reaction. Appearance is monadic, reaction is dyadic, and observation is triadic (Cf. Fig. 5.4).

In social psychology, in particular developmental psychology, there is also much talk about dyads and triads, and about some things being dyadic and other triadic (cf. Tomasello, 1999). Thus, interactions, engagements, eye gaze, and so on, are said

Fig. 5.4 The basic meaning of the Peircean triad

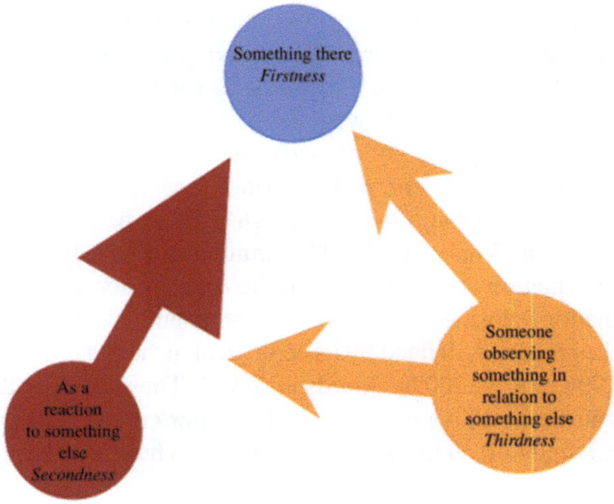

to by either dyadic or triadic. This terminology would seem to have originated in the sociology of Georg Simmel (Cf. Simmel, 1971). Dyads and triads are to Simmel groups of two or three individuals, respectively. Units, not relationships are counted. Between two individuals there may be any number of relationships, just as there may be between three individuals. When, in contemporary articles, we read about a "mother-child dyad", etc., this is clearly what is meant. In general, translated into the terminology of Sonesson (2000b), a dyadic situation seems to be taken to consist of Ego and Alter (another person) or Ego and Alius (a thing or a person treated as a thing), whereas a triad includes all three types. Even more specifically, the triad tends to involve child, caretaker and a referent.

Other uses are more explicitly relational: dyadic is opposed to triadic as the relation of a subject to an object, or another subject is opposed to the relation of a subject both to another subject and another object. Thus, on one hand, there is "dyadic eye gaze: looking at object or person", and on the other hand there is "triadic eye gaze: looking back and forth between object and person" (Cf. Bates, 1979). A more complex interpretation would suppose that a dyadic relation is a relation between two individuals, while a triadic relation is a relation to the relation between two individuals. This is similar to what Peirce seems to mean, according to the interpretation given above. It should be noted that such a relation to the relation between Alter and Alius is not the same thing as two relations, to Alter on the one hand, and to Alius on the other. However, in practice, the only way to know that somebody is attending to the relationship between two individuals may be to observe him or her looking first at one individual and than at the other. Perhaps we would even need to go further, introducing relations between relations as well as relation between such relations.

Clearly social psychology, in spite (or because) of being a much more practical concern that Peircean philosophy, is as unclear about what is dyadic and triadic as Peirce. Basically, however, it seems that what is involved in dyadic relations, in both cases, is a subject taking cognizance of the world, and in the triadic relations, somebody being aware of what the first subject is doing.[6] Typically, in social psychology, this is the caretaker observing the child's perceptual interchange with the world. In other words, it involves Ego and Alter interacting with reference to Alius.

Understood in this way, Peircean semiosis (which we should no longer restrict to being a sign) is not properly speaking "communicative", in Donald's sense, but certainly "public" or, perhaps better "spectacular". It is available to others. Yet, for it to be available, it is not enough for it to be present, but it must be accessible to attention. What is needed is a community (not only a single caretaker) for which this information is available – and the capacity for attending, without which the information is lost, as it is on so many other animals than man, as soon as it goes beyond the properties defined by its ecological niche (Cf. Gurwitsch, 1957; Sonesson, 1989, 1996, 2007a, 2007b; Arvidson, 2006). Thus the capacity for attending freely to the outside world – going beyond the *Umwelt* to the *Lebenswelt* –, may well be the first missing link on the way from animals to human beings.

References

Arvidson, S. (2006). *The sphere of attention: Context and margin*. London: Kluwer.

Bates, E. (1979). *The emergence of symbols*. New York: Academic.

Bentele, G. (1984). *Zeichen und Entwicklung. Vorüberlegungen zu einer genetischen Semiotik*. Tübingen: Narr.

Daddesio, T. C. (1995). *Of minds and symbols*. Berlin and New York: Mouton de Gruyter.

Deacon, T. (1997). *The symbolic species*. New York: Norton.

Deely, J. (2001). *Four ages of understanding*. Toronto: University of Toronto Press.

Donald, M. (1991). *Origins of the modern mind*. Cambridge, MA: Harvard University Press.

Donald, M. (2001). *A mind so rare*. New York: Norton.

Gallagher, S. (2005). *How the body shapes the mind*. Oxford: Clarendon Press.

Gurwitsch, A. (1957). *Théorie du champ de la conscience*. Bruges: Desclée de Brouver.

Hribar, A., Call, J., & Sonesson, G. (in press). From sign to action. Studies in chimpanzee pictorial competence. *Semiotica*.

Husserl, E. (1939). *Erfahrung und Urteil*. Prag: Academia Verlagsbuchhandlung.

Jappy, T. (2000). Iconicity, hypoiconicity. In J. Quiroz & R. Gudwin (Eds.), *The digital encyclopaedia of Charles S. Peirce*. Retrieved September 2, 2011, from http://www.digitalpeirce.fee.unicamp.br/jappy/hypjap.htm

Mertz, E., & Parmentier, R. J. (Eds.). (1985). *Semiotic mediation: Sociocultural and psychological perspectives*. Orlando, FL: Academic.

Piaget, J. (1945). *La formation du symbole chez l'enfant*. Neuchatel: Delachaux & Niestlé.

Piaget, J. (1967). *La psychologie de l'intelligence*. Paris: Armand Colin.

Piaget, J. (1970). *Epistémologie des sciences de l'homme*. Paris: Gallimard.

Peirce, C. (1931–58). *Collected Papers I–VIII*. C. Hartshorn, P. Weiss, & A. Burks (Eds.). Cambridge, MA: Harvard University Press (Quoted in the text as CP).

Simmel, G. (1971). In D. Levine (Ed.), *On individuality and social forms: Selected writings*. Chicago: University of Chicago Press.

Sonesson, G. (1989). *Pictorial concepts*. Lund: Aris/Lund University Press.

Sonesson, G. (1992a). *Bildbetydelser*. Lund: Studentlitteratur.

[6] Or something: The mind is not necessarily a subject to Peirce, but he does admit that there is no way of explaining it, at least at present, than by reference to a subject.

Sonesson, G. (1992b). The semiotic function and the genesis of pictorial meaning. In E. Tarasti (Ed.), *Center/periphery in representations and institutions. Imatra, Finland* (July 16–21, 1990, pp. 211–156). Imatra: Acta Semiotica Fennica.

Sonesson, G. (1996). An essay concerning images. From rhetoric to semiotics by way of ecological physics. *Semiotica, 109*(1/2), 41–140.

Sonesson, G. (1998). Icon — Iconicity — Index — Indexicality, entries. In P. Bouissac in collaboration with G. Sonesson, P. Thibault, & T. Threadgold (Eds.), *Encyclopedia of semiotics* (pp. 293–297, 206–311). New York: Oxford University Press.

Sonesson, G. (2000a). Action becomes Art. "Performance" in the Context of Theatre, Play, Ritual – and life. *VISIO, 5*(2), 105–122.

Sonesson, G. (2000b). Ego meets Alter: The meaning of otherness in cultural semiotics. *Semiotica, 128–3/4*, 537–559.

Sonesson, G. (2001). From semiosis to ecology. *VISIO, 6*(2–3), 85–110.

Sonesson, G. (2006). The meaning of meaning in biology and cognitive science. A semiotic reconstruction. Semiotiké. *Trudy po znakovym sistemam/Sign system studies, 34*, 135–213.

Sonesson, G. (2007a). From the meaning of embodiment to the embodiment of meaning. In T. Ziemke, J. Zlatev, & R. Frank (Eds.), *Body, language, and mind* (pp. 85–28). Berlin and New York: Mouton de Gruyter.

Sonesson. G. (2007b). The extensions of man revisited. From primary to tertiary embodiment. In J. Krois, M. Rosengren, A. Steidle & D. Westerkamp (Eds.), *Embodiment in cognition and culture* (pp. 27–56). Amsterdam and Philadelphia, PA: Benjamins.

Sonesson, G. (2009). The view from Husserl's Lectern: Considerations on the role of phenomenology in cognitive semiotics. *Cybernetics and Human Knowing, 16*(3–4), 107–148.

Sonesson, G. (2010a). Here comes the semiotic species: Reflections on the semiotic turn in the cognitive sciences. In B. Wagoner (Ed.), *Symbolic transformations* (pp. 38–58). London: Routledge.

Sonesson, G. (2010b). Semiosis and the elusive final interpretant of understanding. *Semiotica, 178–1/2*, 511–624.

Sonesson, G. (in press). *From iconicity to pictorality. Iconicity revisited/L'iconicité révisité*. Paris: L'Harmattan.

Sonesson, G., & Zlatev, J. (in press). Overall theoretical summary of the SEDSU project. In C. Sinha, G. Sonesson, & J. Zlatev (Eds.), *Signing up to be human.*

Tomasello, M. (1999). *The cultural origins of human cognition.* Cambridge, MA: Harvard University Press.

Tomasello, M. (2008). *Origins of human communication.* Cambridge, MA: MIT Press.

Part II
The Prehistoric and Comparative Connection

Chapter 6
The Natural History of Intentionality. A Biosemiotic Approach

Jesper Hoffmeyer

Abstract Our lives cannot but implant the knowledge in our souls that the mind is one thing and the world is another. Out of this separation arises the problem of intentionality, that our minds necesarily occupy themselves with things in the world, or that mind processes are always "about" something. In the scholastic tradition from Thomas Aquinas this "aboutness" is still seen as an immaterial or *intentional direct union* between the knower and the known. To know about things, e.g. a storm or a flower, implies that these things exist in the mind of the knower as *intentional beings,* and the nature of this kind of being is that of a relation or interface. This understanding is radically different from the cognitive theories that came to dominate in the course of the scientific revolution. According to Descartes the exterior world is grasped through the mechanical work of the senses, which then required some intermediate entity, a concept or an idea, to stand between the outside world (reality) and the mind. Henceforward the mind lost its direct access to the world, and logically enough this line of thought ended up in the conception that we can never understand the world as it is in itself. The idea of *intentional being* was taken up once again by Franz Brentano in 1874, who claimed that "Mental phenomena . . . are those phenomena which contain an object intentionally within themselves". To Brentano – and the phenomenological tradition he thus initiated – *mind should be seen as real, irreducibly intentional, and inexplicable naturalistically*. Philosphers of the analytic tradition rejected this whole notion claiming that *whatever is real is nonintentional and explicable naturalistically.* Unnoticed by most thinkers a third position was suggested by Charles Peirce, who agreed with Brentano that *mind is real and irreducibly intentional* but in the same time maintained, contra Brentano, that *mind is explicable naturalistically.* This chapter takes the semiotic realism of Charles Peirce as a starting point and discusses a biosemiotic approach to the problem of intentionality. Intentionality is seen as implicit to semiosis (sign processes) and semiosis and life is seen as co-existant. The needs of all living beings for expressing a degree of anticipatory capacity is seen as an evolutionary lever for the development of species with increased semiotic freedom. Human intentionlity is not therefore unique in the world but must be understood as a peculiar and highly

J. Hoffmeyer (✉)
Biological Institute, University of Copenhagen, Copenhagen DK 2200, Denmark
e-mail: jhoffmeyer@me.com

T. Schilhab et al. (eds.), *The Symbolic Species Evolved*, Biosemiotics 6,
DOI 10.1007/978-94-007-2336-8_6, © Springer Science+Business Media B.V. 2012

sophisticated instantiation of a general semiotics of nature. Biosemiotics offers a way to explicate intentionality naturalistically.

6.1 Anthropomorphism

In the 1890s the Russian physiologist Ivan Pavlov studied gastric function in dogs when he made the striking observation that his experimental animals began to salivate whenever an assistant entered the doors. Pavlov realized that the salivary response was not due to an automatic physiological process, and started his now famous series of experiments to study this "psychic excretion" as he called it. Pavlov's experiments probably more than any other observations served to corroborate the firm belief held by the scientific society that living beings were in fact "mere machines". Nearly a century of behaviorist psychology has helped cementing this conception.

The machine metaphor for organismic life dates back of course to René Descartes' mechanical biology from the 17. century. Descartes' own image of the organism was the clockwork, but as technology developed other machines took its place. In the 19th century the steam engine became the preferred metaphor, and in modern times the computer is the unchallenged candidate. There is indeed a striking likeness between machines and living creatures in that both exhibit goal-directed activity. But as Terrence Deacon has observed, "whereas machines ... exhibit derived functionality and intention by virtue of a kind of teleological parasitism on human teleology, living functions of the body and mind are intrinsically teleological" (Deacon, 2007, 2008). Thus the functionality of organisms may well be said to depend on delicate machinery, but this machinery has not been created by human minds and it therefore requires explanation of another sort than the explanation needed for the existence of real machines. In fact, the very popularity among scientists of the machine metaphor paradoxically discloses a strong scientific intuition, that organismic life does indeed exhibit intentionality, even though the machine metaphor was obviously meant to do away with exactly this presumed antiscientific idea.

The opposition among scientists and (most) philosophers to the idea of intentionality in animals – or worse plants, fungi, and bacteria – is probably due to the general taboo against anthropomorphisms in science. The automatic rejection by modern science of all theories carrying even the faintest trace of anthropomorphism (a rejection reminding one of the *horror vacui* of an earlier epoch) is however itself deserving of critical study (see Favareau, 2007). As Karl Popper once remarked, if we are talking about the *nose* of a dog, we are also anthropomorphizing the dog, but we are doing so for good reasons, because the nose of the dog and the nose of the human individual are homologous organs, i.e., their structural and functional similarities are accounted for by the well-established fact of common ancestry. Likewise, claimed Popper, we are well-justified in speaking about *knowledge* in animals to the extent that homology implies that animal brains and human brains are evolutionarily related organs performing related functions (Popper, 1990, p. 30). In fact, any claim to the effect that human beings are the only animals to possess intentionality would require additional theories to explain why other mammals should be so fundamentally different from us. No satisfactory theories pertaining to such an effect is known to this author.[1]

Needless to say, the widespread belief – not least among pet owners – in the presence of sophisticated human-like psychological intelligence in animals (and sometimes even in plants) hardly satisfies the slightest critical scrutiny, and anthropomorphism clearly poses a challenge that must be confronted and dealt with in any concrete case. But so does the opposite danger, the danger of anthropocentrism: "the *reading humanness out of nature*", as the American philosopher Maxine Sheets-Johnstone has put it: "By such an act, nonhuman creaturely life is interpreted in ways that consistently exalt the measure of humans: humans become special creations" (Sheets-Johnstone, 2009, p. 125). The religious overtones are hardly accidental in this derogative characterization: considering the rationalist commitments underlying the taboo against anthropomorphism it is indeed remarkable that the godlike status hereby implicitly ascribed to the human being goes unnoticed. Like every other species in the world the human species is a product of evolution, and it is not reasonable to think that a world that has managed to create a human species would be deprived of all and every trace of human faculties. We therefore refuse to let the fear for anthropomorphisms deter us from considering the occurrence of natural intentionality.

6.2 Semiotic Realism

Gregory Bateson in the book "Angels fear" (posthumously edited by his daughter Mary Catherine Bateson) observed that "if we had continual awareness of our image-making process, our images would cease being credible," and he goes on to assert that "The links between sense and motion are indispensable to living, but the links depend always on presuppositions that are commonly either absolutely inaccessible to consciousness, or momentarily left unexamined in the immediacy of action" (Bateson & Bateson, 1987, pp. 96–97). Accordingly one might say that

[1] There is a third possibility of course, a possibility that has been adopted by philosophers such as Daniel Dennett, who recommended our taking of "the intentional stance" (Dennett, 1987). Briefly stated, this view holds that we cannot understand the life of other humans (or of animals) without describing those lives as guided by, or woven into, intentionality. This does not mean that these creatures possess intentionality as a real property – rather, the thesis states only that we cannot understand these creatures unless we pretend that they do. I must confess that this position reminds me of the evermore complex (and increasingly less likely) sets of epicycles that Ptolemaic astronomers had to introduce into their explanations of the planetary orbits in order to uphold the belief in the geocentric system. Rather than seeking shelter in such powerless conceptions about what, for all of us without exception, is the deepest and most real content of our lives – i.e., the fact that such life is being experienced – we shall suggest that it is instead the ingrained belief in animals as machine-like robots that ought to be given up.

living in a very deep sense depends on faith, faith in what our senses tell us – even though we have no idea of how the senses actually managed to tell us this or that; and even though we know quite well, that our senses may be telling us a wrong story. The inaccessibility to consciousness of the brain activities that leads us to perception is obvious alone from the fact that most nerve cells in the brain do not connect to the outside world or even to sense organs. They only "know" the local surround of other brain cells and body fluids. And yet these cells are our only means for grasping the outside world (Roepstorff, 2004, p. 149).

The kind of faith we are talking about here might perhaps be called *animal faith*, it is a faith that comes to us unconsciously and is hard to escape. The human species nevertheless is the only species in this world that may – to some degree at least – manage to escape it. We – or most of us – know that we may err; immersed as we are in a *linguistic lebenswelt* we are forced to distinguish between the reality of self-subsisting *things* and the more immediately known and equally real *objects* of our experience (Deely, 2001, p. 8). Our lives cannot but implant the knowledge in our souls that the mind is one thing and the world is another. Language gives us the capacity to reorganize our interior cognitive and affective states in ways that are not tied to the biological constitution of the human organism.

This partial loss, or transcending, of brute "animal faith" is the source for a deep-rooted existential skepticism that clings to the human condition and which makes faith itself, *human faith*, a turning point for much of social life. In the philosophy of science it has caused persistent warnings against naïve realism. As Francis Bacon famously said in 1620: "The human understanding is like a false mirror, which, receiving rays irregularly, distorts and discolors the nature of things by mingling its own nature with it" (*Novum Organum*).

Yet, if human knowledge did not actually help us – as it helped our remote ancestors – in the everyday dealings with the natural and social conditions of our lives, why would we have evolved brains and language in the first place, and why would we have prospered as a species? The rejection of scientific realism may seem tempting from an intellectual point of view, but it doesn't accord well with empirical evidence, i.e., the success story (at least as seen from an evolutionary point of view) of human civilization on our planet.

Nominalism nevertheless has dominated philosophy of science for centuries, not the least due to the broad acceptance of the Kantian claim that we can never reach an understanding of the *Ding-an-sich* (the thing in itself) but will always only grasp the "Ding-für-mich" (the thing for me). Scientists, accordingly, generally feel justified to understand their work as dealing with "data", not with reality, a concept that is relegated to "metaphysical speculation." The nominalist doctrine implies that the relation between existing things is not itself part of what exists, but should instead be regarded as a mere construction in the mind of the observer. That a mammalian upper arm bone fits into the shoulder joint of the same animal is, according to this view, not part of reality, for *fits into* is a purely *relational concept*, and relations are not real things in themselves, only things like arm and shoulder bones are. The fitting-in *relation*, correspondingly, does not – to a strict nominalist – refer to any independently existing reality (Deely, 1994; Hoffmeyer, 2008). The claim in other words is, that our all too human habit of connecting things into relational systems misrepresents a reality that does not contain any such relations.

But again, why, one must ask, has the human species evolved this habit of placing things in relationships? Do we really have to believe in the Kantian dictum on the *Ding-an-sich* and the nominalist skepticism it logically entails? After all, modern philosophy might have led us astray. Perhaps relational order is indeed part of a mind-independent reality that humans do right in emphasizing in their descriptions of the natural order.

This was the position taken by the American scientist and philosopher Charles Sanders Peirce (1839–1914). Peirce developed a *semiotic realism* where knowledge was seen quite generally as based upon sign action, *semiosis*. The way the outside world of an organism and its inside world are connected is not, in this view, by way of something from the outside entering into the inside, but by the formation in the organism of a sign relation connecting it to the outside world through an interpretative act. We shall return to this philosophy, but we must first recreate the historical context for Peircean semiotics[2] which unfortunately predates the scientific revolution and has therefore by and large become extinguished from the curricula of the educational system.[3] We must, in other words, look back to the Latin thinkers of medieval times, the period of scholasticism characterized by an attempt to reconcile ancient classical philosophy (notably Plato and later Aristotle) with Christianity.

From the point of view of science the most influential figure in this period was the Italian Dominican priest Thomas Aquinas (1225–1275) who held that the study of "the Book of Nature" was a necessary way to understanding "the Book of God", the Bible, and *vice versa*. The metaphor of the book of nature dates back to the church father Augustin (354–430) who saw the created world as a message to us from the Creator. Augustin was also the originator of the general category of a *sign* (signum) to be understood as something awakening us to infer something else (Deely, 2001),[4] and for Augustin this meant that God gave us signs to show us his intentions. A storm or a flower was not just a storm or a flower since God would have had a purpose by creating storms and flowers – or idiots for that matter. Our linguistic categories were not then – as the nominalists claimed – just inventions of the human mind, for they served as necessary tools to disclose God's purposes with his creation. Thomas Aquinas' teachings were very influential in the scholastic period and

---

[2] Semiotics unfortunately is still best known as a branch of linguistics due to the influential work of the Swiss linguist Ferdinand de Saussure (1916). In the Saussurean tradition, which should better be termed *semiology* (Saussure's own original term) to avoid confusing it with Peircean semiotics, human language is taken as the primary model for semiosic activity and if semiosis is admitted to take place in the animal world at all, it is seen as a degenerate version of human semiosis. Modern semiotics, however, following the semiotic understanding of Peirce, considers human language as just one peculiar instantiation of a much broader semiotics pertaining to evolution at large.

[3] History is written by winners.

[4] Augustine defined the sign as "something, that besides the impressions it conveys to the sense, make something else come into cognition" (Bains, 2006, p. 40).

his position on the necessity of studying the Book of Nature was one important stepping stone on the route to the scientific revolutions of the 16th and 17th centuries. Particularly important in this context was also his insistence on God's reliability, that God, in his benevolence, would not have created nature as an unruly and lawless place. Among the nominalists many protested against this theses that might be seen as an undue limitation of God's freedom: God must be free to create whatever kind of world he might want, they claimed. But in the end the Thomist understanding prevailed – not the least, one may suspect, because the conception of a capricious and unpredictable God would imply, as Luther and Calvin both saw, that God might choose to punish the pious and reward evildoers. The belief in an orderly nature was of course a necessary precondition for the birth of the natural sciences, for had nature not to some extent been conceived as predictable there would have been no natural laws to study in the first place and the attempt to seek systematic knowledge about nature could hardly be expected to pay off.

For Thomas Aquinas himself and the Thomist tradition there is an emphasis on realism in which there is an immaterial or *intentional direct union* between the knower and the known (Bains, 2006). To know about things, e.g. a storm or a flower, implies that these things exist in the mind of the knower as *intentional beings,* and the nature of this kind of being is that of a relation or interface. This understanding is radically different from the cognitive theories that came to dominate in the course of the scientific revolution, where "intentional being" was seen rather as an intermediary "obstacle posited between the knower and the known", an obstacle "*that would first be known reflexively* before the thing was known" (ibid, 43, my emphasis). In Aristotle form and matter were seen as different aspects of things, and in cognition the soul or mind would take on the form of the thing perceived without receiving its matter. So, according to Aristotle, when I think about or perceive a flower, my mind receives the form of the flower. Aquinas instead argues that the flower has a different existence in nature (*esse naturale*) and in thought (*esse intentionale*) (Bains, 2006, p. 44).

The crux of the matter is the direct union between knower and known; the concept of intentionality is descriptive of this *relation* between the mind and the things cognized. We are aware not of the idea or concept but rather of that which it represents – its object. The idea or concept does not stand in between the cognizing organism and the thing (physical or mental), rather the idea or the concept is a formal sign, (an interpretant in the later terminology of Peirce), i.e. "that by which – or rather that on the basis of which – we know, . . . not that which we know . . ." (ibid, 50).

The nature of this relation would be the theme of the next several centuries of scholastic thought culminating in the semiotic philosophy of John (João) Poinsot from Coimbra (1589–1644) that has only recently been dug out of near oblivion thanks to the efforts of John Deely (2007). There is not space to delve upon the subtleties of Poinsot's thinking. Let me here just with Paul Bains summarize: "Poinsot sought to conserve the possibility of an adequate "correspondence" or coherence between thought and thing, but he also laid the semiotic foundation for an image of thought that could not be reduced to that perspective. Relations are truly *between* things: rhizomes or *interbeings*, to use Deleuze and Guattari's terms, or the "Being of the between" to use Heidegger's characterization of Dasein." (ibid, p. 51).

Poinsot was a contemporary of Galileo and at this point of history scholasticism was more or less swept away from center stage by the aggressive new philosophy of the scientific revolution. As Bains remorsefully observes: "modern western philosophy (particularly from Descartes and onwards) chooses to dispense with the doctrine [of intentional being] and embrace the aporias of a "classical" metaphysics of representation in which what the mind knows directly is its own products, positing a beneficent God to make our "objective" ideas conform to the world" (ibid, p. 45). Several hundred years had to pass before Charles Peirce in the late 19th Century took up again the line of thought from the Latin thinkers and developed it to a full blown theory of semiotic realism.

"Modern philosophy", writes Bains "began once *the idea* came to be considered the immediate object of knowledge rather than an interface, or relation" (ibid, p. 51). According to Descartes the exterior world is grasped through the mechanical work of the senses, which then required some intermediate entity, a concept or an idea, to stand between the outside world (reality) and the mind. Henceforward the mind lost its direct access to the world. Humans do not usually react automatically upon sense stimuli but "translate" them to a conceptual world, which then serves as the substrate for thoughts and action. What is at stake is the nature of this "translation": How can a material process be converted into a concept through a purely mechanical processes? It cannot of course, and therefore there is no escape from Descartes' *res cogitans* in post-Cartesian philosophy, with the implication, that realism was essentially impossible from the very beginning of modern philosophy. For as long as thought is imprisoned in its own solipsistic res cogitans there is no way to measure it against the world. The only way to transcend this dualism, we shall claim, is to see organisms as connected to their world in a relational semiotic network rather than through the mechanics of their sensory organs.

6.3 Intentionality

The modern concept of intentionality in philosophy goes back to the German philosopher Franz Brentano (1838–1917) who in 1874 proposed intentionality as the one "positive attribute" that holds true of all mental phenomena: "Mental phenomena ... are those phenomena which contains an object intentionally within themselves" (Brentano, 1874/1973, pp. 88–89, cit. from Short, 2007, p. 6). Brentano was himself, as Deely notes, a Dominican priest and was well read in the literature of scholasticism to which he explicitly refers. (Deely, 2007, p. 4). The problem we are concerned with here is how it can be that some things in this world are "about" some other things. Thoughts, hopes, desires etc. are always about something else, and we distinguish them according to what they are about. Lifeless things, on the other hand, such as stones or clouds are not – to the best of our knowledge – about anything else. The term intentionality was meant to catch this strange property of *aboutness*.

We can summarize the Brentano thesis in three points: 1) Only mental phenomena exhibits intentionality, 2) Intentionality is an irreducible feature of mental phenomena, 3) Since no physical phenomena could exhibit it, mental phenomena could not be a species of physical phenomena. Brentano's pupil Edmund Husserl (1859–1938) saw the identification of intentionality with the mental as a fundamental principle whereupon he founded a new science: *phenomenology*. Phenomenology was the science of the mental and naturalistic explanation was excluded in principle.

From the point of view of analytical philosophy a major problem with Brentano's concept of intentionality is the claim that mental phenomena contains an object inside themselves even when that object does not exist. If for instance we think of a unicorn or of Santa Claus, then in both cases Brentano's thesis implies that our thoughts contains a non-existent object. As Short says: "one fears the unreal and desires the impossible" (Short, 2007, p. 7). But how can something be an object without existing? W. V. O. Quine (1908–2000), for instance, would admit that intentional idioms are irreducible but would claim that they do not denote anything real, and like most contemporary philosophers of mind he held a) that reality is physical, b) that physicalist language is wholly free of intentional idioms (ibid, p. 13).

So, confronting the Brentano/Husserl position, that *mind is real, irreducibly intentional, and inexplicable naturalistically* we have the predominant position of analytical philosohy, that *whatever is real is nonintentional and explicable naturalistically.* A fruitful discussions between two so fundamentally different positions is hardly possible and yet for decades these are the two positions between which we have had to choose. Unknown to the great majority of contemporary thinkers, however, a third position was suggested more than hundred years ago by Peirce: that *mind is real, irreducibly intentional, and yet explicable naturalistically.*

As the reader will expect by now, this is the position we will take here and explore in the rest of this chapter. Peirce was against dualisms of any kind, an option he saw as closing the door for deeper understanding, and he would not accept that mental life should evade naturalistic explanation. But neither could he accept the physicalism of his own time which he saw as constrained by a much too narrow understanding of what is meant by "the physical". Instead the way he crossed the mind-body dualism was by extending the concept of intentionality beyond the confines of human cognition, and he did so by grounding intentionality in a very generalized understanding of sign action, *semiosis,* which he explained thus: "by "semiosis" I mean . . . an action, or influence, which is, or involves, a cooperation of *three* subjects, such as a sign, its object, and its interpretant, this tri-relative influence not being in any way resolvable into actions between pairs" (Peirce, 1931–1958, p. 5.484). On the one hand this triadic notion of semiosis, which we shall discuss below in more detail, entails intentionality since to the interpreter (the system in which the interpretant is formed) the sign obviously is "about" something, and on the other hand Peirce did not conceive of the interpreter as being necessarily a human person. Peirce explicitly referred to this point in the following passage, from a letter to Lady Welby, written in 1908: "It is clearly indispensable to start with an accurate and broad analysis of the nature of a Sign. I define a sign as a thing which is so determined by something else, called its Object, and so determines an effect upon a person, which effect I call its interpretant, that the latter is thereby mediately determined by the former. My insertion of "upon a person" is a sop to Cerberus, because I despair of making my own broader conception understood.[5]" (Peirce, 1908, pp. 80–81).

6.4 Biosemiotics

Biosemiotics is an approach to the understanding of living systems that takes sign processes or *semiosis* to be constitutive for life. Customarily the field has been divided into two main areas, 1) *endosemiosis*, meaning semiotic processes taking place inside an organism, as for instance in signal transduction (mediating extracellular signals to intracellular activity) or secretion of hormones; 2) *exosemiosis* meaning semiotic processes taking place between organisms, as for instance the dance of cranes or mating rituals of water mites. From a theoretical point of view however such a division is less satisfying since many, if not most, processes in nature connect semiotic processes of both kinds. In sexual reproduction for instance hundreds of endosemiotic processes (e.g. complex schemes of recognition processes between egg and sperm cells, hormonal regulations etc.) are mixed up in the sophisticated exosemiotics of mating behavior.

By making the sign fundamental to living systems biosemiotics undoubtedly will arouse fear of vitalism in the minds of many biologists. Signs do not belong to the habitual tool set of scientific theory and may be felt to allude to uncontrollable subjectivist aspects of life. In biology, vitalism refers to the belief that the functions of living organisms must be explained through the action of peculiar *vital forces*, that do not in any way influence inanimate nature. Biosemiotics rejects appeal to such forces. Sign processes are neither forces nor things; rather, they are processual relations that, as shown below, organize many activities. The causality of signs thus differs from the causality of forces. Indeed, while signs are frequently misunderstood or ignored, forces always exert their power with merciless efficiency. Biosemiotics is not a new version of vitalism (Hoffmeyer, 2010).

But the concept of *semiosis* indeed brings a novel element to the scientific tool set for, by definition, a sign-process requires an interpretative agency. This new element, moreover, may be felt to jar with the hegemonic ontology of mainstream science. From a biologist's perspective, however, it can hardly be controversial to attribute agency to living systems. Indeed, the mechanism of natural selection can only work as, in Darwin's terms, organisms "strive" (Darwin, 1971 [1859], p.71: cf. Swenson, 1989; Hoffmeyer, 2009a) for resources such as those provided by food, water, shelter, mating partners and escape from predators. Without such strife there could be no

[5] Peirce had no illusions that his contemporaries would accept his own broader conception whereby nature teemed with beings, for example, bees, that could stand in the place of persons as sites for the establishment of interpretants.

competition, and without competition there could be no selection. Although rarely stated, natural selection does not magically provide a mechanistic explanation the agency of living beings. At a deep level, mainstream biology needs a theory of living agency as much as does biosemiotics.

A promising avenue towards such a theory comes from studying the thermodynamics of irreversible systems and complex system dynamics. These approaches open up a non-deterministic world in which bottom-up processes engage in intricate interactions with their top-down counterparts (Laughlin, 2005; Kauffman, 2008). The evolutionary roots of agency, function and semiosis may well extend back to the extinct prebiotic systems that, about 4 billion years ago, gave rise to the first life forms. While realistic modeling of such systems may indicate how such a process arose, for our purposes we push back the origin question to a "threshold zone." Under the threshold we find neither semiosis, function or agency and above these properties are indeed exhibited by the system. Important work is presently done in this area by several groups (Kauffman & Clayton, 2005; Deacon, 2006).

The first important thing to say about the sign, as understood by modern semiotics, is that the sign is a process: the sign does not exist apart from the process through which it exerts its effect. When we say e.g. that smoke is a sign of fire we think of the sign simply as identical to the smoke itself, but obviously for the sign to have any effect in the world qua sign it must be understood (or misunderstood). A baby sensing smoke would not turn its head to search for something burning, but even the baby might make an *icon*[6] of the smoke to the extent that the smoke reminds her of something like it that she had previously experienced. To adult people the smoke normally acts as an *index,* it indicates the eventual occurrence of a fire even when such a fire cannot be seen. This interpretation may of course be false, as would usually – but not always – be the case in a theater room. *Signs, in other words, refer to something else by eliciting an interpretative process in an organism.*

There is no need in this context to go into the complicated questions of the taxonomy of signs; suffice it to be aware that human language constitutes a very peculiar sign system based on the advanced category of signs called symbols. A few mammalian species may in exceptional cases access the world of symbolicity, but the systematic symbolicity of human language is unique to our species (Deacon, 1997). Considering the uniqueness of human language it is perhaps not so strange that semiotics has traditionally been conceived as a branch of linguistics. But the belief in a sharp demarcation line between human and prehuman nature, corresponding to the realm of necessity and the realm of free choice respectively, makes no sense in the light of evolutionary theory and our growing understanding of animal cognitive skills (see e.g. Bekoff & Pierce, 2009)[7]. And worse yet, it makes us misunderstand

[6] In Peirce's icon, index, symbol trichotomy the icon is a sign that refers to an object because of a supposed "likeness"; an index refers to an object because of a causal or correlative relation; the symbol refers to the object via a convention (often historically based as when the word "cheval" refers to an animal that in the English-speaking world may be referred to by the word "horse").

[7] According to Peirce "All thinking is by signs, and the brutes use signs. But they perhaps rarely think of them as signs. To do so is manifestly a second step in the use of language. Brutes use

what is truly unique about the human being. For we are certainly not alone in the world in being sign users, all organisms must to some limited extent be capable of reading cues in their surroundings in order to survive. In fact, semiosis is exactly what distinguishes life from non-life (Hoffmeyer, 2009a). What is unique about human beings is the kind of semiosis we engage in: While we share the capacity for interpreting iconic and indexical signs with all living creatures, we alone are bound to live our lives embedded in a symbolic world, a world of language (Deacon, 1997).

Charles Peirce defined the sign as a triadic relation connecting the *sign* vehicle (the smoke in our example above) with an *object* (the fire) through the mediation of an *interpretant* (conscious – or instinctive – sensomotoric activity) (Fig. 6.1):

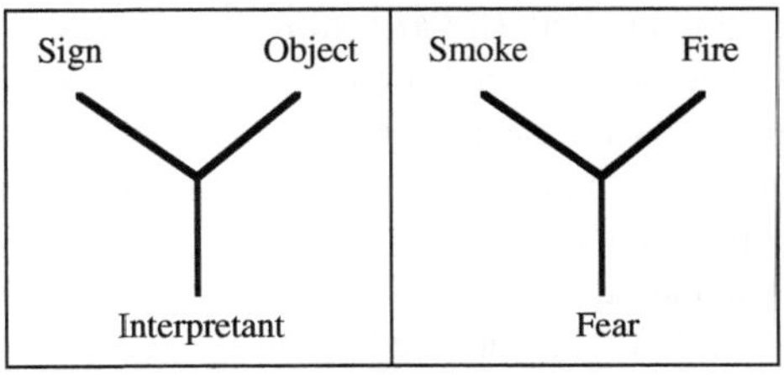

Fig. 6.1 The Peircean concept of a sign as a triadic relation connecting a sign vehicle with an object through the formation of an interpretant in a receptive system.[8] In the *right part* is shown how smoke may act as a sign that evoke a sense of fear by making us aware of the risk of burning

Thus, when a deer senses smoke it is immediately "seized by alarm" (the interpretant) and flees away. The deer may or may not understand that smoke signifies fire but it certainly knows that smoke signifies danger (the object). Likewise, when a macrophage (a cell from the immune system) lets HIV virus into its interior, this is caused by the cell falsely interpreting the virus as belonging to the body itself. In achieving this the virus has acted as an *icon* for one of the normal components involved in the immunoresponse reaction chain. Semiotics cannot restrict itself to deal with human language, but must encompass all kinds of sign systems as they unfold in time and space throughout organismic life on our planet.

A sign is not necessarily linked to a communicative context. Most sign processes in this world are not only unconscious but also unintended in the sense that the sign was not produced for the sake of interpretation. Most people for instance don't want to blush when embarrassed but cannot help doing so, and much to our

language, and seem to exercise some little control over it. But they certainly do not carry this control to anything like the same grade that we do. They do not criticize their thought logically." ("Consequences of Critical Common-Sensism", c. 1905, (Peirce, 1931–1958, p. 5.534)).

[8] It may seem contradictory that "sign" is put in as one element in the sign-relation. As explained in the text the sign always presupposes the whole triadic relation and technically speaking the term "representamen" or at least "sign vehicle" should have been used instead of sign. However, since everyday language uses the term "sign" as equivalent to the representamen as such, I have chosen to stick with it.

annoyance observers readily interpret the result. All mammalian animals unwittingly emit butyric acid with the sweat, but for the tick the butyric acid acts as a signal that causes her to abandon her post (on top of the blade of grass/bush) and fall blindly downward toward her prey. Or, at another level, an antelope has been hurt and thereby draws the attention of hunting lions that selects it out for its slightly awkward movements.

I have suggested the term *semethic interaction* for this kind of co-evolution whereby "habits become signs" in the sense that individuals of one species have acquired the capacity to interpret certain regular activity patterns (habits) characteristic for individuals of another species, which then eventually may release new kinds of regular behavioral patterns in the first species etc. As an example we can take the case of the large blue butterfly *Maculinea arion* where the female lays her eggs in thyme plants. The larvae spend their first three weeks on thyme flowers on which they feed until they have reached the last larval instar. They then drop to the ground, where they produce a mixture of volatile chemicals that mimics the smell of larvae of the red ant species *Myrmica sabuleti*. The patrolling worker ants mistake the larvae for their own and carry the caterpillars into the ant nests. Once there, the caterpillars change their diet and start feeding on eggs and larvae of the ants until they pupate. They undergo metamorphosis in the ant colony, surfacing as butterflies (Gilbert & Epel, 2009, p. 86). Here the female butterfly profits from the ants' habit of locating their nests on well grazed grassland with plenty of thyme plants so that she will "know" where to put her eggs (presumably a parameter connected to the thyme plant is interpreted as a sign for oviposition). The caterpillars furthermore are capable of fooling the ants by interfering with the ants' own signaling system. That this is indeed the case is proven by the fact that if the height of the grass exceeds 4 cm *M. sabuleti* will disappear and another ant species, *M. scabrinodis*, will replace it with fatal consequences for the caterpillar, since this ant species will not misinterpret the volatile chemicals produced by the caterpillar as a message of belonging in the colony.

Now, even if successful the semethic interaction between butterfly and ant may still fall victim to a new kind of dangerous parasitism. As Gilbert and Epel explains: "It seems these caterpillars are the sole food source for the larvae of several species of wasps of the genus *ichneumon*. A female wasp can detect not only the ant colonies but also the presence of butterfly larvae within them. She enters only colonies where caterpillars are present; once there, the wasp emits pheromones that cause the ants to fight among themselves while she goes about laying a single egg in each butterfly larva. Each wasp egg hatches into a larva that eats the caterpillar as it begins pupation. Eleven months later, the pupal case is shed and there emerges not a butterfly but an adult wasp" (ibid).

Many more examples on these webs of habits-signs-habits... are given in my book *Biosemiotics. An Examination into the signs of life and the life of signs* (Hoffmeyer, 2008). Among biochemists, there is a rule of thumb saying that whenever nature keeps a store of energy (e.g., food) there will also always be a species that makes its living by consuming it. I shall suggest a quite similar rule of thumb by saying that there never occurs a regularity or a habit in nature that has not *become a sign* for some other organism or species. Admittedly, this rule may be less well investigated (so far!) than the biochemical rule, but it does catch an important semiotic aspect of the evolutionary process, for due to the mechanism of semethic interactions, the species of this world have become woven into a fine-meshed global web of semiotic relations. These semiotic relations, more than anything else, are responsible for the ongoing stability of Earth's ecological and biogeographical patterns. At the individual level as well as at the level of ecosystems all interaction patterns are controlled through semiotic relations – more or less in the same way the traffic in a city is controlled through signals. This relational network can be looked upon as an internal semiotic scaffold.

Biosemiotics then is not so much about *communication* as it is about *signification,* the many processes whereby organisms ontogenetically or phylogenetically have learned to ascribe meaning to whatever regularities around them that may be useful as trigger mechanisms. Biosemiosis therefore does not fit into the traditional scheme from communication science of a sender and a receiver connected through a channel, for to the extent there are clear-cut senders and receivers at all (hardly the normal situation) the channel is itself part of the message as interpreted by the receiving system. Semiotic causation is based on a "trigger-mechanism" whereby the interpreting system creates the interpretive response by its own means, not by any external intervention – apart form whatever "trigger-event" it has learned to select. For this reason *there is no need for physical compatibility between the sign and the activity it releases*, the sound of a bell may get children running, a few molecules hitting the antenna of a wasp may release flying behavior, and the absence of light stimulation of the right eye retina of a kitten during a critical period may cause the cat to be functionally blind on the right eye for the rest of its life. The evolution of a semiotically based regulation of organismic activity opens for a versatility of means and a complexity of interactive entities that had been impossible if the controlling agency relied exclusively on efficient causality, good old biochemistry.

6.5 Relative Being and Semiotic Freedom

A sign is a triadic relation that organizes a process whereby an event or entity inside or outside of an organism (or a lineage seen as a *supraindividual* historical organism) is reacted upon. A bird sees an unexpected shape on the stem of a birch tree, interpret the shape as "food" and changes its flight for the catch. This process is totally different from the case where the moth is smashed into pieces by a falling heavy object. The moth will die in both cases, and also in the last case do we have a causative relation, gravity causing the heavy object and the Earth to approach. But this is a dyadic (cause-and-effect) relation in no need of semiotic explanation; the bird on the other hand is led by an interpretative act, that could not be described or explained in the absence of the particular triadic relation involved. A hardliner reductionist may object that the feeding behavior of the bird is indeed explainable through an "infinite" web of efficient causative events. Rather than argue about such speculations, I shall point out that for all practical purposes the explanation for the bird's behavior is a triadic relation, and to insist upon explaining it through an infinitely long series of dyadic relations will bring us nowhere.

The point is this: There is no reason to assume that interpretative events are not perfectly physically caused processes. But interpretative processes are special because *they are organized according to semiotic dynamics*. If we try to separate them out of this context we will just end up having myriads of dyadic (cause-effect) processes without any discernable pattern, and we would not have a chance to elucidate the real dynamics at work in the situation. In reality, though, it may seem as if we manage quite well by describing everything in terms of dyadic processes, but this is because we instinctively understand the intentionality of the organisms we are studying. We are therefore capable, to some extent at least, to "guess" on what is the real dynamics at play. The success of this strategy may persuade us to think that dyadic relations are all we need, when in reality this need has been "falsely" satisfied through a loan from our own intentionality. Biosemiotics is precisely needed in order "to make explicit those assumptions imported into biology by such unanalyzed teleological concepts as function, adaptation, information, code, signal, cue, etc., and to provide a theoretical grounding for these concepts" (Kull, Deacon, Emmeche, Hoffmeyer, & Stjernfelt, 2009, p. 170).

Most relations in this world are of course quite uninteresting, as for instance the relation one might hypothetically draw between my big toe and any of planet Jupiter's 63 moons. We have no reason to believe that any of these particular relations will ever get to have any causative influence upon events on Earth or elsewhere in the universe and have no reason to ascribe any reality to them. Other relations are important enough, as for instance the relation between prolonged day length and approaching summer solstice. This is a purely dyadic relation that may be precisely predicted at any given latitude thousands of years forward in time. The relation is important because – among other things – it assures that trees may "know" approximately when to burst into leaves. Still, since this biological use of the relation is not intrinsic to the relation it does not count as an ontological relation. The term "know" here is put inside quotation marks to indicate that this is not knowledge in the human sense of this word, but it *is* knowledge in the sense that trees have "learned" to use this relation as a sign (actually, degree-days rather than day length is the releasing factor here, but the two are of course closely connected). Each single beech tree that bursts into leaves does not "know" why it does so, but as a species the beech trees have learned through evolution to size up on the time when this parameter has reached a threshold level.

Knowledge in the biological sense of the term, as we have used it here, necessarily depends on predictability, and the mechanism behind all learning is the creation of a triadic relations on the basis of stable dyadic relations. The predator, for instance, goes for any prey animal that moves awkwardly because it "knows" that clumsy behavior signifies easy catch. In other words, the predator converts the dyadic relation between slow-rate flight and clumsiness into a triadic relation in which an interpretant is produced on the top of the simple dyadic relation, and this interpretant makes it hunt the prey animal that moves clumsily. A bird, on the other hand, may predict that if it moves clumsily, e.g. by pretending to have a broken wing, the fox will try to catch it instead of it's youngs. This way the bird may often manage to lure the fox away from the nest only to fly away when the fox has lost trace of the nest itself.

The moment we turn from the physical world to the organic world relations tend to take on a reality of their own. Returning to the already mentioned relation between the shoulder of an animal and the upper arm bone we are dealing with a highly sophisticated version of an ordinary ball and socket relation. But when we follow the evolution of this particular relation in animals since it first occurred hundreds of millions of years ago as fins of fishes, it is obvious that the relation as such has been the focus of natural selection, and not only the bones. Whatever change selection may have favored in the evolution of these two bones, the relation between them would have to remain fit, for if one of the two bones changed without a corresponding change in the other bone the individual would be crippled and could not be expected to leave offspring. The adaptive function of the arm-shoulder joint is an *intrinsic property* of the relation as such and it would be stretching to claim, as the nominalists do, that this relation is not part of reality but only of our own minds. This type of relation has been termed *categorial relations*.

Sign relations, or ontological relations, are different since they do not depend upon the (mind-independent) existence of that which they relate. Thus most people in the western hemisphere will think of Santa Claus if they are shown an image of a white-bearded man dressed in red clothes walking in a snow covered landscape, but this does not mean that these people believe in Santa Claus. As we said above, the sign is a *pure relation*: it is neither material nor mental, it simply consists in the process of producing a connection between events or entities in such a way that one event or entity becomes related to another event or entity according to the needs of an organism. The sign relation is both established and goes extinct in the very action of the sign. Eventually it may endure for some time of course, but this persistence depends on the perpetual repetition of the same – or a similar – sign process.

Anticipation – in the broadest sense of the term – is what living systems do for a living. Contrary to lifeless systems organisms do not passively sit (or flow) and wait for things to happen, they actively search for the resources they need and actively protect themselves against a range of possible dangers.[9] All of this presupposes some kind of anticipation where present cues are used to tell about future conditions in some sense or other.

There are of course many strategies a species can "choose" for coping with the challenge of change and the evolution of increased anticipatory talent is only one among them. It is however a very decisive strategy that may even have changed the dynamics of the later stages of evolution on our planet. This is because such a strategy possess an intrinsic potential to drive forward a growth in the capacity I have called *semiotic freedom*, or *interpretance*. Semiotic freedom may be defined

[9] Plant movements may not seem of much, but if you increase the time scale it may actually look quite impressive when played at video. Plants move by growth (beneath and above the earth), by off-shoots and runners, and by spreading their seeds.

as the capacity of a system (a cell, organism, species etc.) to distinguish relevant sensible parameters in its surroundings or its own interior states and use them to produce signification and meaning.[10] An increase in semiotic freedom implies an increased capacity for responding to a variety of signs through the formation of (locally) "meaningful" interpretants (Hoffmeyer, 2008; Hoffmeyer, 2009a). The term freedom in this context should be taken to mean: underdetermined by natural lawfulness.

The appearance on our planet of biosemiosis opened a new agenda for the evolutionary process by providing entities with the agential property presupposed for Darwinian "striving" and thus for natural selection. For billions of years the semiotic freedom of agents remained low, and a bacterium, for instance, cannot itself chose to *not* swim upstream in a nutrient gradient. Therefore, at this stage of evolution semiotic freedom is primarily exhibited at the level of the lineage (the species as an evolving unit).[11] I suggested the term *evolutionary intentionality* for this kind of intentionality (Hoffmeyer, 1996b). Only gradually would emerge a more advanced stage of biosemiosis, in which semiosic activity was no longer a property of the lineage but also, and importantly so, a property of individual organisms.

This "individualization" of semiotic freedom, i.e., its displacement from the level of the species to the level of the individual, would have initiated a change in the dynamics of the evolutionary process. Patterns of interactive behavior now became increasingly regulated or released by semiotic means, and this would have induced a new kind of flexibility upon inter- and intraspecific interactions. Innovations more and more came to depend on semiotically organized cooperative patterns at all levels from single organisms and species to whole ecological settings. In fact, as I have suggested elsewhere, natural selection from now on would more and more follow directions given by the ecosemiotic interaction patterns (called *ecosemiotic motif*s in (Hoffmeyer, 1997)). The more natural systems become scaffolded through semiotic interaction patterns (semiotic scaffolding) the less will be the role played by genetic scaffolding, and the more derivative will the role of natural selection become. Natural selection will now favor such genetic adjustments that might support already established semiotic interaction patterns, but will not itself to the same extent mark out the direction of change. As a consequence the individual rather than its genes become the main evolutionary agent, and the concrete life history of individuals will increasingly determine their behavior. By implication learning, interpretance and semiotic freedom will be more and more

---

[10] Originally I defined semiotic freedom as "the depth of meaning that an individual or species is capable of communicating" (Hoffmeyer, 1993, p. 109, 1996a, p. 61), but the essence of this ability is interpretation rather than communication, although the two aspects are of course closely connected.

[11] Even at this level one cannot rule out individual semiotic freedom right away though. A bacterium is a hugely complex and well tuned system of proteins and other components and although learning processes do probably not directly play a role at this level the bacterium is capable of changing its behavior by the active uptake of foreign DNA from bacteriohages.

important parameters in the games played out in the evolutionary theater. Or, in other words, a self-sustaining dynamics leading to increased semiotic freedom is set in motion.

6.6 Human Intentionality

Very late in organic evolution a further potentiation of semiosic capacity took place through the appearance of human beings that from the first beginnings were embedded in a linguistic Lebenswelt, based on the particular ability of this species to understand symbolic linguistic referencing (Deacon, 1997). Due, not the least, to the indefatigable efforts of the late Thomas Sebeok it has now gradually become accepted that human semiotic capacity is only one – although radical – further refinement of a biosemiotic capacity that has unfolded itself on Earth through nearly 4 billion years (Sebeok, 1979; Sebeok & Umiker-Sebeok, 1992). The semiosic difference between the human animal and other living systems is staggering indeed, but, as John Deely has repeatedly pointed out, by far the most important dimension of this difference is that humans know the difference between signs and things, while animals don't (Deely, 2001). In our attempts to cope with this discomforting knowledge the winning strategy in modern time has been the de facto institution of a dualism between mind and body,[12] a dualism that left biology and medicine with only half of the human person, the so-called body, and which therefore has not managed to understand what a human being is, or what health is (Hoffmeyer, 2010).

Human beings are persons and persons cannot be divided into one part, the body, that must be treated somatically, and another part, the mind, that must be treated psychologically. This is where the biosemiotic approach may help out, because biosemiotics sees meaning and signification (sema) as inherent to the body proper (soma) and not as something separated out to non-descript locations in the brain or mind. Whatever the mind is it is also body, not body in the physical sense this word has got in present day biology or medical science, but body in a semiotic sense of the word, a body that is inherently engaged in communicative processes that serve to coordinate the activities of the cells, tissues and organs inside the body as well as to exchange integrating messages across hierarchically distinct levels. Seen in this light *the mental system or mind is simply the interface through which a human organism manages its coupling up to the surrounding web of things, natural or social*. The mind, thus, is not a thing and has no more distinct location (in the brain?) than has the electronically mediated processes whereby a changing pattern of pixels is at each moment shaped on the TV-screen. The outsourcing of the body-world interface into a distinct disembodied field, the mind, to be studied by a separate science, psychology, was of course a necessary step as a compensation for the imprisonment by natural science of the diseased person into a body that could

[12] Often unwittingly disguised as materialistic monism which, however, in a deeper analysis can be shown to presuppose dualism (Searle,1992).

no longer – qua body – integrate itself into the world in a meaningful way. But not only was this outsourcing based on unequal balance in which the medical expertise was the unquestioned highest authority in matters of therapy. It also forced psychology into an impossible role as caretaker of functions that were neither corporeal nor social but mysteriously suspended in a no-mans-land produced by lack of any substance apart from the unbearable idea of Cartesian res cogitans.

I must emphasize that this criticism is not directed towards the many excellent practitioners inside the respective areas of medicine and psychology. Ingenious ways have been invented to overcome the absence of a unified theory of the human person, and present day psychological and medical approaches to health are invaluable and absolutely necessary stepping stones for a further development of a healthy health strategy. Nevertheless biology and medicine must reintegrate the body's interface with its external world, the mind, into its core theory of a human organism. And the way to do this is by admitting semiotics into its basic tool set. "A sign is not just something for the mind to interpret, but something for the body to interpret, and the body is itself of the same stuff as dreams are made of, significative biomolecular processes – our dreams are constantly nourished by the semiotic processes going on in the bodymind" (Hoffmeyer, 2010).[13]

With the birth of this animal, the human being, the natural history of intentionality seems to have reached a threshold level, where the social and cultural environment attained an autonomous kind of creativity that irreducibly interacts with, and largely – but never completely – determines the horizon inside which the personal intentionality of human beings exhibits itself. And unlike biological creativity (organic evolution) the history of cultural creativity is deeply dependent on semiotic scaffolding right from the beginning. Language itself is of course a powerful semiotic scaffolding tool, allowing for oral transmission of cultural experiences in time (from generation to generation) and space (from group to group). But a range of additional and increasingly sophisticated scaffolding devices follows the development of human civilizations, primarily in the form of technical practices and art. Sculptures, paintings and, in time, written texts support the transmission of social skills and the myths that makes the world meaningful to people and serves as a much needed memory store, necessary because most people through most of human history were illiterate and had no access to other kinds of external memory stores.

The cathedrals of the middle ages, the invention of the printing press, the radio, films, TV, computer networks and the internet are some of the major semiotic scaffolding tools that supported the route to the modern world. No need to go in more detail. Let me rather end this little sketch by observing that for each new step in the development of this endless series of still more powerful semiotic scaffolding tools, the semiotic freedom of individuals took on new dimensions, because each of these steps trivialized insights that earlier generations had had to spend their cognitive

[13] The term bodymind was introducced by immunologist Candace Pert and co-workers, (Pert, Ruff, Weber, & Herkenham, 1985), and discussed in a semiotic context in Hoffmeyer (1996a).

resources to acquire or learn, where now in the new generation semiotic scaffolding made the insight more or less part of an "inherited" skill (Hoffmeyer, 2009b). Air pictures, to take just one example, now instantly show us the geography that earlier generations had spent so much energy in calculating through laboriously acquired techniques. As a result each new generations tend to possess more powerful semiotic scaffolding systems on which to base their own contribution to the changing world, the semiotic freedom get ever greater horizons to work on.

Finally, one may suppose that already from the earliest modest steps on the route to speech its function as a means for semiotic scaffolding would feed back into the development of the human brain itself. The gradual appearance in our remote ancestor's cognitive system of a referencing system based upon sound signs, speech, would have opened a new kind of social intelligence, an intelligence derived from the capacity of the social group as a whole to learn through the exchange of experiences, skills, empathy, sensitivities, fantasy and inventiveness in a public or shared process. In the framework of this new communal functionality it must increasingly have become a criterion for individual success that one could contribute in talented ways to this emerging social intelligence. The establishment of a communal or social intelligence might have further influenced selection of individual intelligence for instance by favoring the development of brains talented for social and linguistic competences. Therefore one might suggest that the very special way the human brain has been organized is as much a result of the social life made possible by speech, as the social life is a result of highly the developed intelligence of humans (cf. Deacon, 1997). But most probably the input went both ways: from the social-semiotic level to the genetic level, and from the genetic level to the social-semiotic level.

References

Bains, P. (2006). *The primacy of semiosis. An ontology of relations*. Toronto, ON: Toronto University Press.

Bateson, G. & Bateson, M. C. (1987). *Angels fear. Towards an epistemology of the sacred*. New York: Macmillan.

Bekoff, M., & Pierce, J. (2009). *Wild justice. The moral lives of animals*. Chicago and London: University of Chicago Press.

Brentano, F. (1874/1973). *Psychology from an empirical standpoint*. New York: Humanities Press.

Darwin, C. (1971 [1859]). On *the origin of species by means of natural selection or the preservation of favored races in the struggle for life*. London: J.M. Dent & Sons.

Deacon, T. (1997). *The symbolic species*. New York: Norton.

Deacon, T. (2006). Reciprocal linkage between self-organizing processes is sufficient for self-reproduction and evolvability. *Theoretical Biology, 1*, 136–149.

Deacon, T. (2007). Shannon-Boltzmann-Darwin: Redefining information. Part 1. *Cognitive Semiotics, 1*, 123–148.

Deacon, T. (2008). Shannon-Boltzmann-Darwin: Redefining information. Part 2. *Cognitive semiotics, 2*, 167–194.

Deely, J. (1994). How does semiosis effect Renvoi? *The American Journal of Semiotics, 11*, 11–61.

Deely, J. (2001). *Four ages of understanding. The first postmodern survey of philosophy from ancient times to the turn of the twenty-first century*. Toronto, ON: Toronto University Press.

Deely, J. (2007). *Intentionality and semiotics*. Scranton, PA: Scranton University Press.
Dennett, D. C. (1987). *The intentional stance*. Cambridge, MA: MIT Press/Bradford Books.
Favareau, D. (2007). The evolutionary history of biosemiotics. In M. Barbieri (Ed.), *Introduction to biosemiotics* (pp. 1–67). Dordrecht: Springer.
Gilbert, S. F., & Epel, D. (2009). *Ecological developmental biology. Integrating epigenetics, medicine, and evolution*. Sunderland, MA: Sinauer.
Hoffmeyer, J. (1993). *En Snegl på Vejen. Om Betydningens Naturhistorie*. København: Rosinante/Munksgaard.
Hoffmeyer, J. (1996a). *Signs of meaning in the universe*. Bloomington, IN: Indiana University Press.
Hoffmeyer, J. (1996b). Evolutionary intentionality. In E. Pessa, A. Montesanto, & E. P. Penna (Eds.), *Third European conference on systems science* (pp. 699–703). Edzioni Kappa.
Hoffmeyer, J. (1997). Biosemiotics: Towards a new synthesis in biology. *European Journal for Semiotic Studies, 9*, 355–376.
Hoffmeyer, J. (2008). *Biosemiotics. An examination into the signs of life and the life of signs*. Scranton and London: University of Scranton Press.
Hoffmeyer, J. (2009a). Semiotics of nature. In P. Cobley (Ed.), *The routledge companion to semiotics* (pp. 29–42). London and New York: Routledge.
Hoffmeyer, J. (2009b). *Tro på tvivl. Kritik af religiøs og videnskabelig ufornuft*. København: Ries Forlag.
Hoffmeyer, J. (2010). A biosemiotic approach to health. In S. Cowley, et al. (Eds.), *Signifying bodies. Biosemiosis, interaction and health*. Braga: The Faculty of Philosophy of Braga, Portuguese Catholic University.
Kauffman, S. (2008). *Reinventing the sacred. A new view of science, reason, and religion*. New York: Basic Books.
Kauffman, S., & Clayton, P. (2005). On emergence, agency, and organization. *Biology and Philosophy, 21*, 501–521.
Kull, K., Deacon, T., Emmeche, C., Hoffmeyer, J., & Stjernfelt, F. (2009). Theses on biosemiotics: The Saka convention. *Biological Theory, 4*, 167–173.
Laughlin, R. (2005). *A different universe. Reinventing physics from the bottom down*. New York: Basic Books.
Peirce, C. S. (1908). A letter to Lady Welby dated Dec. 23. In C. S. Hardwick & J. Cook (Eds.), *Semiotics and significs: The correspondence between Charles S. Peirce and Victoria Lady Welby* 1977. Bloomington, IN: Indiana University Press.
Peirce, C. S. (1931–1958). Collected Papers of Charles Sanders Peirce. Vols. 1–4, C. Hartstone & P. Weiss (Eds.), Vols. 7–8 A. W. Burks (Ed.). Cambridge, MA: Harvard University Press.
Pert, C. B., Ruff, M. R., Weber, R. J., & Herkenham, M. (1985). Neuropeptides and their receptors: A psychosomatic network. *The Journal of Immunology, 135*, 820s–826s.
Popper, K. (1990). *A world of propensities*. Bristol: Thoemmes Antiquarian Books.
Roepstorff, A. (2004). Cellular neurosemiotics: Outline of an interpretive framework. *Theorie der Biologie, 6*, 133–154.
Saussure, F. de (1916). *Cours de linguistique générale*. Paris: Payot.
Searle, J. R. (1992). *The rediscovery of mind*. Cambridge, MA: MIT Press.
Sebeok, T. A. (1979). *The sign & its masters*. Austin, TX: University of Texas Press.
Sebeok, T. A., & Umiker-Sebeok, J. (1992). *Biosemiotics: The semiotic web 1991*. Berlin: Mouton de Gruyter.
Sheets-Johnstone, M. (2009). *The corporeal turn. An interdisciplinary reader*. Charlottesville, VA: Imprint Academic.
Short, T. L. (2007). *Peirce's theory of signs*. Cambridge, MA: Cambridge University Press
Swenson, R. (1989). Emergent attractors and the law of maximum entropy production. *Systems Research, 6*, 187–197.

Chapter 7
The Evolution of Learning to Communicate: Avian Model for the Missing Link

Irene M. Pepperberg

Abstract Exclusively primate-centric models for the study of the evolution of communication, although reasonable considering the close phylogenetic relationships between present day human and nonhuman primates, overlook parallel or convergent evolution and the possibility that birds—with their advanced cognitive and communicative abilities—can provide models for the evolution of communication, particularly for vocal learning. Through similar evolutionary pressures and parallel exploitation of ecological niches, similar communicative abilities likely evolved, and birds are among the few nonhuman species to learn their vocal communication system. Even the neuroanatomical structures subserving vocal behavior in birds and humans are now evaluated for similarity. Thus, I suggest that examining avian subjects, particularly their learning and use of various vocal systems, will shed light on the evolution of learned vocal communication.

7.1 Introduction

Given the close phylogenetic relationships among present-day humans and apes, models for communication and language evolution not surprisingly focus on the primate lineage (e.g., Deacon, 1997) with a prominent role reserved for common ancestors or "missing links". A plethora of books and articles suggest possible evolutionary pathways (recently, Smith, Smith, & Ferrer i Cancho, 2008; Bickerton, 2010; Slocombe, Waller, & Liebal, 2011), including those involving mirror neurons (MNs; Arbib, 2005, 2008; Fogassi & Gallese, 2002; Gallese & Lakoff, 2005). Clearly, human language evolved from something simpler, but primate-centric models overlook parallel or convergent evolution, the likelihood that similar (albeit non-identical) communicative abilities evolved in different species, and that birds, because of their advanced cognitive, social, and communicative abilities (e.g., Emery & Clayton, 2004; Pepperberg, 2007), might be superior models for the evolution of communication, particularly for *vocal learning*, possibly even language (Pepperberg, 2011, in press). Parallels between birdsong and human language (e.g.,

I.M. Pepperberg (✉)
Department of Psychology, Harvard University, Cambridge, MA 02138, USA
e-mail: impepper@media.mit.edu

T. Schilhab et al. (eds.), *The Symbolic Species Evolved*, Biosemiotics 6,
DOI 10.1007/978-94-007-2336-8_7, © Springer Science+Business Media B.V. 2012

issues of adequate input, presence of babbling or practice periods, learning appropriate context for specific vocalizations; Marler, 1970, 1973; Nottebohm, 1970; Byers & Kroodsma, 1992), once commonly cited but often ignored at present, are still valid; neuroanatomical structures subserving vocal behavior in birds and humans are now evaluated for possible homologies (Jarvis et al., 2005; Fitch & Mietchen, in press; cf. Person, Gale, Farries, & Perkel, 2008). Thus, I suggest that birds—their vocal learning, use of various communication systems, and possible "missing link" species between those that do and do not learn song—will shed light on the evolution of vocal communication (Pepperberg, 2007, in press).

7.2 Nonhuman Primate Models for Language Evolution

Some precursors of what was likely early human vocal communication exist in present-day nonhuman primates—e.g., alarm calls (vervets: Strusaker, 1967; Seyfarth, Cheney, & Marler, 1980; Diana and putty-nosed monkeys: Arnold & Zuberbühler, 2006); differential food calling (tamarins: Roush & Snowdon, 2001)—suggesting some form of reference and even combinatorial ability (albeit far simpler than for humans). But vocal *learning* is all but absent in nonhuman primates.[1] So how did learned vocal human language evolve?

One proposal was Hewes' (1973) motor theory,[2] in which voluntary use of manual signals as a means of communication (albeit gestural) arose fairly early in the hominid line—a sensible hypothesis, given modern apes' communicative use of gestures in highly nuanced, contextually-related, culturally-distinct ways (Pollick & de Waal, 2007) and their acquisition of some intentional, referential ASL signs (e.g., Gardner & Gardner, 1969). Initial associations of these manual gestures with innate cries, calls or other movements (e.g., sucking, feeding) could have arisen and become more tightly connected if the combinations enhanced communication; eventually these precursor non-speech movements could then become *articulatory gestures* adapted for communicative intent (e.g., Fogassi & Ferrari, 2004; Studdert-Kennedy, 2005)—the *hidden* constrictions and releases later subsumed by the human vocal tract. But for vocal communication to have evolved as we know it, the brain also had to have transferred voluntary control from manual to vocal gestures and been able to represent someone else's speech as hidden motor articulatory behavior (Liberman & Mattingly, 1985; Vihman, 1993).

That is, two additional steps were necessary. Corballis (1989, 1991, 2003, 2008) suggested that the left hemisphere took control of voluntary manual communicative gestures—which are often lateralized in modern apes (see Hopkins & Cantalupo, 2008)—and that this laterality and voluntary behavior were preserved when manual

[1] Critical rearing experiments have yet to test claims for some form of vocal dialect learning in apes (Crockford, Herbinger, Vigilant, & Boesch, 2004) and marmosets (de la Torre & Snowdon 2009).

[2] A similar proposal exists for birds (Williams & Nottebohm, 1985), but more on that later.

gestures became associated with facial motions. Next, representation of others' speech as motor articulatory behavior would be assisted if, for example, your articulatory system responded to my voice as if you were talking; enter the fortuitous discovery of mirror neurons (MNs), which involve exactly that kind of parity (e.g., Arbib, 2005). MNs also have an inhibitory component, allowing you to choose whether or not to repeat my utterance (Baldissera, Cavallari, Craighero, & Fadiga, 2001). Too, MNs are found in both the language-related Broca's area in humans and a Broca's homologue, F5, in monkeys. The monkey MN system reacts to grasping, mouthing, and related actions (e.g., Fadiga, Fogassi, Pavesi, & Rizzolatti, 1995; Fogassi & Gallese, 2002), being activated during both production and perception of gestures but, interestingly, not vocalizations (Jürgens, 1998). What if, instead of a major brain reorganization to shift from voluntary control of manual to vocal communicative gestures in the hominid line, all that actually happened was the evolution of Broca's area from the monkey-like F5 MN system (Arbib, 2005; Rizzolatti & Arbib, 1998), resulting in an early hominid mirror system as a neural "missing link" between our nonhuman ancestors' communication abilities and modern human language? Arbib (2008), for example, argues for expansion of the projection from F5 that controls vocal folds to one that could control the tongue and lips.[3]

Recently, an additional intriguing correlation between gesture and vocal communication has been observed. Several scientists (e.g., Wilson, Braida, & Reed, 2010; Ro, 2011) have presented data suggesting strong connections between areas of the brain that process hearing and touch. Possibly areas sensitive both to tactile gesture and concomitant calls or cries might early on have processed how one experiences one's own utterances/movements; such areas might later have been rewired so as to additionally process aspects of both types of gestures in others in an MN-like manner.

But these theories do not explain evolution of vocal *learning*, which is basic to language (Pepperberg, 2007, in press). Vocal behavior can be under voluntary control, but be unlearned and quite distinct from language, even if mediated by an MN system—e.g., monkeys choose which alarm call to use and learn when to use it, but the sounds are innately specified. Understanding connections among learning, voluntary control, and MNs requires discussing a form of learning implicated in many aspects of vocal communication—imitation. Initially, MNs were thought to underlie imitation, because when individuals see an action, their MN system enables them to recognize it (through a form of resonance) and to configure their own body parts so as to replicate (imitate) the action if they so choose (even if initially only roughly; Fogassi & Ferrari, 2004; Vauclair, 2004). But monkeys, despite having an MN system, don't precisely imitate (Visalberghi & Fragaszy, 1990, 2002). In fact, monkey MNs cannot respond to or replicate novel actions, only to those already in their repertoire (Chaminade, Meary, Orliaguet, & Decety, 2001; Rizzolatti, Fogassi, &

[3] Recent studies show that various language-related functions (e.g., mapping of auditory sound representations onto motor representations for producing speech; mapping speech sounds onto word concepts) may require parallel processing of widely distributed brain areas (e.g., Holt & Lotto, 2008, Poeppel & Monahan, 2008), but these may also be tied to gesture recognition.

Gallese, 2001). Human MNs, in contrast, seem able to parse a novel behavior into a set of actions that can be *approximated* by variants of actions or collections of actions already in the repertoire (see Arbib, 2005, cf. Dinstein, Hasson, Rubin, & Heeger, 2007; Dinstein, Thomas, Behmann, & Heeger, 2008), thus assisting in imitation of the behavior. Might evolution of MN systems be implicated in the evolution of vocal as well as physical imitation?

The answer is yes, but only if we propose the existence of various levels of imitation and, likely, various types of MNs related to these levels of imitation and learning (Fogassi & Ferrari, 2004), for different species *and* along evolutionary and developmental pathways (Pepperberg, 2005a, 2005b, 2007, in press). Ideas about these possible intermediate forms are presented elsewhere (Pepperberg, 2005a, 2005b, 2007); the implication (e.g., Arbib, 2005, 2008) is that an ancestral MN system, intermediate between that of present humans and nonhuman primates, enabled a simple vocal communication system via imitative learning. This hypothesis is untestable, however, as we lack fossil evidence[4] for appropriate language-ready or proto-linguistic brain structures in this so-called "missing link" (discussion in Pepperberg, in press).

7.3 Avian Models for Language Evolution

But what if we look instead at other creatures that engage in vocal learning, for whom a gestural theory of communicative evolution has also been proposed (Williams & Nottebohm, 1985), and with likely present-day "missing links"—that is, birds? An avian model makes neurobiological sense: Researchers (e.g., Jarvis et al., 2005) argue that avian and mammalian brain areas have a common pallial precursor and that many birds have large cortical-like structures, likely with an MN system for vocal learners that functions in ways similar to that of humans (Bauer et al., 2008; Prather, Peters, Nowicki, & Mooney, 2008). Thus birdsong evolution also likely involved an intermediate MN system, and, unlike the primate line, an intermediate "missing link" species, in terms of vocal learning, may still exist. Specifically, the many parallels between avian and primate species, from those with little in the way of learned vocal communication to those having many traits in common with humans, might extend to those possible missing links, providing an intriguing avian model (Pepperberg, 2011, in press).

7.3.1 Avian Vocal Nonlearners

These birds' communicative behavior is primarily genetically determined (note De Kort & ten Cate, 2001), generally consisting of only a few distinct sounds that can be

[4] We can, of course, draw inferences about brain morphology of early hominins and their descendents from endocasts (note Holloway, Sherwood, Hof, & Rilling, 2009), but such data cannot provide conclusive information concerning brain structures and their specific *interconnections*.

repeated but are rarely combined. Chickens' unlearned alarm calls (Evans, Evans, & Marler, 1993) provide an avian parallel to those of vervets (Cheney & Seyfarth, 1990), in that both species learn appropriate contexts for use of their different calls. Passerine birds such as flycatchers have more complex, but still unlearned, vocal communication systems (e.g., more flexibility in the context of use and meaning) that parallel those of apes: In addition to calls, flycatchers have relatively simple innate songs, consisting of just a few notes, but learn from social interactions how meaning is altered by context; they also combine actions and vocalizations to extend meaning (Smith & Smith, 1992, 1996; Leger, 2005), much like apes (Pollick & de Waal, 2007). Thus, flycatchers who signal different aggression levels by altering the number of repetitions of their single song or vary flight patterns or body postures while singing (i.e., engage in a form of rule-governed behavior that could be interpreted as a very simple combinatory syntax; Smith & Smith, 1992, 1996) could be viewed as living models of an early hominin who might have mixed grunts and gestures to serve a similar purpose (Pepperberg, 2007, 2011, in press; cf. Bickerton, 2003, 2010).

What about parallels in brain structures? In avian and primate species discussed above, brain nuclei obviously must exist to control the physical production of vocalizations; brain centers for vocal *learning* are, however, lacking (Kroodsma & Konishi, 1991; Jürgens cited in Arbib, 2008). Because communication must involve parity for both sender and receiver (Smith, 1997), and an MN system purportedly facilitates this parity (e.g., Arbib, 2005), all these species likely have a simple MN system that codes relationships among another agent's action (e.g., adults' calling), the context of the action (e.g., presence of a particular predator or competitor), and the ability to replicate the action—that is, allowing for choice in *whether* to execute the action (i.e., control over inhibitory neurons so that calls are not emitted in the absence of a receiver) but with strong limitations as to exactly *what* vocal action is possible (Pepperberg, in press).

7.3.2 Avian Vocal Learners

For many avian species, "song" may mean from one to hundreds of songs, songs of a few notes to those of considerable length and complexity, but vocalizations that are not innately specified and that must be *learned*. Birdsong is a simpler communication *system* than human language, but important parallels exist between *vocal learning* in songbirds and humans (e.g., Marler, 1970; Baptista, 1983, 1988; Kroodsma, 1988; Jarvis et al., 2005; Gentner, Fenn, Margoliash, & Nusbaum, 2006). Both birds and humans have (a) a sensitive period during which exposure to the adult system allows development to proceed most rapidly, although acquisition is indeed possible beyond this period, particularly if social interaction is involved[5];

[5] For example: Studies by Baptista (1983, 1988) and his colleagues on white-crowned sparrows showed that the song-learning period described by Marler (1970) for birds that were tape-tutored in social isolation could be doubled if the birds were exposed to live interacting tutors.

(b) a babbling or practice stage wherein juveniles experiment with sounds that will ultimately become part of their repertoire; (c) a need to learn not just what to produce but to understand the appropriate context in which to produce specific vocalizations; (d) the ability to process hierarchically structured vocal sequences, a precursor to grammatical syntax; and (e) lateralized brain structures devoted to acquisition, storage, and production of vocalizations.

The behavioral correlates are obviously of considerable import, but the last point—avian and mammalian brain structures, responsible for vocal learning, now thought to be derived from the same pallial structures (Jarvis et al., 2005)—is central to the use of birds as models for vocal learning (Pepperberg, in press). Of course, the direct correlations between human Broca's area and monkey F5 are unlikely to exist for avian brains—possibly the bits of brain corresponding to specific mammalian language/articulatory gesture centers are apportioned across several song centers in the avian brain (Reiner, pers. comm., April 19, 2005; see Jarvis, 2004). Nevertheless, recent studies strengthen the avian-human correlations, particularly with respect to a possible MN system (Prather et al., 2008; cf. Person et al., 2008).

Some form of avian MN system corresponding to that in humans seems likely. Interestingly, for the song sparrow, Prather et al. (2008) found HVC_X neurons (a population of neurons in the songbird HVC, the higher vocal center, that innervates Area X, important to song learning and perception) that display nearly identical patterns of activity when a bird sings *and* hears the same sequence of notes. The authors refrain from claiming that these are indeed MNs, but the involved brain areas correlate with those involved in human vocal behavior and MNs. Other possible MNs, found in an area (CLM, the caudolateral mesopallium) that links into HVC (Bauer et al., 2008; Keller & Hahnlose, 2009), may provide some additional insights into how articulatory motions are refined in the course of vocal learning for both birds and humans.

7.3.3 The "Missing Link": Evolutionary Pathways, Avian Species

But how do parallels between avian nonlearners, monkeys and apes, and avian vocal learners and humans, provide a model for the *evolution* of vocal learning (Pepperberg, in press)? Because the different forms of avian and primate communication described above not only are reflected in different but parallel neuroanatomical systems (Nottebohm, 1980; Kroodsma & Konishi, 1991; Jarvis & Mello, 2000; Jarvis et al., 2005), but the avian system also provides a model for how a fully-developed vocal learning system could have evolved from pre-existing (nonlearning) motor pathways (e.g., Farries, 2001; Perkel, 2004), via addition and subtraction of certain projections between brain nuclei (e.g., Farries, 2004; Feenders et al., 2008). Too, beaks are often used in ways similar to primate forelimbs; motor control of the beak resides in areas separate from, but near to, the neural song system (Wild, Arends, & Zeigler, 1985); and these areas relate to those controlling human jaw movements (Wild, 1997). By examining avian brain areas coopted for the evolution of song learning and song decoding, we might find parallel areas co-opted in mammals for the evolution of language (Pepperberg, in press). Lieberman (2000) argues that human communication structures evolved from the reptilian brain; certainly so did those of birds (Medina & Reiner, 2000).

Notably, both songbirds and humans have physiological and anatomical features that evolved to produce and process rapid sound sequences (Stevens, 1998; Williams, 1989; Lieberman, 1991; Carr & Soares, 2002; Margoliash, 2003), representations that relate to stored templates of vocalizations (Kuhl, Tsao, & Liu, 2003; Phan, Pytte, & Vicario, 2006), and some form of rule-governed, syntax-like behavior (see Gentner et al., 2006). Parrots, quail, nonhuman mammals and humans parse phonological space similarly, adjusting for auditory context effects (Kuhl, 1981; Kluender, Diehl, & Killeen, 1987; Patterson & Pepperberg, 1994, 1998; Lotto, Kluender, & Holt, 1997; Pepperberg, 2007, in press). Such data suggest phonology evolved to use existent auditory sensitivities basic not just to humans or even mammals, but at least to vertebrates (e.g., Dent, Brittan-Powell, Dooling, & Pierce, 1997; cf. Locke, 1997).

Birds also may be models for mechanisms of primate co-development of gestural and vocal combinations (Pepperberg, 2011). Young children almost simultaneously acquire the ability to combine objects (e.g., spoon-into-cup) and phonological/grammatical units (e.g., "more+X" type of emergent syntax; Greenfield, Nelson, & Salzman, 1972). Greenfield (1991) posited that control of such parallel development initially resides in a single neural structure (roughly Broca's area) that differentiates as a child's brain matures into specialized areas for, respectively, physical combinations versus language and that such competence was a critical aspect of language (i.e., human) development. Subsequent research on both physical combinatorial behavior in nonhuman primates (Johnson-Pynn, Fragaszy, Hirsh, Brakke, & Greenfield, 1999) and combinatorial communicative acts by apes (*Pan paniscus, P. troglodytes*) trained in a human-based code (Greenfield & Savage-Rumbaugh, 1990, 1991) showed that apes' combination of physical objects and also labels (e.g., "more tickle") are similar to, if simpler than, those of young children, but in monkeys such behavior develops only with intensive training and to a much more limited extent. Greenfield (1991) then proposed that nonhuman primate behavior derives from a homologous structure just predating the evolutionary divergence of apes and hominids (see Deacon, 1992). Such arguments support Hewes' (1973) thesis. Notably, however, some bird species also show co-occurrences of hierarchical vocal and physical combinations. Grey parrots trained to communicate with humans using English speech have the same spontaneously co-occurring vocal and physical combinatory behavior as children and apes (Pepperberg & Shive, 2001). Male marsh wrens (*Cistothorus palustris*) form complicated woven nests (Leonard & Picman, 1987) as they construct/memorize hierarchies of neighbors' song repertoires to order their own responses serially (i.e., reorder or recombine their songs in new ways) to best defend their territories; Kroodsma, 1979). Whatever neural structures are involved, parallel physical and vocal combinatory behavior is not limited to primates.

But the real use of birds as evolutionary models hinges on possible present-day "missing links": two avian species that apparently straddle the vocal learning-nonlearning divide and might be models for missing hominin ancestors that did the same (Pepperberg, 2011, in press). One of these avian species, the three-wattled bellbird (*Procnias tricarunculata*), a close relative of flycatchers, is supposedly a suboscine (at least it has, until now, been classified as such)—the technical term for birds with innate song—but seems to learn its songs (Snow, 1973); the evidence is that males have dialects, can be bilingual with respect to these dialects (at least for several years), and that a close relative, the bare-throated bellbird (*P. nudicollis*), learns allospecific song (that of another species; Kroodsma, 2005). DNA samples of the three-wattled bellbird prove that the different dialects are from one, not different closely-related, species (Saranathan, Hamilton, Powell, Kroodsma, & Prum, 2007). Adding these facts to the knowledge that some bellbirds don't begin to sound like (or even look like) adults until they are four or five years old (Kroodsma, 2005), we find a pattern that is highly unusual, both for suboscines and "normal" song-learning species—known as oscines. Even oscines that continue learning songs over their lifetimes usually have a recognizable, characteristic song their first year as an adult. And, when bellbirds alter their songs in adulthood, they don't seem to change their overall dialect but rather shift frequency (pitch) over the years[6]; apparently older males shift, forcing younger ones to shift as well or lose status (and possibly mating chances) within the group (Kroodsma, 2005). The bellbirds' learning abilities thus seem similar to oscines, except for the extraordinarily long juvenile stage and the fact that, as noted above they are technically classified as suboscines. We might expect that they, like oscines, have specific brain areas devoted to song learning, but no studies have yet been performed. Given their unusually prolonged babbling stage, we might also expect a brain that is "differently" equipped for learning, but, again, no experiments have yet been done. Might bellbirds' behavior be explained by a vocal MN system that is primitive compared to that of the ocsines? An MN system that, as a consequence, is slow to mature, slow to take the bellbird beyond the babbling stage (Pepperberg, 2007, 2011, in press)?

Such hypotheses support the use of bellbirds as models for vocal learning in a "missing-link" hominin species (or multiple species) that bridged the gap between *Homo sapiens*' and our nonhuman primate ancestors' communication, that is, as a model for an intermediary MN system mediating the first elements of vocal learning (Pepperberg, 2011, in press).[7] Such a model could help us to determine what is innate and what is learned. Most likely a continuum, rather than a sharp break, existed between innate and learned, with certain communicative elements shifting as flexibility provided evolutionary advantages. Conceivably, the same evolutionary pressures that led from the innate, relatively simple song of true suboscines to the

[6] Other features of the song have also changed over the years (Kroodsma, pers. comm., September 2005), but the change in frequency is the most obvious (Kroodsma, 2005).

[7] The bellbird is endangered, but conceivably data might be obtained in the future from captive birds in a noninvasive manner.

fairly simple but slowly learned song of the bellbird to the amazing complexity of, for example, the brown thrasher's hundreds of songs were exerted on the nonhuman-to-hominid line (Pepperberg, in press). If so, these evolutionary pressures were likely exerted on a MN system, such that the complexity of the MN system and the complexity of the behavior involved evolved in parallel, synergistically supporting the next evolutionary stage (Pepperberg, 2007, in press). Articulatory gestures grounded in feeding behavior and contact calls/cries that can be co-opted for other uses were as likely in birds (see Homberger, 1986) as primates; possibly MN systems shifted in the same manner (Pepperberg, in press).

7.4 Conclusions

In sum, although the above arguments are strongest if avian and human communicative abilities did indeed evolve convergently—adapting independently in association with similar environmental pressures—a common core of skills nonetheless likely underlies complex cognitive and communicative behavior across species, even if their specific skills manifest somewhat differently (Pepperberg, in press). The main point, in any case, is that because birds—like humans (but few other mammals)—*learn* their vocal communication systems, they can provide models for both acquisition and use of vocal behavior. Marler (1973) suggested this possibility decades ago, albeit in a more limited construct; later studies on the effects of social interaction on vocal learning served to strengthen the potential use of avian models (review in Pepperberg, 2004). Specifically, few theses concerning language origins focus on the evolution of *vocal learning* as the basis for sophisticated communicative skills, yet vocal learning (and its interconnections with social interaction) is a central issue—not only because humans must learn to communicate in the vocal mode, but also because it is one of the most transparent of modes for study (i.e., what is learned is obvious; Pepperberg, 1999, 2004) and allows for cultural evolution and adaptation to novel circumstances. Although we no longer have access to the precursor neuroanatomy that gave rise to current human language abilities, parallels between the acquisition, development, and use of current human and some avian communication systems suggest that parallels likely existed in their evolutionary histories. As a consequence, species such as the bellbird could be a model for the missing human precursor (Pepperberg, in press).

This chapter (along with other, more detailed presentations; e.g., Pepperberg, in press) is meant to suggest lines of research, not definitively answer the difficult questions about origins of communicative abilities and language. Present-day humans can only guess at the concatenation of the many cultural, social, and neuroanatomical changes likely responsible. But maybe the bare-bones model presented here (given space limitations) will stimulate studies using avian models.

Acknowledgments Manuscript preparation was supported by donors to *The Alex Foundation*. Ideas described herein were developed partly during a Bunting Fellowship at the Radcliffe Institute. Significant portions of this manuscript have been adapted from Pepperberg (2011, in press).

References

Arbib, M. A. (2005). From monkey-like action recognition to human language: An evolutionary framework for neurolinguistics. *Behavioral & Brain Sciences, 28*, 105–167.
Arbib, M. A. (2008). From grasp to language: Embodied concepts and the challenge of abstraction. *Journal of Physiology-Paris, 102*, 4–20.
Arnold, K., & Zuberbühler, K. (2006). Semantic combinations of primate calls. *Nature, 441*, 303.
Baldissera, F., Cavallari, P., Craighero, L., & Fadiga, L. (2001). Modulation of spinal excitability during observation of hand actions in humans. *European Journal of Neuroscience, 13*, 190–194.
Baptista, L. F. (1983). Song learning. In A. H. Brush & G. A. Clark, Jr. (Eds.), *Perspectives in ornithology* (pp. 500–506). Cambridge: Cambridge University Press.
Baptista, L. F. (1988). Song learning in white-crowned sparrows (*Zonotrichia leucophrys*): Sensitive phases and stimulus filtering revisited. *Proceedings of the 100th Deutsche Ornithologische Gesellschaft: Current topics in avian biology, Bonn.*
Bauer, E. E., Coleman, J. J., Roberts, T. F., Roy, A., Prather, J. F., & Mooney, R. (2008). A synaptic basis for auditory-vocal integration in the songbird. *Journal of Neuroscience, 28*, 1509–1522.
Bickerton, D. (2003). Symbol and structure: A comprehensive framework for language evolution. In M. H. Christiansen & S. Kirby (Eds.), *Language evolution* (pp. 77–93). Oxford: Oxford.
Bickerton, D. (2010). *Adam's tongue*. New York: Hill and Wang.
Byers B. E., & Kroodsma, D. E. (1992), Development of two song categories by chestnut-sided warblers. *Animal Behaviour, 44*, 799–810.
Carr, C. E., & Soares, D. (2002). Evolutionary convergence and shared computational principles in the auditory system. *Brain, Behavior & Evolution, 59*, 294–311.
Chaminade, T., Meary, D., Orliaguet, J.-P., & Decety, J. (2001). Is perceptual anticipation a motor simulation? A PET study. *Brain Imaging, 12*, 3669–3674.
Cheney, D. L., & Seyfarth, R. M. (1990). *How monkeys see the world*. Chicago: University of Chicago Press.
Corballis, M. C. (1989). Laterality and human evolution. *Psychological Review, 96*, 492–505.
Corballis, M. C. (1991). *The lopsided ape: evolution of the generative mind*. Oxford: Oxford.
Corballis, M. C. (2003). From mouth to hand: Gesture, speech, and the evolution of right-handedness. *Behavioral & Brain Sciences, 26*, 199–260.
Corballis, M. C. (2008). Of mice and men—and lopsided birds. *Cortex, 44*, 3–7.
Crockford, C., Herbinger, I., Vigilant, L., & Boesch, C. (2004). Wild chimpanzees produce group-specific calls: A case for vocal learning? *Ethology, 110*, 221–243.
Deacon, T. W. (1992), Brain-language coevolution. In J. A. Hawkins & M. Gel-Man (Eds.), *The evolution of human languages* (Vol. 10, pp. 49–83). Redwood City, CA: Addison-Wesley.
Deacon, T. W. (1997). *The symbolic species: The co-evolution of language and the brain*. New York: Norton.
De Kort, S. R., & ten Cate, C. (2001). Response to interspecific vocalizations is affected by degree of phylogenetic relatedness in Streptopelia doves. *Animal Behaviour, 61*, 239–247.
de la Torre, S., & Snowdon, C. T. (2009). Dialects in pygmy marmosets? Population variation in call structure. *American Journal of Primatology, 71*, 1–10.
Dent, M. L., Brittan-Powell, E. F., Dooling, R. J., & Pierce, A. (1997). Perception of synthetic /ba/-/wa/ speech continuum by budgerigars (*Melopsittacus undulatus*). *Journal of the Acoustical Society of America, 102*, 1891–1897.
Dinstein, I., Hasson, U., Rubin, N., & Heeger, D. J. (2007). Brain areas selective for both observed and executed movements. *Journal of Neurophysiology, 98*, 1415–1427.
Dinstein, I., Thomas, M., Behmann, M., & Heeger, D. J. (2008). A mirror up to nature. *Current Biology, 18*, R13–R18.
Emery, N. J., & Clayton, N. S. (2004). The mentality of crows: convergent evolution of intelligence in corvids and apes. *Science, 306*, 1903–1907.
Evans, C. S., Evans, L., & Marler, P. (1993). On the meaning of alarm calls: Functional reference in an avian vocal system. *Animal Behaviour, 46*, 23–38.

Fadiga, L., Fogassi, L., Pavesi, G., Rizzolatti, G. (1995). Motor facilitation during action observation: A magnetic simulation study. *Journal of Neurophysiology, 73*, 2608–2611.

Farries, M. A. (2001). The oscine song system considered in the context of the avian brain: Lessons learned from comparative neurobiology. *Brain, Behavior, & Evolution, 58*, 80–100.

Farries, M. A. (2004). The avian song system in comparative perspective. *Annals of the New York Academy of Sciences, 1016*, 61–76.

Feenders, G., Liedvogel, M., Rivas, J., Zapka, M., Horita, H., Hara, E., et al. (2008). Molecular mapping of movement-associated areas in the avian brain: A motor theory for vocal learning. *PLoS One, 3*(3): e1768. doi:10.1371/journal.pone.0001768

Fitch, W. T., & Mietchen, D. (in press) Convergence and deep homology in the evolution of spoken language. In J. J. Bolhuis & M. Everaet (Eds.), *Birdsong, speech, and language: Converging mechanisms*. Cambridge, MA: MIT Press.

Fogassi, L., & Ferrari, P. F. (2004). Mirror neurons, gestures, and language evolution. *Interaction Studies, 5*, 345–363.

Fogassi, L., & Gallese, V. (2002). The neural correlates of action understanding in non-human primates. In M. I. Stamenov (Ed.), *Mirror neurons and the evolution of brain and language* (pp. 21–43). Philadelphia, PA: John Benjamins.

Gallese, V., & Lakoff, G. (2005). The brain's concepts: The role of the sensory-motor system in reason and language. *Cognitive Neuropsychology, 22*, 455–479.

Gardner, R. A., & Gardner, B. T. (1969). Teaching sign language to a chimpanzee. *Science, 165*, 664–672.

Gentner, T. Q., Fenn, K. J., Margoliash, D., & Nusbaum, H. C. (2006). Recursive syntactic pattern learning by songbirds. *Nature, 440*, 1204–1207.

Greenfield, P. M. (1991). Language, tools and brain: The ontogeny and phylogeny of hierarchically organized sequential behavior. *Behavioral & Brain Sciences, 14*, 531–595.

Greenfield, P. M., & Savage-Rumbaugh, E. S. (1990). Grammatical combination in *Pan paniscus*: Processes of learning and invention in the evolution and development of language. In S. T. Parker & K. R. Gibson (Eds.), *'Language' and intelligence in monkeys and apes: Comparative developmental perspectives* (pp. 540–578). New York: Cambridge.

Greenfield, P. M., & Savage-Rumbaugh, E. S. (1991). Imitation, grammatical development, and the invention of protogrammar by an ape. In N. A. Krasnegor, D. M. Rumbaugh, R. L. Schiefelbusch, & M. Studdert-Kennedy (Eds.), *Biological and behavioral determinants of language development* (pp. 235–258). Hillsdale, NJ: Erlbaum.

Greenfield, P. M., Nelson, K., & Salzman, E. (1972). The development of rulebound strategies for manipulating seriated nesting cups: A parallel between action and grammar. *Cognitive Psychology, 3*, 291–310.

Hewes, G. W. (1973). Primate communication and the gestural origin of language. *Current Anthropology, 33*, 65–84.

Holloway, R. L., Sherwood, C. C., Hof, P. R., & Rilling, J. K. (2009). Evolution of the brain in humans—paleoneurology. In M. D. Binder, N. Hirokawa, U. Windhorst, & M. C. Hirsch (Eds.), *Encyclopedia of neuroscience, Part 5* (pp. 1326–1334). New York: Springer.

Holt, L. L., & Lotto, A. J. (2008). Speech perception withini an auditory cognitive science framework. *Current Directions in Psychological Science, 17*, 42–46.

Homberger, D. G. (1986). *The lingual apparatus of the African grey parrot, Psittacus erithacus Linne (Aves: Psittacidae) Description and theoretical mechanical analysis. Ornithological Monographs, No. 39*. Washington, DC: The American Ornithologists' Union.

Hopkins, W. D., & Cantalupo, C. (2008). Theoretical speculations on the evolutionary origins of hemispheric specialization. *Current Directions in Psychological Science, 17*, 233–237.

Jarvis, E. (2004). Learned birdsong and the neurobiology of human language. *Annals of the New York Academy of Sciences, 1016*, 749–777.

Jarvis, J. D., Güntürkün, O., Bruce, L., Csillag, A., Karten, H., Kuenzel, W., et al. (2005). Avian brains and a new understanding of vertebrate evolution. *Nature Reviews Neuroscience, 6*, 151–159.

Jarvis, J. D., & Mello, C. V. (2000). Molecular mapping of brain areas involved in parrot vocal communication. *Journal of Comparative Neurology, 419*, 1–31.

Johnson Pynn, J., Fragaszy, D. M., Hirsh, E. M., Brakke, K. E., & Greenfield, P. M. (1999). Strategies used to combine seriated cups by chimpanzees (*Pan troglodytes*), bonobos (*Pan paniscus*), and capuchins (*Cebus apella*). *Journal of Comparative Psychology, 113*, 137–48.

Jürgens, U. (1998). Neuronal control of mammalian vocalization, with special reference to the squirrel monkey. *Naturwissenschaften, 85*, 376–388.

Keller, G. B., & Hahnlose, R. H. R. (2009). Neural processing of auditory feedback during vocal practice in a songbird. *Nature, 457*, 187–190.

Kluender, K. R., Diehl, R. L., & Killeen, P. R. (1987). Japanese quail can learn phonetic categories. *Science, 237*, 1195–1197.

Kroodsma, D. E. (1979). Vocal dueling among male marsh wrens: Evidence for ritualized expressions of dominance/subordinance. *Auk, 96*, 506–515.

Kroodsma, D. E. (1988). Song types and their use: Developmental flexibility of the male blue-winged warbler. *Ethology, 79*, 235–247.

Kroodsma, D. E. (2005). *The singing life of birds* (pp. 96–101). New York: Houghton Mifflin.

Kroodsma, D. E., & Konishi, M. (1991). A suboscine bird (Eastern phoebe, *Sayornis phoebe*) develops normal song without auditory feedback. *Animal Behaviour, 42*, 477–487.

Kuhl, P. K. (1981). Discrimination of speech by nonhuman animals: Basic auditory sensitivities conducive to the perception of speech-sound categories. *Journal of the Acoustical Society of America, 70*, 340–349.

Kuhl, P., Tsao, F. M., & Liu, H. M. (2003). Foreign-language experience in infancy: Effects of short-term exposure and social interaction on phonetic learning. *Proceedings of the National Academy of Sciences, USA, 100*, 9096–9101.

Leger, D. W. (2005). First documentation of combinatorial song syntax in a suboscine passerine species. *Condor, 107*, 765–774.

Leonard, M. L., & Picman, J. (1987). The adaptive significance of multiple nest building by male marsh wrens. *Animal Behaviour, 35*, 271–77.

Liberman, A. M., & Mattingly, I. G. (1985). The motor theory of speech perception revised. *Cognition, 21*, 1–36.

Lieberman, P. (1991). Preadaptation, natural selection, and function. *Language & Communication, 11*, 63–65.

Lieberman, P. (2000). *Human language and our reptilian brain*. Cambridge, MA: Harvard.

Locke, J. L. (1997). A theory of neurolinguistic development. *Brain and Language, 58*, 265–326.

Lotto, A. J., Kluender, K. R., & Holt, L. L. (1997). Perceptual compensation for coarticulation by Japanese quail (*Coturnix coturnix japonica*). *Journal of the Acoustical Society of America, 102*, 1134–1140.

Margoliash, D. (2003). Offline learning and the role of autogenous speech: New suggestions from birdsong research. *Speech Communication, 41*, 165–178.

Marler, P. (1970). A comparative approach to vocal learning: Song development in white crowned sparrows. *Journal of Comparative & Physiological Psychology, 71*, 1–25.

Marler, P. (1973). Speech development and bird song: Are there any parallels? In G. A. Miller (Ed.), *Communication, language, and meaning*. (pp. 73–83). New York: Basic Books.

Medina, L., & Reiner, A. (2000). Do birds possess homologues of mammalian primary visual, somatosensory and motor cortices? *Trends in Neurosciences, 23*, 1–12.

Nottebohm, F. (1970). Ontogeny of bird song. *Science, 167*, 950–956.

Nottebohm, F. (1980). Brain pathways for vocal learning in birds: A review of the first ten years. *Progress in Psychobiology and Physiological Psychology, 9*, 85–124.

Patterson, D. K., & Pepperberg, I. M. (1994). A comparative study of human and parrot phonation: Acoustic and articulatory correlates of vowels. *Journal of the Acoustical Society of America, 96*, 634–648.

Patterson, D. K., & Pepperberg, I. M. (1998). A comparative study of human and Grey parrot phonation: Acoustic and articulatory correlates of stop consonants. *Journal of the Acoustical Society of America, 103*, 2197–2213.

Pepperberg, I. M. (1999). *The Alex studies*. Cambridge, MA: Harvard University Press

Pepperberg, I. M. (2004). The evolution of communication from an avian perspective. In D. K. Oller & U. Griebel (Eds.), *Evolution of communication systems: A comparative approach* (pp. 171–192). Cambridge, MA: MIT Press.

Pepperberg, I. M. (2005a). Evolution of language from an avian perspective. In M. Tallerman (Ed.), *Language origins: Perspectives on evolution* (pp. 239–261). Oxford: Oxford.

Pepperberg, I. M. (2005b). Insights into vocal imitation in Grey parrots (*Psittacus erithacus*). In S. Hurley & N. Chader (Eds.), *Perspectives on imitation: From mirror neurons to memes* (Vol. 1, pp. 243–262). Cambridge, MA: MIT Press.

Pepperberg, I. M. (2007). Emergence of linguistic communication: Studies on Grey parrots. In C. Lyon, C. L. Nehaniv, & A. Cangelosi (Eds.), *Emergence of communication and language* (pp. 355–386). London: Springer.

Pepperberg, I. M. (2011). Evolution of communication and language: Insights from parrots and songbirds. In M. Tallerman & K. Gibson (Eds.), *Oxford handbook of language evolution* (pp. 109–119). London: Oxford.

Pepperberg, I. M. (in press). Evolution of vocal communication: an avian model. In J. J. Bolhuis & M. Everaet (Eds.), *Birdsong, speech, and language: Converging mechanisms*. Cambridge, MA: MIT Press.

Pepperberg, I. M., & Shive, H. (2001). Simultaneous development of vocal and physical object combinations by a Grey Parrot (*Psittacus erithacus*): Bottle caps, lids, and labels. *Journal of Comparative Psychology, 115*, 376–384.

Perkel, D. J. (2004). Origin of the anterior forebrain pathway. *Annals of the New York Academy of Sciences, 1016*, 736–748.

Person, A. L., Gale, S. D., Farries, M. A., & Perkel, D. J. (2008). Organization of the songbird basal ganglia, including Area X. *Journal of Comparative Neurology, 508*, 840–866.

Phan, M. L., Pytte, C. L., & Vicario, D. S. (2006). Early auditory experience generates long lasting memories that may subserve vocal learning in songbirds. *Proceedings of the National Academy of Sciences USA, 103*, 1088–1093.

Poeppel, D., & Monahan, P. J. (2008). Speech perception: Cognitive foundations and cortical implementation. *Current Directions in Psychological Science, 17*, 80–85.

Pollick, A. S., & de Waal, F. (2007). Ape gestures and language evolution. *Proceedings of the National Academy of Sciences USA, 104*, 8184–8189.

Prather, J. F., Peters, S., Nowicki, S., Mooney, R. (2008). Precise auditory–vocal mirroring in neurons for learned vocal communication. *Nature, 451*, 305–310.

Rizzolatti, G., & Arbib, M. (1998). Language within our grasp. *Trends in Neuroscience, 21*, 188–194.

Rizzolatti, G., Fogassi, L., Gallese, V. (2001). Neurophysiological mechanisms underlying the understanding and imitation of actions. *Nature Review Neurology, 2*, 661–670.

Ro, T. (May, 2011). *Feeling sounds: Auditory influences on touch perception*. Paper presented at the 161st meeting of the Acoustical Society of America, Seattle, WA.

Roush, R. S., & Snowdon, C. T. (2001). Food transfer and development of feeding behavior and food-associated vocalizations in cotton-top tamarins. *Ethology, 107*, 415–429.

Saranathan, V., Hamilton, D., Powell, G. V. N., Kroodsma, D. E., & Prum, R. O. (2007). Genetic evidence supports song-learning in the three-wattled bellbird *Procnias trucarunculata* (Cotingidae). *Molecular Ecology, 16*, 3689–3702.

Seyfarth, R., Cheney, D., & Marler, P. (1980). Monkey responses to three different alarm calls: Evidence for predator classification and semantic communication. *Science, 210*, 801–803.

Slocombe, K., Waller, B. M., & Liebal, K. (2011). The language void: the need for multimodality in primate communication research. *Animal Behaviour, 81*, 919–924.

Smith, W. J. (1997). The behavior of communicating, after twenty years. In D. H. Owings, M. D. Beecher, & N. S. Thompson (Eds.), *Perspectives in ethology* (Vol. 12, pp. 7–53). New York: Plenum.

Smith, W. J., & Smith, A. M. (1992). Behavioral information provided by two song forms of the Eastern kingbird, *T. tyrannus*. *Behaviour, 120*, 90–102.

Smith, W. J., & Smith, A. M. (1996). Information about behavior provided by Louisiana waterthrush, *Seurus motacilla* (Parulinae), songs. *Animal Behaviour, 51*, 785–799.

Smith, A. D. M., Smith, K., & Ferrer i Cancho, R. (Eds.). (2008). *The Evolution of language: Proceedings of the 7th international conference*. London: World Scientific Publishing Company.

Snow, D. W. (1973). Distribution, ecology, and evolution of the bellbirds (*Procnias*, Cotingidae). *Bulletin of the British Museum of Natural History, 25*, 369–391.

Stevens, K. N. (1998). *Acoustic phonetics*. Cambridge, MA: MIT Press.

Studdert-Kennedy, M. (2005). How did language go discrete? In M. Tallerman (Ed.), *Language origins: Perspectives on evolution* (pp. 48–67). Oxford: Oxford University Press.

Strusaker, T. (1967). Auditory communication among vervet monkeys (*Ceropithecus aethiops*). In S. Altmann & K. Gibson (Eds.), *Social communication among primates* (pp. 281–324). Chicago: University of Chicago Press.

Vauclair, J. (2004). Lateralization of communicative signals in nonhuman primates and the hypothesis of the gestural origin of language. *Interaction Studies, 5*, 365–386.

Vihman, M. H. (1993). Variable paths to early word production. *Journal of Phonetics, 21*, 61–82.

Visalberghi, E., & Fragaszy, D. M. (1990). Do monkeys ape? In S. T. Parker & K. R. Gibson (Eds.), *'Language' and intelligence in monkeys and apes* (pp. 247– 273). Cambridge: Cambridge University Press.

Visalberghi, E., & Fragaszy, D. M. (2002). "Do monkeys ape?" Ten years after. In K. Dautenhahn & C. L. Nehaniv (Eds.), *Imitation in animals and artifacts* (pp. 471–499). Cambridge, MA: MIT Press.

Wild, M. (1997). Neural pathways for the control of birdsong production. *Journal of Neurobiology, 33*, 653–670.

Wild, M., Arends, J. J. A., & Zeigler, H. P. (1985) Telencephalic connections of the trigeminal system in the pigeon (*Columba livia*): a trigeminal sensorimotor circuit. *Journal of Comparative Neurology, 234*, 441–464.

Williams, H. (1989). Multiple representations and auditory-motor interactions in the avian song system. *Annals of the New York Academy of Sciences, 563*, 148–164.

Williams, H., & Nottebohm, F. (1985). Auditory responses in avian vocal motor neurons: A motor theory for song perception in birds. *Science, 229*, 279–282.

Wilson, E. C., Braida, L. D., & Reed, C. M. (2010). Perceptual interactions in the loudness of combined auditory and vibrotactile stimuli. *Journal of the Acoustical Society of America, 127*, 3038–3043.

Chapter 8
From Parsing Actions to Understanding Intentions

Richard W. Byrne

Abstract "Having a theory of mind" is often invoked to explain remarkable abilities in social cognition, but in reality this is little more than a re-description of the data, a challenge for theorists to understand what it really means, and how we—and perhaps some other species—evolved those abilities. I argue that these abilities most likely grew out of an understanding of action, and in particular that the key first step was the evolution, by shared ancestors of modern humans and the living great apes, of an ability to parse the skilled actions of others. Parsing organized, goal-directed behaviour allows the intention of the model to be "seen" (as the typical result which terminates action), and the cause-effect relations in the steps of the process to be "seen" (as the sequence and coordination of actions necessary for that result to be achieved). In contrast, reasoning about true intentionality and true causality may depend on possession of language, and be of relatively minor usefulness in everyday interactions.

When we notice someone engaged in activity, we see not only how their body moves and what effects those movements are having on other things, but we also see what it *means*. The meaning of action includes *what is likely to happen next*, as a consequence of what has been done already; and what overall result is to be expected from the activity, in short, *why it is being done*. This description applies to the simplest of organized, purposeful actions but also to what is arguably our most sophisticated cognitive ability, the ability to talk. When we hear someone talking our language, we don't merely register a series of sounds, phonemes, words, phrases, and meanings; we immediately have some understanding of what thoughts have led to their speaking, whereabouts (metaphorically) their speech is going, and what pragmatic effects the speaker might be trying to achieve by it. These observations are so familiar and commonplace that normally we pay them no heed: rather, we only notice when people do things that make no sense to us or say things that seem irrational. Yet our ability to perceive the everyday world of social action as a world of meanings, purposes, intentions and reasons is an extraordinary one.

At the heart of the ability to read meaning in perceived action is *parsing*. A characteristic of skilled action is that, in physical terms, no structure is overt. The

R.W. Byrne (✉)
School of Psychology, University of St Andrews, St Andrews, Fife KY16 9JP, Scotland
e-mail: rwb@st-andrews.ac.uk

T. Schilhab et al. (eds.), *The Symbolic Species Evolved*, Biosemiotics 6,
DOI 10.1007/978-94-007-2336-8_8, © Springer Science+Business Media B.V. 2012

sequence of components is *linear*—although with action that involves both hands or extends to mouth or feet, there may be several partly-linked streams running in parallel, each linear in sequence. But whether driving a car, uttering a sentence, or baking a cake, all that is physically present to be perceived is *smooth, fluid movement*. The absence of "real gaps" between many of the separate words in a spoken sentence is part of every entry-level linguistics course; and just the same is true of manual actions. Once a skilled sequence of actions has been assembled, practising will result in smoother and smoother performance, to the point when underlying structure is not signalled by any detectable interruptions in the sequence. That is the first part of the parsing problem: seeing a linear sequence of fluid behaviour, but perceiving it as segmented into discrete units that correspond to real entities for the actor who is observed.

The second parsing problem concerns the fact that organized, complex behaviour is *hierarchical* in structure. This means that elements lying together in sequence may be closely related logically, because they form part of a module or subroutine or phrase (depending on what sort of behaviour is under discussion); or much less closely related, only lying together by virtue of the organization of some higher order unit of organization. To understand action, and thereby detect the meaning in it, it is crucial to parse its hierarchical structure accurately. The output of the parsing process must go beyond a sequence of discrete units, to get at the underlying *relationships* that we conventionally represent in terms of a bracketed string, a tree-diagram or a phrase-structure grammar. Without that, there would be no systematic way to connect observed behaviour to the purposes that underlie it in the mind of the actor—and thus, to go on to understand the actor's *intentions* and the *cause-and-effect* of how that particular behaviour is efficient for achieving their purposes.

It is the thesis of this chapter that parsing has its evolutionary origins in an unexpected place. Rather than deriving from a selective pressure for more sophisticated vocal communication, a function in which we see the full flowering of parsing ability in modern humans, I argue that parsing was originally part of a *feeding adaptation*; and that these abilities, evolved for efficient feeding, were themselves based on earlier evolution of abilities in *social behaviour reading*.

After briefly considering primate vocal communication, I will first sketch the evidence that a segmentation system, one that can parse a smooth behavioural performance into separate but meaningful units of action, is present in monkeys—and probably in many other species even more distant from us on an evolutionary timescale. The main biological function of action segmentation in those species is most likely the estimation of current behavioural dispositions in conspecifics and the prediction of their likely actions in the immediate future. Among the primates, it seems, only in great apes did rather special abilities of *hierarchical* parsing and anticipatory planning develop, and I will suggest that these special capacities may be parasitic on that earlier segmentation system—but are not dependent on any prior ability to understand intentions or causality. In non-human great apes hierarchical parsing seems only to be only found within the *manual skill domain,* where it functions in the wild by allowing more efficient feeding; and there are plausible ecological reasons why enhanced feeding abilities should have evolved specifically in the great apes. Under the artificial conditions of human rearing, hierarchical parsing and anticipatory planning give rise to a wide range of richly complex behaviours, and can be deliberately co-opted into human-derived communication systems such as American Sign Language.

Given such abilities in living apes, in the manual-spatial domain, it is only a small step to speculate that in one of our own early ancestors these hierarchical parsing skills became available also in the vocal-auditory domain. Linguistic syntax is thereby seen as evolutionarily derived from hierarchical behaviour parsing. Further implications may be drawn out. As emphasized, behaviour parsing is not dependent on prior causal-intentional understanding; however, it could have been a crucial step on the way to achieving this level of mental representation—an essential precursor to human cognition, and still a necessary part of the process of representing phenomena as causal-intentional structures. Moreover, the fact that so much can in principle be achieved *without* involving that level of mental representation—parsing of behavioural structure, social learning of complex skills by program-level imitation, and so on—opens the door to a heretical thought. Could it be that the prevalence of causal-intentional interpretation of our social world is illusory, a consequence of retrospective contemplation? Certainly, when we choose to ponder causation and attribution, or when we are asked to justify our actions by others, as adult humans we are well able to construct causal-intentional theories that make sense. But perhaps the cut-and-thrust of everyday social action and interaction does not need this mentalizing, or would indeed be slowed or disrupted by it (Apperly & Butterfill, 2009; Bargh & Chartrand, 1999), and we should look elsewhere for the evolutionary functions of theory of mind and causal reasoning.

8.1 Primate Vocal Communication—Primitive Speech?

Extensive study for many years has focused on primate vocalizations, driven partly by theoretical interest in language origins and partly by the availability of sound-manipulation technology. We now know that the potential for flexibility in the production of calls by primates is very limited.

No primate can copy another's sounds, in the way that many birds and some cetaceans can do (Janik & Slater, 1997). Even vocal dialects are nearly unknown in primates, except in cases where human influence may have unintentionally conditioned a local variation (Green, 1975; Mitani, Hasegawa, Gros-Louis, Marler, & Byrne, 1992). "Nearly", because there is recent evidence that zoo communities of chimpanzees develop characteristic group dialects (Auser & Wrangham, 1987), and adjacent communities in the wild have been found to differ more in their vocalizations than do more distant ones—just as a dialect in human communities can serve to identify group membership and label an out-group (Crockford, Herbinger, Vigilant, & Boesch, 2004). Even in these cases, the modifications are small ones, to calls which are biologically fixed in form. Young primates of many species have often been reared out of any auditory contact with conspecifics: nevertheless, they all develop a normal repertoire of vocalizations. Learning does play a role in the normal development of calling, but this is *contextual learning* not production learning (Janik & Slater, 1997): primates learn the appropriate circumstances in which to call, rather than learning the calls themselves. The famous case of predator-specific alarm calls in vervet monkeys shows this process in action (Seyfarth & Cheney, 1986). The referential specificity of these calls is to a limited extent innate, but whereas a young vervet will initially make an "eagle alarm" to a wide range of flying things (even a large, falling leaf on one occasion), as it matures calling is restricted to large broad-winged birds, then specifically to raptorial species, and finally the call is given almost exclusively to the martial eagle *Polemaetus bellicosus*, a vervet's main aerial predator.

Most non-human primates have a vocal repertoire of more-or-less discrete calls, but also show some graded variation, most extensively in the chimpanzee and gorilla (Marler & Tenaza, 1977). Animals have been found to perceive human speech categorically (Kuhl, 1982), and primate calls which sound like a smoothly varying continuum to the human ear have been shown to be composed of several circumstance-specific and function-specific calls (Gouzoules, Gouzoules, & Marler, 1984). However, nothing remotely like the multiple levels of patterning and syntactic structuring found in human speech has been detected in any primate vocal system. The closest to hierarchical organization is the recent discovery that one call can modify another and so qualify its degree of definiteness, as if adding "maybe" to its meaning (Zuberbuhler, 2002). This is a far cry from the generative, productive nature of everyday human speech, and theories that try to make direct connection between primate vocal communication and language have a large gap to fill—with pure speculation. For these reasons, I now turn instead to the manual skills of human and non-human primates.

8.2 Segmentation of the Action Stream

When we approach a range of problems, from car maintenance to public speaking, we do so with a pre-existing repertoire of motor actions ready to deploy. Some of these "elements" of action are no doubt innate; and many others are constructed by trial-and-error exploration of previous similar situations; and a significant part of our repertoire of actions is learned by noticing other people's behaviour (and listening to their speech). By all three routes of acquisition, we meet the many novel problems in adulthood prepared with a rich vocabulary of elements of action which we can permute and organize into creative solutions, as well as deploy effortlessly in response to more familiar demands.

The stream of action that we observe, however, does not come with ready-made gaps that correspond to logically distinct elements. This has been classically noted to apply to speech, where a sound-gap is more likely to be part of a plosive consonant than to signal a new word, but in fact the point applies to all skilled behaviour. Thus, with motor action, the physical stimulus that confronts us is smooth and fluid, not segmented. How are we nevertheless able to pick out functional elements in the smooth and apparently unbroken flow of action?

To be used as building blocks in effective motor planning, elements of action discerned in another's behaviour must meet one simple principle: each element should *already* be within the repertoire of the observer (Byrne, 2003). In contrast, the "size" of an element is irrelevant: I propose that people are able to "see" (pick out) within a stream of action any element which is already present as a pattern in their personal repertoire. For different observers, or at different times in the life of a single observer, one particular movement of a single finger or an elaborate sequence of bimanual movements might both properly be seen as single elements. When we watch a relatively unfamiliar process being performed, the *level* at which we notice elements will be low, perhaps that of finger movements; whereas when we watch a slight variant of an already familiar activity, the basic elements that we notice might themselves be high-level, complex processes. Most commonly perhaps, the level at which observed behaviour matches parts of our existing repertoire would be neither of these, but rather consist of simple and highly-practised movements that produce visible effects on environmental objects: that is, simple, goal-directed movements. Such elements may be particularly easy to delimit because they are marked by a characteristic pattern of acceleration and deceleration, just like the cadence of syllables in a sentence. Consistent with this idea, people are able to pick out the basic elements of action, even when the stimulus is experimentally reduced to fluorescent spots on the joints (Baldwin, Andersson, Saffran, & Meyer, 2008; Loucks & Baldwin, 2009). Is it plausible that this means of segmentation is a primitive part of the human cognitive system? A digression into recent neuropsychological studies of monkeys suggests that it is.

Non-human primates have been shown able to pick out, in the behaviour of others they observe, actions that are already in their own repertoire. A system of single neurons has been identified in the premotor cortex of rhesus monkeys *Macaca mulatta* (Gallese, Fadiga, Fogassi, & Rizzolatti, 1996; Rizzolatti, Fadiga, Fogassi, & Gallese, 1996, 2002), each of which responds to a simple manual action, and responds equally whether the monkey makes the action or sees another do it. The cardinal properties of these *mirror neurons* are (1) they detect goal-directed movements that are in the observing monkey's own repertoire, and (2) they generalize over whether the movement is performed by the monkey itself or by another agent. It is unlikely that mirror neurons have any role in imitation for monkeys, simply because monkeys have repeatedly failed to show evidence of imitative capacity (Visalberghi & Fragaszy, 1990). Rather, Rizzolatti and his collaborators relate the evolutionary origin of mirror neurons to monkey social sophistication: i.e. they suggest that the system functions in revealing the demeanour and likely next actions of conspecifics, by reference to those actions the observing monkey might itself have done (Rizzolatti et al., 2002).

These units have sometimes been described as "monkey see, monkey do" cells, and in a very restricted sense this is accurate. Much of what is described as imitation in experimental studies of non-human primates involves provoking a subject to repeat an action that is in its repertoire, upon seeing another perform the same action (e.g. Whiten, Custance, Gomez, Teixidor, & Bard, 1996; Custance, Whiten, & Fredman, 1999; Bugnyar & Huber, 1997; see Byrne, 2002a for discussion). However, in these studies nothing new is being learned: this sort of imitation has been argued to be better described as *response facilitation* (Byrne, 2002a, 2002b; Rizzolatti et al., 2002). In response facilitation, as opposed to any more general sense of imitation, a pre-existing response is facilitated (i.e. made more available) by seeing it done, and this causes a higher probability of the response occurring subsequently (Byrne, 1994; Byrne & Russon, 1998). Response facilitation is closely related to stimulus enhancement (Galef, 1988; Spence, 1937), and they may indeed be two manifestations of the same phenomenon: *priming* of neural correlates (Byrne, 1994, 1998b, 2005a). On this view, priming neural correlates of some aspect of the social situation or the environment results in stimulus enhancement; whereas priming neural correlates of an action pattern within the current repertoire results in response facilitation. The mirror neuron system provides a possible neural instantiation for imitation in this restricted sense of response facilitation, but not for imitative learning of new skills. (This does not mean the mirror system is innate and fixed; indeed, at least for the human equivalent, there is evidence that it can be affected by learning: Catmur, Walsh, & Heyes, 2007.)

Despite these reservations, a segmentation system, based on elements of action that the observer can already perform, would be a very useful starting point for more elaborate forms of imitation—and that is what I have proposed underlies great ape imitation (Byrne, 2003). By responding to precisely those movement patterns that correspond to potential actions in the viewer's repertoire, segmentation by response facilitation, operating by means of mirror neurons, has the power in principle to convert automatically a continuous flow of observed movements into a *string* of recognized, familiar actions. If seeing a string of familiar actions also allows construction of links between them, then "action-level" imitation occurs (Byrne, 2002a, Byrne & Russon, 1998). In action-level imitation, a linear sequence of actions is copied without recognition of any higher-order organization that may be present: the organization is "flat". Chimpanzees have been reported to copy the order of actions, even though the sequence was entirely arbitrary and unrelated to success (Whiten, 1998), and a detailed learning model has been developed to describe action-level imitation in animals (Heyes & Ray, 2000).

If it were beneficial to copy arbitrary, random actions or behaviour that is genuinely linear in structure (e.g. the "fixed action patterns" described by early ethologists), action-level imitation might be useful. However, most human action, and arguably also much of the behaviour of non-human great apes, is planned: with a hierarchical, not linear organization. The question is, can this planning be "seen" by an ape, in the behaviour of another? More generally, can a bottom-up, mechanistic analysis go beyond action-level imitation to explain how behavioural organization can also be parsed and thereby copied, i.e. *program-level imitation* (Byrne & Russon, 1998)? If so, then the evolution of behavioural parsing has implications far beyond imitation itself.

8.3 Parsing Hierarchical Structures of Behaviour

It is no coincidence that a theory of how program-level imitation might be achieved should have been developed to explain manual behaviour, specifically in great apes (Byrne, 2003). Most animals simply do not learn sufficiently complex patterns of behaviour for imitative learning to be detectable in observational data from them, nor would they have much need for the ability to learn by imitation (Byrne, 2002a). The primitive 5-fingered primate hand (Napier, 1961) is highly effective as a manipulator and in many species shows some opposability. In great apes, however, the hand shows a considerably augmented range of aptitudes compared even to those of monkeys. For example, in the mountain gorilla (Byrne, Corp, & Byrne, 2001b), everyday food preparation typically involves using the two hands in different but complementary roles (i.e. *manual role differentiation*: Elliott & Connolly, 1984). The resulting "asymmetric bimanual co-ordination" is augmented by the gorilla's ability to control individual digits of the hand independently (i.e. *digit role differentiation*: Byrne et al., 2001b). This allows items to be held in part of the hand while other digits can carry out other activities; for instance, part-processed food can be accumulated in the hand, while part of the food-processing routine is iteratively repeated to build up a larger handful of food. This remarkable dexterity allows mountain gorillas to deal with plants that are physically defended by an array of spines, stings and hard casings (Byrne, 2001). In the process, they display a huge repertoire of functionally distinct elements of action (i.e. single actions that produce clear changes to the plant substrate; for instance, thistle-processing alone requires 72 such elements). Although attention has been drawn away from the chimpanzee's general manual skills by the anthropological emphasis on tool use, when chimpanzee plant-processing has been studied qualitatively, similar abilities are found to those of gorillas (Corp & Byrne, 2002a, 2002b). (Note that the sophisticated tool-making of early hominins also relied on manual dexterity: Byrne, 2004, 2005b). With animals of such dexterity, manual behaviour is sufficiently rich for complex organizations of learnt behaviour to be detectable by researchers. Moreover, the manual tasks confronted by great apes are challenging, so it would certainly pay the apes to be able to learn new skills by imitation of others.

The evidence that great apes do indeed learn skills by imitation comes from observational data rather than experiment, since no useful experimental test of program-level imitation in animals has yet been devised. Although the evidence is therefore oblique, cumulatively it is fairly impressive (Byrne, 2002a, 2005a). First, there is the very fact that young great apes learn complex, hierarchically structured routines of manual behaviour (some of them essential to survival in adulthood) in just a few years before their weaning, in contrast to monkeys where there is no evidence of anything comparable. Evidence of complexity is strongest for the mountain gorilla, where 5-stage sequential processes have been described (Byrne, 1999c, Byrne & Byrne, 1993; Byrne, Corp, & Byrne, 2001a), but also clear in chimpanzees, both in tool-using tasks (Boesch & Boesch, 1990; Goodall, 1986; Matsuzawa, 2001; Matsuzawa & Yamakoshi, 1996) and in dealing with complicated plant foods (Corp & Byrne, 2002a; Stokes & Byrne, 2001). The fact that orangutans sometimes also use tools to deal with complex plant defences (Fox, Sitompul, & Van Schaik, 1999) suggests that they have similar abilities, and this is confirmed by studies of young orangutans' efforts to deal with the vicious spines of certain palm trees (Russon, 1998). Far more studies have been carried out on the foraging behaviour of monkeys than that of apes; yet no comparable evidence has come to light. Second, in a detailed analysis of variation in the skills of adult mountain gorillas, it was striking that minor details (grip type, exact fingers employed, hand preference, extent of movement) varied idiosyncratically between individuals, even between mother and offspring, whereas the overall "program-level" organization of each technique was remarkably standardized in the local population (Byrne & Byrne, 1993). If idiosyncrasy is characteristic of trial and error learning, such standardization of techniques needs explaining. There are two possibilities: either the affordances of the gorilla's hands, combined with the physical form of the plant defences, define a clear gradient of optimization and thus with practice every gorilla will inevitably acquire the same method; or, observational learning is involved, and some aspects of the skills are passed on culturally. The third line of evidence is specifically relevant to this issue, as it involves the study of animals disabled by crippling snare wounds. Snares are not set to catch gorillas, but young individuals may suffer injury because of their explorative behaviour (Stokes, Quiatt, & Reynolds, 1999). If the standardized pattern of an adult technique is a product of affordances, then in an animal with severely maimed hands a quite different technique should result from the same trial and error experience. Yet in both chimpanzees and gorillas, disabled individuals acquire the same organization of behaviour as the able-bodied, and instead work around their difficulties by modifying the low-level details of implementation (Byrne & Stokes, 2002; Stokes & Byrne, 2001). This favours the hypothesis that the standard technique is a culturally transmitted pattern. Finally, one anecdotal observation supports the case that great apes can only learn certain aspects of their complex feeding skills by observation. When processing stinging nettles, one single adult in the study population differed in technique: the female Picasso did not fold bundles of leaves, so was presumably often stung on her lips (Byrne, 1999a). Picasso had transferred into the study area from lower altitude, where nettles do not grow. Because adult gorillas feed alone and out of sight of others in dense herbage, mountain gorillas' only opportunity for observational learning of plant processing comes in infancy. It seems most likely that a lack of opportunity to observe accounts for Picasso's incomplete technique, and intriguingly her juvenile was the only other gorilla in the study population to lack that particular element of the skill.

8.4 Imitation Without Intentionality

In the face of this evidence, I therefore developed a theory of how great apes could learn the program-level structure of behaviour by imitation, aiming to avoid any assumption that the animals had prior understanding of purpose or intention (Byrne, 1999b, 2003). This "behaviour parsing" model is based instead on the *statistical regularities* present within the variability of multiple performances of the same skilled sequence of action.

Every execution of a motor act, however familiar and well-practised it is, will differ slightly from others. Nevertheless, this variation is constrained—because if certain characteristics are missing or stray too far from their canonical form the act will fail to achieve its purpose. Watching a single performance will not betray these underlying constraints, but the statistical regularities of a repeated, goal-directed action can serve to reveal the organizational structure that lies behind it. Unweaned great apes spend most of each day within a few feet of their mothers, and (since their main nutrition still comes from milk) they have almost full-time leisure to watch any nearby activities, as well as learn about the structure of the local environment by their own exploration. For instance, by the time a young gorilla first begins to handle a plant like a nettle, at the rather late age of about two years because the stinging hairs discourage earlier attempts, it will have watched many hundreds of nettle plants being expertly processed by its mother.

Consider how a young gorilla might learn from statistical regularities of observed behaviour how to process stinging nettles (Fig. 8.1).

Its mother's behaviour will be perceived as a string of discrete elements, where each of these actions is a familiar one that it can already perform. At this time, the young ape's repertoire of familiar elements of action derives from: (i) its innate manual capacities; (ii) from many hours of playing with environmental objects, such as plants and discarded debris of the mother's feeding; and (iii) from its own experience of feeding on other plants, perhaps ones simpler to process than nettles. Suppose that it also has some way of focusing on those particular sequences of its mother's action that are relevant to eating nettles; perhaps it has explored nettle plants and found that they are painful, yet puzzlingly the mother seems to enjoy interacting with them, making her nettle-interactions intrinsically interesting. (Some such mechanism to focus learning on relevant action sequences seems to be essential for any "bottom up" model of motor learning.) Because motor behaviour is intrinsically variable, and plants also vary from individual to individual, the string of elements that the young gorilla sees when watching its mother eat nettles will differ each time. However, her starting point will always be a growing, intact nettle stem, and—because she is expert at this task—her final stage will always be the same, popping a neatly folded package of nettle leaves into the mouth. In between these points, variation will be particularly associated with non-critical parts of the performance, and certain aspects must necessarily be the same—or else, the result simply will not be success. With repeated watching, and a mind that tends automatically to extract regularities in behaviour that varies over time, a pattern will gradually begin to become apparent. The mother always makes a sweeping movement of one hand, held around a nettle stem which is sometimes held in the other hand even though the plant is still attached to the ground, and this leaves a leafless stem protruding from the ground; she always makes a twisting movement of the hands against each other, and immediately drops a number of leaf-petioles (which she does not eat) onto the ground; she always uses one hand to fold a bundle of leaf-blades protruding from

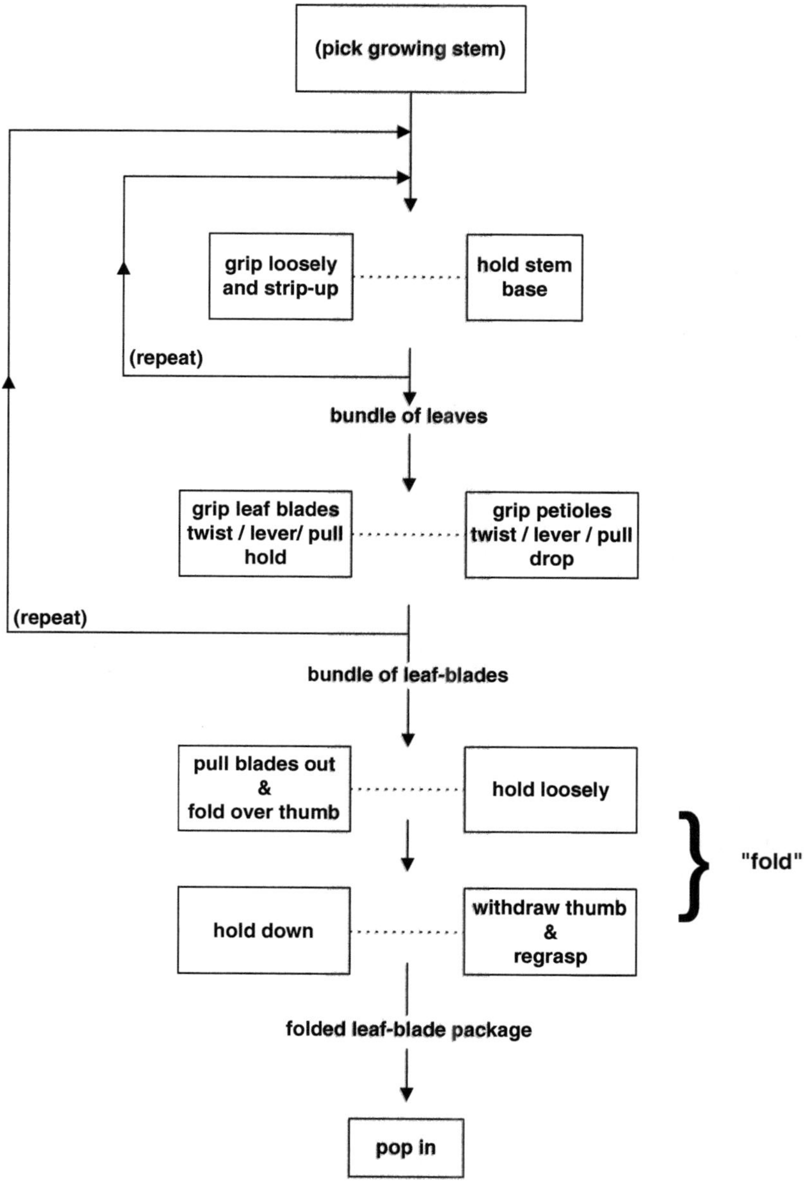

Fig. 8.1 Flow-chart for a typical adult gorilla processing nettle *Laportea alatipes* leaves. The action starts at the top, with selection of a growing nettle to eat, and works downwards. Processes are shown in rectangles; those which are optional, depending on the state of the plant itself, are shown in brackets. As with conventional flow-charts, diamonds represent choice points, with the alternative options shown by the directed links leading from each diamond. Unlike the single linear process of most flow-charts, the diagram represents the actions of both *left* and *right hand*: actions which are significantly lateralized to the *left hand* are shown on the *left* of the figure, and vice versa for the *right hand*. Some of these actions are nevertheless co-ordinated together, though the two actions are different: these cases of asymmetric bimanual co-ordination are shown with *dotted lines* connecting the separate processes

the other hand, and holds down this folded bundle with her thumb. Moreover, these stages always occur in exactly the same order each time.

Statistical regularities, in behaviour that is repeatedly observed, thereby mark out the minimal set of *essential actions* from the many others that occur during nettle eating but which are not crucial to success, and reveal the *correct order* in which they must be arranged. (The ability of human babies as young as eight months to detect statistical regularities in spoken strings of nonsense words shows that just such sensitivity to repeated orderings is active early in human development: Saffran, Aslin, & Newport, 1996.). The usefulness of detecting regularities applies not only to the linear sequence of movements of each hand, but also the hands' operation together: stages that crucially depend on the hands' close temporal and spatial co-ordination while doing different jobs will recur in every string, while other coincidental conjunctions will not.

Other statistical regularities derive from modular organization and hierarchical organization. Whenever the operation of removing debris is performed (by opening the hand that holds nettle leaf-blades, and delicately picking out debris with the other hand), it occurs at the same place in the sequence. Also, on some occasions but not others, a section of the program sequence may be repeated twice or several times. For instance, the process of <pull a nettle plant into range, strip leaves from its stem in a bimanually co-ordinated movement, then detach and drop the leaf-petioles> may be repeated several times before the mother continues to remove debris and fold the leaf-blades before eating. Subsections of the string of actions that are marked out in this way may be single elements, or as in this example a string of several elements. Both omission and repetition signal that some parts of the string are more tightly bound together than others, i.e. that they function as *modules*. Optional stages, like cleaning debris, occur between but not within modules. Moreover, repetition of a sub-string gives evidence of a module used hierarchically as a *subroutine*, for example, iteration to accumulate a larger handful.

Further clues to modular structure are likely to be given by the distribution of pauses (occurring between but not within modules), and the possibility of smooth recovery from interruptions that occur between modules. Gorillas often pause for several seconds during the processing of a handful of plant material, in order to monitor the movements and actions of other individuals. Finally, a different module entirely may be substituted for part of the usual sequence (e.g. if one hand is required for postural support, then a normally bimanual process may need to be performed unimanually), and if this module is recognized as an already-familiar sequence its substitution again reveals structure; eventually, it may be that a taxonomy of substitutable methods is built up.

All these statistical regularities are precisely what enabled us, the researchers, to discover the hierarchical nature of nettle processing by adult gorillas (Byrne & Byrne, 1993). The behaviour parsing model proposes that the same information can be extracted and used *by the apes themselves*, and that this ability is what enables a young ape to perceive and copy the sequential, bimanually co-ordinated, hierarchical organization of complex skills from repeated watching of another.

Behaviour parsing enables the underlying hierarchical organization of planned behaviour to be picked out—under certain circumstances. The first caveat, from what we know of living apes in the wild, is that it is entirely possible that non-human apes' capacity to parse behaviour is limited to the visible domain of manual and bodily actions, and thus not available in the auditory domain. The bonobo Kanzi's apparent ability to parse human speech, when he responds correctly to words whose referent depends on the syntactical organization of a relative clause within a sentence (Savage-Rumbaugh et al., 1993), may cause this qualification to be relaxed, at least for extensively human-reared apes. For the moment, however, I will assume that living apes under natural conditions, and our own earliest ancestors, had no such ability. The great ape forte is evidently the manual domain, as convincingly demonstrated in the hundreds of ASL signs acquired by participants in "ape language" experiments (see chapters in Gardner, Gardner, & Van Cantfort, 1989). In contrast, modern humans are routinely able to parse vocal material.

The second limitation, from the way the model works, is that "multiple independent looks" are necessary. A single view of skilled behaviour that is unfamiliar in its organization will not result in a useful parsing, so seeing *multiple* samples of efficient behaviour is required. The samples must be *independent*, so that there is information about the variance within the strings of perceived elements; that is because only by having sensitivity to the relative variability of elements can behaviour parsing locate the key (unvarying) elements. Thus, viewing a film-clip of the same segment of skilled behaviour would not serve to allow unfamiliar behaviour to be parsed. Note that, although we may well substantially overrate our everyday abilities (Bargh & Chartrand, 1999), modern humans do not seem to be subject to this limitation. Gergely, Bekkering, and Kiraly (2002) show that babies over sixteen months old are able to pick out for imitation the key elements of behaviour demonstrated only once, according to simple rationality criteria; behaviour parsing alone could not explain these data. Before the critical age, I predict that babies are still able to show program-level imitation, but will not at that point be able to select out specifically rational features of the process to copy. In circumstances not requiring acquisition of new behavioural organization, there is also some evidence for similar selectivity in imitation without multiple views of the behaviour in both chimpanzees and domestic dogs (Buttleman, Carpenter, Call, & Tomasello, 2007; Horner & Whiten, 2005; Range, Viranyi, & Huber, 2007). Thus, the behaviour parsing model can only be part of the eventual answer of how human imitative abilities evolved.

8.5 Why Great Apes?

To be precise, why should it have been only this one taxon of primate that developed the rather special ability to parse a segmented stream of action into a hierarchically-organized structure—and thereby acquire novel, complex skills by imitative learning? At present, the social brain or Machiavellian intelligence hypothesis is widely accepted as the most plausible explanation for the origin of primate intelligence (Brothers, 1990; Byrne & Whiten, 1988; Dunbar, 1998; Humphrey, 1976; Jolly, 1966; Whiten & Byrne, 1997). However, when it comes to accounting for cognitive differences *between* monkeys and apes, it will not do. According to the social brain hypothesis, the root cause of intellectual advance is social complexity. Because the ancestors of modern haplorhine primates (monkeys and apes) needed to live in increasingly large social groups, yet individuals of each species were thereby put in direct competition for resources with other group members, a selection pressure resulted that favoured increased social intelligence and a concomitant enlargement in neocortex volume (Byrne, 1996). Thus, today, we find that primates living in larger groups have larger brains (Barton & Dunbar, 1997; Dunbar, 1992), and are more likely to employ subtle means of social manipulation such as deception (Byrne & Corp, 2004). While this fits nicely with the differences among living species of varied brain sizes, and gives a good account of the evolutionary origins of the large-brained haplorhines, it does not distinguish between monkeys and apes. There is no systematic difference in the causal variable: the great apes simply do not live in larger social groups than do many monkey species, which have much smaller brains and show little sign of the sophisticated cognition of apes.

This means that serious attention must be paid to alternative, ecological selection pressures that might have promoted intelligence, at least for this special case (Byrne, 1997): for instance, is there an ecological challenge that affects great apes more than monkeys? Because of the anatomical differences between monkeys and apes, the answer is yes. Great apes are systematically larger than monkeys, and since they are adapted to brachiation (hanging below branches on long, powerful arms) costs of long-distance travel are much greater for them than for monkeys. However, apes are all specialists in easy-to-digest plant material (fruit or soft leaves) which is ephemerally available and patchily distributed, so apes they must regularly travel to find their food. Almost everywhere they live, great apes share the forest with Old World monkeys—which are not only smaller and more efficient in long-range travel, but happen to have gut adaptations enabling them to eat fruit when slightly less ripe, or leaves when slightly tougher, than can apes. Monkeys, in short, are in direct niche competition with great apes and possess all the aces: how have living apes survived at all? The explanation becomes clear when the details of their diet are examined: chimpanzees make tools to extract social insects from their nests, and to break open hard nuts; gorillas, and to a lesser extent chimpanzees, use elaborate, multi-stage routines to deal with plant defences; orangutans use complex, indirect routes to reach defended arboreal food, and sometimes make tools to gain access to bees' nests or defended plant food. In each case, "clever" methods of food extraction are used to gain access to foods which monkeys would be unable to reach. Thus, it becomes plausible that the Miocene ancestors of the living great apes (whom we share) may have adapted cognitively, in ways that would enable a broader range of food types to be exploited: and I propose that learning new skills by behaviour parsing was just this adaptation.

8.6 Parsing to "See" Intentions: The Origin of Mime and Gestural Language?

If this behaviour parsing model is correct, human language and speech evolved in a species that was *already* able to parse hierarchically organized behaviour—which might be no coincidence. Moreover, this ability to "see below the surface" of behaviour, and detect the logical organization that produced it, has implications for other cognitive activities. Indeed, the ability to learn new skills by imitation may be seen as just part of a fundamental process of interpreting or understanding complex behaviour.

It was important in the development of the behaviour parsing model that processing should start from observed behaviour and require no prior understanding of the physical cause-and-effect of the actions upon objects in the world, nor the intentions or other mental states of the demonstrator. However, we know from common experience that these more abstract representations form regular parts of how adult humans understand and discuss the world: so their evolutionary origin must be explained. Behaviour parsing might be a *necessary* step on the road to seeing the world in an intentional-causal way.

Consider causation. Since a perceptual parsing of complex action will (in many cases) be applied to actions-upon-objects in the world, changes in the physical world will become linked to the sequence of action—statistically. Of course, there is more to cause than correlation, but it can be questioned whether that matters for everyday purposes, or for evolution. Reliable correlation might be described as a "Pretty Good Cause", and only physicists dealing with the fundamentals of matter may need to go much beyond it. The fact is that most things are seen as likely to happen to the extent that they, or things very like them, have happened before under the same circumstances. The sun will rise tomorrow morning because it has been doing so for a long time at rather regular and statistically predictable intervals; not flawless logic, but good enough. Any parent who has tried to answer a series of "Why?" questions from a young child will know how soon one gets out of one's depth with causation: ok, so day and night are caused by the Earth going round the Sun, but why does it do that? In fact, probing deeper into the physics of most everyday situations helps little with everyday living, and does not provide a very satisfying advance on *cause-as-correlation.* In contrast, behaviour parsing picks out the correlational structure of a changing environment quite well.

How could behaviour parsing help us with intentionality? The perceived organization of behaviour that results from the parsing process will inevitably be set in a real-world context of achievement of valuable ends, just because the individuals observed engaged in skilful action will only be doing so for biologically sufficient reasons. Often, demonstrators will be close associates or relatives of the observers, confronting much the same problems as them. Thus, associating a particular organizational structure with the typical *result* of its performance is in many cases a relatively trivial task: the *point* of achieving that particular result is something the observer probably already understands. Intended purpose is indicated by the usual result of successful performance. ("Unsuccessful" is of course also identified statistically, here, on the basis of visible behaviour. It corresponds to those occasions when the action needs to be re-done, rather than moving on to another action.) This means that, in principle, behaviour parsing makes it possible to compute the prior intention of the other individual: by recognizing a behaviour pattern that would, if the observing self performed it, achieve a comprehensible goal for the self. Any animal capable of program-level imitation should therefore also be able to detect at least some intentions of others from their behaviour, in cases where they have been able to gain the necessary prior experience of that behaviour. And indeed, the living great apes do show some aspects of theory of mind (Byrne, 1995, 1998a, 2000; Cartmill & Byrne, 2007; Tomasello, Call, & Hare, 2003), although it seems likely that these fall short of the full mentalizing abilities of five year old children (Astington, Harris, & Olson, 1988; Perner & Wimmer, 1985; Wellman, 1990). As in the case of causation, the intentions extracted by behaviour parsing are intentions in a weak sense of the term: rather than an imagined mental state, intentions of these kinds need be no more than proper results of the normal behaviour sequence. But similarly, this sense of intention may be good enough for most everyday purposes: animals sensitive to *intentions-as-results* will not be able to conceive of false belief and deliberate trickery, but they will be able to pick out the purposes of many everyday social actions.

Animals with behaviour parsing abilities, as indexed by their ability to imitate at program-level, might still be rather limited in understanding—with causation reduced to correlation, and intentions reduced to expected results. However, combined with the delicate and sophisticated manual control of action that we find in all the living apes, even this limited kind of understanding should be sufficient for communication by means of *gesture*. Natural gestural communication in non-human apes is a rather neglected topic, but current evidence shows that in captivity both chimpanzees and gorillas develop gestures not seen in the wild, and use them intentionally in dyadic communication (Genty & Byrne, 2010; Genty, Breuer, Hobaiter, & Byrne, 2009; Hobaiter & Byrne, 2010, 2011; Pika, Liebal, & Tomasello, 2003; Tanner, 1998; Tanner & Byrne, 1996, 1999; Tomasello, George, Kruger, Farrar, & Evans, 1985; Tomasello, Gust, & Frost, 1989). Moreover, the ability of living great apes to extend their gestural repertoires when helped by humans has been amply demonstrated in the various "ape sign language" projects: whatever is believed of their linguistic sophistication, there is no doubt that those chimpanzees, gorillas and orangutans have learned many new manual gestures.

8.7 Tailpiece: A Heretical Thought

Those who conduct behavioural experiments or analyse observational data from the field, in order to discover whether any animal has the ability to represent the mental states of others, become acutely aware that their task is a difficult one because simpler mechanisms can generate richly complex behaviour. In particular, this chapter has argued that an understanding of planned behaviour, in terms of hierarchically organized structure that can be copied, with causality approximated by correlation and purpose by normal results, can result from a mechanistic process of behavioural analysis that need not involve any "mentalizing" about the actual mental states of the observed party. Thus, great apes show program-level imitation, but might still not possess theory of mind and causal understanding. But what about humans?

Of course, humans can and do represent causes and intentions: we explain (away) our actions, on grounds of our beliefs, false or otherwise; we teach our children by explaining that one thing causes another or that some people have different beliefs to ourselves, and so on. But do these retrospective, verbal accounts actually correspond to causal mental states that generate our behaviour when we are not explaining anything? We are always reluctant to accept how much of our behaviour is an automatic and fast product of mental processes of which we are unaware (Bargh & Chartrand, 1999), but I think this should be seriously considered for the case of theory of mind.

There are two possibilities. On the one hand, it may be that calculations about others' mental states *are* causal, and that the normal process of automatization with practice simply renders them faster and more efficient, to the point when they can only be made conscious by "off-line" deliberation. But the heretical alternative is that rather different, mechanistic but unconscious processes—analogous to those that allow us to parse behaviour—actually cause most of our everyday social behaviour and interactions with the world of objects, and mentalizing is a secondary process (and see Apperly & Butterfill, 2009 for a related discussion). On this view, mentalizing has different functions: these include teaching, when we explain processes or people to a child, and prevaricating, when we retrospectively construe our behaviour in a way very different to what we know to be accurate. Any such process of verbal (mis)construal is certainly a function of language ability, and so must be recent in human evolution; but it may be that the behavioural capacities that we attribute to "theory of mind" were all present at an earlier stage in human evolution, and are perhaps even shared with non-human great apes, though they cannot explain and discuss their actions as we can.

References

Apperly, I., & Butterfill, S. A. (2009). Do humans have two systems to track beliefs and belief-like states? *Psychological Review, 116*, 953–970.

Astington, J. W., Harris, P., & Olson, D. R. (1988). *Developing theories of mind*. Cambridge: University Press.

Auser, M. D., & Wrangham, R. W. (1987). Manipulation of food calls in captive chimpanzees. *Folia Primatologia, 48*, 207–210.

Baldwin, D., Andersson, A., Saffran, J., & Meyer, M. (2008). Segmenting dynamic human action via statistical structure. *Cognition, 106*, 1382–1407.

Bargh, J. A., & Chartrand, T. L. (1999). The unbearable automaticity of being. *American Psychologist, 54*, 462–479.

Barton, R. A., & Dunbar, R. I. M. (1997). Evolution of the social brain. In A. Whiten & R. W. Byrne (Eds.), *Machiavellian intelligence II: Extensions and evaluations* (pp. 240–263). Cambridge: Cambridge University Press.

Boesch, C., & Boesch, H. (1990). Tool use and tool making in wild chimpanzees. *Folia Primatologica, 54*, 86–99.

Brothers, L. (1990). The social brain: A project for integrating primate behavior and neurophysiology in a new domain. *Concepts in Neuroscience, 1*, 27–51.

Bugnyar, T., & Huber, L. (1997). Push or pull: An experimental study of imitation in marmosets. *Animal Behaviour, 54*, 817–831.

Buttleman, D., Carpenter, M., Call, J., & Tomasello, M. (2007). Enculturated chimpanzees imitate rationally. *Developmental Science, 10*, F31–F38.

Byrne, R. W. (1994). The evolution of intelligence. In P. J. B. Slater & T. R. Halliday (Eds.), *Behaviour and evolution*. Cambridge: Cambridge University Press.

Byrne, R. W. (1995). *The thinking ape: Evolutionary origins of intelligence*. Oxford: Oxford University Press.

Byrne, R. W. (1996). Machiavellian intelligence. *Evolutionary Anthropology, 5*, 172–180.

Byrne, R. W. (1997). The technical intelligence hypothesis: an additional evolutionary stimulus to intelligence? In A. Whiten & R. W. Byrne (Eds.), *Machiavellian intelligence II: extensions and evaluations* (pp. 289–311). Cambridge: Cambridge University Press.

Byrne, R. W. (1998a). Cognition in great apes. In A. D. Milner (Ed.), *Brain and cognition in monkeys, apes and man* (pp. 228–244). Oxford: Oxford University Press.

Byrne, R. W. (1998b). Imitation: The contributions of priming and program-level copying. In S. Braten (Ed.), *Intersubjective communication and emotion in early ontogeny* (pp. 228–244). Cambridge: Cambridge University Press.

Byrne, R. W. (1999a). Cognition in great ape ecology. Skill-learning ability opens up foraging opportunities. *Symposia of the Zoological Society of London, 72*, 333–350.

Byrne, R. W. (1999b). Imitation without intentionality. Using string parsing to copy the organization of behaviour. *Animal Cognition, 2*, 63–72.

Byrne, R. W. (1999c). Object manipulation and skill organization in the complex food preparation of mountain gorillas. In S. T. Parker, R. W. Mitchell, & H. L. Miles (Eds.), *The mentality of gorillas and orangutans* (pp. 147–159). Cambridge: Cambridge University Press.

Byrne, R. W. (2000). The evolution of primate cognition. *Cognitive Science, 24*, 543–570.

Byrne, R. W. (2001). Clever hands: The food processing skills of mountain gorillas. In M. M. Robbins, P. Sicotte, & K. J. Stewart,(Eds.), *Mountain gorillas. Three decades of research at Karisoke* (pp. 293–313). Cambridge: Cambridge University Press.

Byrne, R. W. (2002a). Imitation of complex novel actions: What does the evidence from animals mean? *Advances in the Study of Behavior, 31*, 77–105.

Byrne, R. W. (2002b). Seeing actions as hierarchically organized structures. Great ape manual skills. In A. Meltzoff & W. Prinz (Eds.), *The imitative mind: Development, evolution, and brain bases* (pp. 122–140). Cambridge: Cambridge University Press.

Byrne, R. W. (2003). Imitation as behaviour parsing. *Philosophical Transactions of the Royal Society of London (B), 358*, 529–536.

Byrne, R. W. (2004). The manual skills and cognition that lie behind hominid tool use. In A. E. Russon & D. R. Begun (Eds.), *Evolutionary origins of great ape intelligence* (pp. 31–44). Cambridge: Cambridge University Press.

Byrne, R. W. (2005a). Detecting, understanding, and explaining animal imitation. In S. Hurley & N. Chater (Eds.), *Perspectives on imitation: from mirror neurons to memes* (pp. 255–282). Cambridge, MA: MIT Press.

Byrne, R. W. (2005b). The maker not the tool: The cognitive significance of great ape manual skills. In V. Roux & B. Bril (Eds.), *Knapping stone. A uniquely hominid behaviour?* (pp. 159–169). Macdonald Institute Monographs.

Byrne, R. W., & Byrne, J. M. E. (1993). Complex leaf-gathering skills of mountain gorillas (*Gorilla g. beringei*): Variability and standardization. *American Journal of Primatology,* 31, 241–261.

Byrne, R. W., & Corp, N. (2004). Neocortex size predicts deception rate in primates. *Proceedings of the Royal Society of London: Biology, 271*, 1693–1699.
Byrne, R. W., Corp, N., & Byrne, J. M. E. (2001a). Estimating the complexity of animal behaviour: How mountain gorillas eat thistles. *Behaviour, 138*, 525–557.
Byrne, R. W., Corp, N., & Byrne, J. M. E. (2001b). Manual dexterity in the gorilla: Bimanual and digit role differentiation in a natural task. *Animal Cognition, 4*, 347–361.
Byrne, R. W., & Russon, A. E. (1998). Learning by imitation: A hierarchical approach. *Behavioral and Brain Sciences, 21*, 667–721.
Byrne, R. W., & Stokes, E. J. (2002). Effects of manual disability on feeding skills in gorillas and chimpanzees: A cognitive analysis. *International Journal of Primatology, 23*, 539–554.
Byrne, R. W., & Whiten, A. (1988). *Machiavellian Intelligence: Social expertise and the evolution of intellect in monkeys, apes and humans*. Oxford: Clarendon Press.
Cartmill, E. A., & Byrne, R. W. (2007). Orangutans modify their gestural signalling according to their audience's comprehension. *Current Biology, 17*, 1345–1348.
Catmur, C., Walsh, V., & Heyes, C. (2007). Sensorimotor learning configures the human mirror system. *Current Biology, 17*, 1527–1531.
Corp, N., & Byrne, R. W. (2002a) Leaf processing of wild chimpanzees: Physically defended leaves reveal complex manual skills. *Ethology, 108*, 1–24.
Corp, N., & Byrne, R. W. (2002b) The ontogeny of manual skill in wild chimpanzees: Evidence from feeding on the fruit of *Saba florida*. *Behaviour, 139*, 137–168.
Crockford, C., Herbinger, I., Vigilant, L., & Boesch, C. (2004). Wild chimpanzees produce group-specific calls: A case for vocal learning? *Ethology, 110*, 221–243.
Custance, D., Whiten, A., & Fredman, T. (1999). Social learning of an artificial fruit task in capuchin monkeys (*Cebus apella*). *Journal of Comparative Psychology, 113*(1), 13–23.
Dunbar, R. I. M. (1992). Neocortex size as a constraint on group size in primates. *Journal of Human Evolution, 20*, 469–493.
Dunbar, R. I. M. (1998). The social brain hypothesis. *Evolutionary Anthropology, 6*, 178–190.
Elliott, J. M., & Connolly, K. J. (1984). A classification of manipulative hand movements. *Developmental Medicine & Child Neurology, 26*, 283–296.
Fox, E., Sitompul, A., & Van Schaik, C. P. (1999). Intelligent tool use in wild Sumatran orangutans. In S. T. Parker, H. L. Miles, & R. W. Mitchell (Eds.), *The mentality of gorillas and orangutans* (pp. 99–116). Cambridge: Cambridge University Press.
Galef, B. G. (1988). Imitation in animals: History, definitions, and interpretation of data from the psychological laboratory. In T. Zentall & B. G. Galef, Jnr. (Eds.), *Social learning: Psychological and biological perspectives* (pp. 3–28). New York: Erlbaum.
Gallese, V., Fadiga, L., Fogassi, L., & Rizzolatti, G. (1996). Action recognition in the premotor cortex. *Brain, 119*, 593–609.
Gardner, R. A., Gardner, B. T., & Van Cantfort, T. E. (1989). *Teaching sign language to chimpanzees*. Albany, NY: SUNY Press.
Genty, E., Breuer, T., Hobaiter, C., & Byrne, R. W. (2009). Gestural communication of the gorilla (*Gorilla gorilla*): Repertoire, intentionality and possible origins. *Animal Cognition, 12*, 527–546.
Genty, E., & Byrne, R. W. (2010). Why do gorillas make sequences of gestures? *Animal Cognition, 13*, 287–301.
Gergely, G., Bekkering, H., & Kiraly, I. (2002). Rational imitation in preverbal infants. *Nature, 415*, 755.
Goodall, J. (1986) *The chimpanzees of Gombe: Patterns of behaviour*. Cambridge, MA: Harvard University Press.
Gouzoules, S., Gouzoules, H., & Marler, P. (1984). Rhesus monkey *(Macaca mulatta)* screams: representational signalling in the recruitment of agonistic aid. *Animal Behaviour, 32*(1), 182.
Green, S. (1975). Dialects in Japanese monkeys: Vocal learning and cultural transmission of locale-specific vocal behaviour? *Zeitschrift fur Tierpschologie, 38*, 304–314.

Heyes, C. M., & Ray, E. D. (2000). What is the significance of imitation in animals? *Advances in the Study of Behavior, 29*, 215–245.

Hobaiter, C., & Byrne, R. W. (2010). The gestural repertoire of the wild chimpanzee. *Animal Cognition.* DOI 10.1007/s10071-011-0409-2.

Hobaiter, C., & Byrne, R. W. (2011). Serial gesturing by wild chimpanzees: its nature and function for communication. *Animal Cognition.* DOI 10.1007/s10071-011-0416-3.

Horner, V., & Whiten, A. (2005). Causal knowledge and imitation/emulation switching in chimpanzees (*Pan troglodytes*) and children (*Homo sapiens*). *Animal Cognition, 8*, 164–181.

Humphrey, N. K. (1976) The social function of intellect. In P. P. G. Bateson & R. A. Hinde (Eds.), *Growing points in ethology* (pp. 303–317). Cambridge: Cambridge University Press.

Janik, V. M., & Slater, P. J. B. (1997). Vocal learning in mammals. *Advances in the Study of Behavior, 26*, 59–99.

Jolly, A. (1966). Lemur social behaviour and primate intelligence. *Science, 153*, 501–506.

Kuhl, P. K. (1982). Discrimination of speech by non-human animals – basic auditory sensitivities conducive to the perception of speech-sound categories. *Journal of the Acoustical Society of America, 70*, 340–349.

Loucks, J., & Baldwin, D. (2009). Sources of information for discriminating dynamic human actions. *Cognition, 111*, 84–97.

Marler, P., & Tenaza, R. (1977). Signalling behaviour of apes with special reference to vocalization. In T. Sebeok (Ed.), *How animals communicate* (pp. 965–1033). Bloomington, IN: Indiana University Press.

Matsuzawa, T. (2001). Primate foundations of human intelligence: A view of tool use in nonhuman primates and fossil hominids. In T. Matsuzawa (Ed.), *Primate origins of human cognition and behaviour*. New York: Springer.

Matsuzawa, T., & Yamakoshi, G. (Eds.). (1996). *Comparisons of chimpanzee material culture between Bossou and Nimba, West Africa*. Cambridge: Cambridge University Press.

Mitani, J. C., Hasegawa, T., Gros-Louis, J., Marler, P., & Byrne, R. (1992). Dialects in wild chimpanzees? *American Journal of Primatology, 27*, 233–243.

Napier, J. R. (1961). Prehensility and opposability in the hands of primates. *Symposia of the zoological society of London, 5*, 115–132.

Perner, J., & Wimmer, H. (1985). 'John thinks that Mary thinks that': Attribution of second-order beliefs by 5- to 10-year old children. *Journal of Experimental Child Psychology, 39*, 437–471.

Pika, S., Liebal, K., & Tomasello, M. (2003). Gestural communication in young gorillas (*Gorilla gorilla*): Gestural repertoire, learning, and use. *American Journal of Primatology, 60*, 95–111.

Range, F., Viranyi, Z., & Huber, L. (2007). Selective imitation in domestic dogs. *Current Biology, 17*, 868–872.

Rizzolatti, G., Fadiga, L., Fogassi, L., & Gallese, V. (1996). Premotor cortex and the recognition of motor actions. *Brain Research, 3*, 131–141.

Rizzolatti, G., Fadiga, L., Fogassi, L., & Gallese, V. (2002). From mirror neurons to imitation: facts and speculations. In A. Meltzoff & W. Prinz (Eds.), *The imitative mind: Development, evolution, and brain bases* (pp. 247–266). Cambridge: Cambridge University Press.

Russon, A. E. (1998). The nature and evolution of intelligence in orangutans (*Pongo pygmaeus*). *Primates, 39*(4), 485–503.

Saffran, J. R., Aslin, R. N., & Newport, E. L. (1996). Statistical learning by 8-month-old infants. *Science, 274*, 1926–1928.

Savage-Rumbaugh, E. S., Murphy, J., Sevcik, R. A., Brakke, K. E., Williams, S. L., & Rumbaugh, D. M. (1993). Language comprehension in ape and child. *Monographs of the Society for Research in Child Development, 58*, 1–252.

Seyfarth, R. M., & Cheney, D. L. (1986). Vocal development in vervet monkeys. *Animal Behaviour, 34*, 1640–1658.

Spence, K. W. (1937). Experimental studies of learning and higher mental processes in infra-human primates. *Psychological Bulletin, 34*, 806–850.

Stokes, E. J., & Byrne, R. W. (2001). Cognitive capacities for behavioural flexibility in wild chimpanzees (*Pan troglodytes*): The effect of snare injury on complex manual food processing. *Animal Cognition, 4*, 11–28.

Stokes, E. J., Quiatt, D., & Reynolds, V. (1999). Snare injuries to chimpanzees (*Pan troglodytes*) at 10 study sites in East and West Africa. *American Journal of Primatology, 49*, 104–105.

Tanner, J. E. (1998). *Gestural communication in a group of zoo-living lowland gorillas*. Unpublished PhD thesis, University of St Andrews, St Andrews.

Tanner, J. E., & Byrne, R. W. (1996). Representation of action through iconic gesture in a captive lowland gorilla. *Current Anthropology, 37*, 162–173.

Tanner, J. E., & Byrne, R. W. (1999) The development of spontaneous gestural communication in a group of zoo-living lowland gorillas. In S. T. Parker, R. W. Mitchell, & H. L. Miles (Eds.), *The mentalities of gorillas and orangutans. Comparative perspectives* (pp. 211–239). Cambridge: Cambridge University Press.

Tomasello, M., Call, J., & Hare, B. (2003). Chimpanzees understand psychological states – the question is which ones and to what extent. *Trends in Cognitive Sciences, 7*, 153–156.

Tomasello, M., George, B., Kruger, A., Farrar, J., & Evans, E. (1985). The development of gestural communication in young chimpanzees. *Journal of Human Evolution, 14*, 175–186.

Tomasello, M., Gust, D., & Frost, T. A. (1989). A longitudinal investigation of gestural communication in young chimpanzees. *Primates, 30*, 35–50.

Visalberghi, E., & Fragaszy, D. M. (1990). Do monkeys ape? In S. T. Parker & K. R. Gibson (Eds.), *Language and intelligence in monkeys and apes* (pp. 247–273). Cambridge: Cambridge University Press.

Wellman, H. M. (1990). *Children's theories of mind*. Bradford: MIT Press.

Whiten, A. (1998). Imitation of the sequential structure of actions by chimpanzees (*Pan troglodytes*). *Journal of Comparative Psychology, 112*, 270–281.

Whiten, A., & Byrne, R. W. (Eds.). (1997). *Machiavellian intelligence II: Extensions and evaluations*. Cambridge: Cambridge University Press.

Whiten, A., Custance, D. M., Gomez, J.-C., Teixidor, P., & Bard, K. A. (1996). Imitative learning of artificial fruit processing in children (*Homo sapiens*) and chimpanzees (*Pan troglodytes*). *Journal of Comparative Psychology, 110*, 3–14.

Zuberbuhler, K. (2002). A syntactic rule in forest monkey communication. *Animal Behaviour, 63*, 293–299.

Chapter 9
Hominid Diversity and 'Ancestor' Myths

Homo, H. sapiens, and Other Taxa from a Phylocladistic Viewpoint

Niels Bonde

Abstract Many of the 'myths' of direct ancestors of 'all hominids' or of *Homo* or of *H. sapiens* and age of these 'ancestors' are shown to be 'false' or based on poor character analyses and/or suboptimal classifications and/or inconsequent choices of names of taxa. Ernst Mayr's devastating influence since 1950 on naming fossil hominids and therefore on comprehending their diversity is obvious. Recently, that is since mid 1990s, many more new taxa of fossil hominids have been found and named, and this has produced a much better appreciation of the prehistoric diversity, and has questioned and put into doubt, if not outright refuted, many of the traditional, often too simplified and adaptationistic scenarios (or 'just-so stories') about human evolution as evidenced by 'direct fossil ancestors'. The most famous of these, 'Lucy', is here named *Afaranthropus* (n. gen.) *antiquus* (Ferguson, 1984).

The diversity, classification, and age of fossils and the delimitation of taxa are obviously relevant also when speculating about the origin of such non-fossiliseable features of modern humans as (self)-consciousness, cognitive abilities, spoken language, early and simple tool making, wearing of clothes etc. As with the origin of such features as 'nakedness', large penis, certain immune systems and blood types, large female breasts, subcutaneous fat, and in fact also upright stance and walking, we can only be completely sure that these features originated somewhere before the last common stem-mother of living humans and after the split from chimpanzees – not a very precise indication: between 7 and 0.1 m.y.

When discussing the more precise level or stage in the phylogenetic tree, it becomes of utmost importance that we talk about the same taxa, and refer to the same groups designated by taxon names like *Homo* and *H. sapiens*. Here, it becomes clear that these names are not at all used in a consistent way, and also that most anthropologists and palaeontologists are not aware that both groups are completely arbitrarily delimited in the time dimension, as demonstrated by our analysis and

N. Bonde (✉)
Institute of Geography & Geology, University of Copenhagen, Øster Voldgade 10, DK 1350 Copenhagen, Denmark

Fur Museum, DK 7884 Fur, Denmark
e-mail: nielsb@geo.ku.dk; niels.bonde@mail.tele.dk

T. Schilhab et al. (eds.), *The Symbolic Species Evolved*, Biosemiotics 6, DOI 10.1007/978-94-007-2336-8_9, © Springer Science+Business Media B.V. 2012

conventions of non-Linnaean nomenclature (see the subtext to the classifications on the homepage [see p. 189]). To discuss a concept even as simple as the 'origin' or age of a taxon, one must make the cladistic distinctions between crown groups, stem groups and total groups, and taxon names attached to 'nodes', 'stems' or somewhere in between, namely to certain synapomorphic features ('key features' or 'adaptations'). Language and its origin will be discussed, and the relationship between early and primitive African languages compared with the tree of recent hominids based on MtDNA:

The complicated phylogenetic tree and the consequential (non-Linnaean) classification of hominids will be demonstrated, so that the diversity of hominids through time can easily be seen.

9.1 Introduction

Historically, the classification and evolutionary history of humans has been as 'minimalistic' as possible, presumably to reflect the uniqueness and unity of humans in a 'politically correct' way, also in respect of religious considerations. So apart from a period during early 20th century, when 'racial distinctions' were almost universally recognized and named, both living and fossil species and subspecies names have been kept at a minimum, grossly underestimating the 'true diversity'. Especially Ernst Mayr's renowned 1950 paper has had a devastating influence as a prohibition on recognizing new fossil hominid species. It is known that it took about four years between 1960 and '64 for L.S.B. Leakey and Ph. Tobias to agree on a new species name for *Homo habilis* due to the unwillingness to create new hominid species – *Australopithecus africanus* (not even necessarily *A. robustus*), *Homo erectus*, *Homo sapiens and H. neanderthalensis* (perhaps only as a subspecies of *sapiens*) then seemed to be a sufficient number of taxa to describe 'human diversity' through time. Later on (1963), Mayr did admit that recognizing the 'robust australopithecines' as a separate species *A. robustus*, was probably justified. He never understood or accepted that in that way both the genus *Australopithecus* and the subfamily *Australopithecinae* were un-natural (paraphyletic) groups. It is also known from Don Johanson (1990) that it took Tim White and him many discussions and consultations with Mayr before they decided to name *Australopithecus afarensis* in 1978 after the find of 'Lucy' in 1975. (Mary Leakey refused to be part of that paper, despite the fact that the type material from Laetoli, Tanzania, was primarily her findings, and White had worked for her – so Yves Coppens 'forced his way' into the paper very late, not to let the Americans run off with all the glory, as the original finds in the Afar region – and Ethiopia – were French [see Coppens, 1975] – see part of that story also in Tattersal [1995, p. 152]).

Admittedly, the four decades before Mayr's 'warning' had seen taxonomic names of fossil hominids being used almost like museum or field numbers with a pletora of generic and specific names. But only fairly recently, in the wake of many spectacular new finds, has it been more generally accepted that there are, in fact, quite a few recognizable human taxa, beginning perhaps with Michael Day's *Guides to Fossil Man* conservative, but gradually more diverse over the four editions from 1965 to 1986.

The aim of the present review is not only to expose the substantial diversity of fossil hominids, but also to disclose the many myths of 'hominid ancestors', especially of *Homo sapiens*, which have been published earlier – and which still occur at high frequency. Most of these postulates are poorly founded from a cladistic viewpoint – difficult as they must be, also from a general methodological viewpoint. Further, the influence this has on certain traditional ideas about the evolutionary history of 'soft' (non-fossiliseable) human features will be traced. Extremely good photos of most of the fossil specimens mentioned below have been published in natural size by Johanson and Edgar (1996), and likewise, continent by continent, by Schwartz and Tattersal (2002, 2003).

9.2 Classificatory Methods of Phylocladistics

Co-working with Bjarne Westergaard (who died 2008), I have developed a modern, non-Linnaean, 'phylocladistic' classification of all fossil and living 'hominids' (2004), taking into account all possible needs of a complete and precise system including taxonomic conventions for clades (monophyletic groups), paraphyletic groups, 'ancestors', metaphyly, uncertainties, doubts, lack of precise knowledge, fragmentary fossils, sequencing, subordination, and age. This is based on a rank-free, 'neutral' unit called LITU (Least Inclusive Taxonomic Unit) as the terminal taxon instead of the species, subspecies etc. of the traditional system (so it is the smallest recognizable group of individuals with some characteristic features – to be named as basal unit in the classification, and (usually) placed at the endpoints of the branches of the phylogenetic tree ('stem tree' – or at endpoints in the cladogram). A LITU is the smallest unit that can be given a differential diagnosis (characterization) and which is of historical (phylogenetic) relevance to be defended by the systematist. A taxon is a named group in the classification.

Hybridisation and other complications in terms of reticulate structures of relationships can be included if necessary. Naming of the taxa employs a single name for each taxon in lower case letters for LITUs and Capitalized for groups of LITUs (double names for hybrids). This is in contrast to the traditional Linnaean system and nomenclature where the basal unit is a species written in small case letters after a name of some genus (Capitalized – perhaps abreviated to just the first upper case letter), into which category any species has to be placed. F.ex. *Homo sapiens* or *H. sapiens* (usually written in *italics*). In the non-Linnaean system, there are no (absolute, Linnaean) ranks or categories (like genera, families etc.), because age of origin (or age span from 'origin' to extinction for fossil groups); can act as both relative and absolute 'rank' at the same time in a precise and relevant way, seen from a phylogenetic viewpoint. This classification is based on the simplest possible phylogeny (parsimony) which maximises, in a cladistic framework, the number of synapomorphies for all taxa (constituting their diagnoses) and optimises the number of characters used as synapomorphies (the mutual, derived features characterizing a taxon/group (Patterson, 1980)). It therefore seems difficult, if not impossible, to find a 'better' system in terms of the informations at hand which are the features analysed cladistically for the relevant (or known) groups of organisms.

The diversity of fossil 'hominids' (here Hominina) of our branch as separated from that of the chimps about 7 m.y. ago, is often grossly underestimated by palaeontologists and anthropologists. Here, we demonstrate that there are at least about 35–40 LITUs that can be distinguished among the known fossils (and several more are likely to be found in the future). And, as shown by DNA research and other molecular or physical anthropological analyses, there is also a historical structure among living humans (the LITU of crown group 'modern' *Sapiens* can apparently be subdivided into historically relevant groups, which in this case will be LITU-s incl. in the 'higher' taxon *Sapiens*).

This is certainly of significance for the analysis of language history which, already for decades, has been done by methodologies very similar, if not identical, to those of cladistics (and further similar to the much older techniques of 'textual critics' analysing the history of old manuscripts (Platnick & Cameron, 1977)). It is probably not just a coincidence that some of these ideas on historical analyses of languages and evolutionary history and the way to illustrate them by 'trees' can be traced back to the two colleagues and friends August Schleicher and Ernst Haeckel in the 1860s (1861 resp. 1866; the former sometimes considered father of historical linguistics, the latter inventor of numerous biological terms, among them 'phylogeny'), both directly inspired by Darwin's *Origin*, which has only one figure, a 'tree-like' diagram (1859).

Concerning nomenclature, the naming of the taxa, we have the cladistic problem of distinguishing so-called 'node-based' crown group taxa and apomorphy based ones from 'stem-based' taxa in a consistent way in the nomenclature for the classifications. I believe the only consequent and 'minimalistic' way of doing this is by naming only terminal taxa/LITUs, clades and crown-groups as far as possible. But what we tried to demonstrate in 2004 (Bonde & Westergaard – check homepage [see p. 189]) was that in practice you need rules also for naming other sorts of groups if you want to cover all known diversity of life. Namely, e.g. paraphyletic, metaphyletic and hybrid groups and 'ancestral' species/LITUs (see also Bonde, 2001). But we were not really consequential (2004) as we named some 'stem-based' groups (Queiroz & Gauthier, 1992), also recommended by the PhyloCode (Cantino & Queiroz, 2004 or newer edition), which is, in fact, unnecessary if not outright impossible.

Such stem-based groups originating at the very split from their living sister groups and comprising the very earliest parts of the said lineages – although they must have existed in the past – can never be satisfactorily and precisely indicated in the practical classification and nomenclature of known organisms. With a group which has fossils as well as recent forms present there will always be a certain 'total group' (a living crown group clade [Jefferies' term] combined with its fossil 'stem-group' [paraphyletic] of Hennig (1966) and together forming a clade – see

Queiroz and Gauthier (1992, 1994)). A total group contains a certain clade and the fossil(s) indicated by at least one synapomorphy to represent the first branch to split off from that clade. This, then, in practice constitutes the closest approximation to a truly 'stem-based' group. But this total group does not include the very earliest parts of the stem-lineage, because if such specimens were really found as fossils they would not show the relevant synapomorphy, or at least not in sufficiently high frequency. Those very earliest 'ancestral' fossils can only be classified as *incerta sedis* within the relevant higher crown group because one cannot show whether they belong just above or just below the relevant split (or 'speciation'), as I have repeatedly argued (1977, 1981, 2001 – see also Bonde & Westergaard, 2004). These supposed and very ideal 'ancestors' (morphotypes in Nelson's (1970) terminology), should they ever be found, which is entirely unlikely, they would form a paraphyletic or metaphyletic group anyway and can by no means be precisely classified (a general feature of organisms with only relatively 'primitive' or plesiomorphic characters). So no groups are in pratice really 'stem-based' (contra Queiroz & Gauthier and the PhyloCode); all groups/clades must be based on the earliest node ('speciation') within the group, unless in rare cases the earliest fossil can be assumed to be truely 'ancestral' and lie on the stem-branch itself. Concerning the positive possibility, in rare cases, of being able to indicate such 'actual ancestors', we disagree with most cladists, while many palaeontologists, especially micro-palaeontologists (who may believe they can dig the phylogeny right out of the ground or rather a drill hole) are willing to consider such possibilities – or even (far too) eager to do so. But surely, if somebody think they can identify 'real ancestors' they must have a possibility to precisely indicate that in a classification by certain conventions. And biologists should not believe that this is a problem only concerning fossils, because neontologists also need sometimes to classify taxa which look exactly like 'primitive ancestors' of other groups (usually their derived sister species) and therefore are, in fact, paraphyletic 'species' with no autapomorphy (e.g. diploid plants giving rise to tetraploids).

9.3 Diversity

The taxonomic diversity of fossil hominids in the restricted sense of being those fossils more closely related phylogenetically to living humans than to chimpanzees, is today realized to be much greater than believed fifty years ago. Many recent fossil discoveries surely contribute to that understanding but still there is, in many textbooks and semi-professional accounts (and a few professional ones), a reluctance to accept this diversity (e.g. Napier, 1975; Leakey, 1981; Winkler & Schweikhardt, 1982; Jones et al., 1992; Stringer & Gamble, 1993; Stringer & McKie, 1996; Lewin, 1999; Benton, 2005). And despite lots of beautiful photos of fossil hominids, the entire diversity is not really mapped by Johanson and Edgar (1996), and books like *Lucy* by Johanson and Edey (1990), and Leakey's *Homo* (1994) are not meant to cover everything. The others mentioned above comprise something like 15–20 taxa of hominids, while Bonde and Westergaard (2004) have 35–40 taxa (LITUs, 'species' or 'subspecies' – some 10 of them traditional chrono-species subdividing branches) as a minimum. But some of the simplifications in the most popular reviews (Tattersal, 1993 for the American Museum exhibit, Gore, 1997; Anon, 2004), and especially in those for kids, (e.g. Thomas, 1994; Ebbesen, 1990), are quite misleading.

But see also Schwartz (2004) about the recent trend of recognizing and naming many more fossil hominid taxa after this had been hampered for 30–40 years by Mayr's 1950 paper. The 'taxonomic activity' slowly increased by the descriptions of *H. habilis* (1964) and *H. ergaster* and 'Lucy' (1975, resp. 1978 – but these new species were discussed as very problematic, or were even neglected), and then it accellerated by the 1990s. Tattersal has rightly warned (1986) that if fossil diversity is not formally named then it will not be discussed – he even claims that subspecific names will be completely forgotten and that we underestimate the 'true' diversity anyway by only having hard parts. One reason is the tradition in palaeontology most often to discuss and to register at the generic level, and only rarely at the specific level (Treatise on Invertebrate Paleontology, Romer's 1966 Vertebrate Paleontology and younger followers; Cooper, 1970; Forey et al., 2004).

Admittedly, the trend in the decades before the mentioned period had been using the taxonomic names rather like museum collection numbers, which had created a plethora of names for what was just variations within the natural taxonomic units or species. Understandably, a certain reduction of the names of 'units' and of small and 'blind' collateral branches on the family tree has to be made for pedagogical reasons in the more popular accounts, but it is much less acceptable in professional reviews of our evolutionary history. It has, in the past, created a totally misleading picture of human phylogeny with profound consequences for ideas about the evolution of specific features both in the skulls and the post-cranium with implications for 'soft parts', functional anatomy, behaviour, and palaeoecology.

The most 'simple' and extreme idea, the so-called 'single species hypothesis' (e.g. Brace & Ashley Montagu, 1965; Wolpoff, 1968, 1980 – 'dogmatic' according to Tattersal, 1995 – later called 'ludicrous'), was upheld in certain anthropological 'schools' even until about thirty-five years ago (see Tattersal, 1995, pp. 127ff). Only the finds of robust australopithecines in the same deposits in Kenya as much more *Homo*-like fossils (Leakey, 1976; Leakey & Walker, 1976) put a final stop to this idea (or nearly so). It visualised the entire evolution of humans since the split from chimps as just one ('biological') species lineage ascending through time – albeit divided into successive 'species' (or chronospecies) for 'practical' or stratigraphic reasons, or because of intuitive ideas of important 'adaptive shifts' (Simpson, 1959, 1961). Similar adaptationistic arguments repeated as late as 1999 by B. Wood & Collard – but after the jubilee year of Darwin we should rather recall his remarks in a letter to T.H. Huxley 'In regard to classification ....I believe it ought ... to be simply genealogical ... I think it ultimately will, ... for it will clear away an *immense amount of rubbish about the value of characters*, ... The time will come, I believe, ... when we shall have very fairly true genealogical trees ...' [quoted by Goodman, 1975, here much abbrieviated with my emphasis]). Perhaps the 'time has come' to stop adaptationistic story telling about 'key characters'?

So finally, in 1975–76, it had been 'proved' that there were at least two human lineages, one a 'robust australopithecine' (*Paranthropus* spp.) and another much more *Homo*-like. The latter much later caller *H. (Pithecanthropus) rudolfensis*, and the age of these at Koobi Fora/'East Rudolph' is between 1½ and 2 m.y. (at the time mistakenly dated to be about 1 m.y. older). The relative ages between the South African cave localities themselves, Sterkfontein, Swartkrans, Kromdrai and Makapansgat was not well known at that time, and neither was their ages as compared to those of East Africa, and the age of the Taung locality for the type *Australopithecus africanus* was entirely uncertain. So phylogenetic (or possible ancestor-descendant relations) were very uncertain, although it was generally assumed that the 'gracile' *A. africanus* (comprising also *Plesianthropus*) was older than the 'robust' forms (Paranthropus), but many only recognized one genus, *Australopithecus*, for them all (also classified as a separate subfamily, Australopithecinae, which Simpson [1945] in his famous classification of mammals included in his family Pongidae for the apes !). Robinson (1956, 1972, and with Broom, 1952) always stressed the latter's name for the 'robust' forms, *Paranthropus*, as a very distinct taxon, but despite the supposed age-relations he preferred some unknown 'robust' australopithecine as ancestral to both *Australopithecus* and *Homo* (1972), in the same spirit as Weidenreich (1946) deriving later humans from 'giants'. The few branches of the phylogenetic tree I showed in 1989 (from a 1987 meeting) was even beyond the tradition at that time (but Groves excellent 1989 book appeared just after that with many more taxa). Figure 9.1 shows the diversity presented in a recent Danish Encyclopedia.

9.4 Australopithecus – An Old Homo

The rather old *Australopithecus africanus* (between 2 and 3 m.y. (Delson, 1988) in S-Africa) was a problem: To some, it was a variant of (or a more primitive species grading into) the 'robust' australopithecines (e.g. Johanson & White, 1979 – an older idea that it was just the females (Brace, 1967) had been abandoned (Birdsel, 1975, pp. 263ff, Tattersal, 1995, pp. 128ff)). To others, it was only a very primitive *Homo* (Robinson, 1956, 1972; Bonde, 1976, 1977, pp. 790–791) based mainly on the postcranial skeleton, but also on some skull features which was repeated by many analyses (Groves, 1991; Tattersal, 1995). This problem remained later because to some, *A. africanus* was clearly related to *Paranthropus* (Rak, 1983; Wood, 1992 wavering), while to Groves (1989, 2001; Olson, 1985; Lockwood & Tobias, 1999; Robinson, 1972) and to myself (1976, 1989, 2001; Bonde & Westergaard, 2004) *africanus* is clearly a *Homo,* based on several synapomorphies, but very primitive. Remember though, that before much was known about the chronology of the South African caves (Taung, Sterkfontein, Makapansgat), and under influence of Weidenreich's *Apes, Giants and Man* (1946), Robinson (1972) also thought that we were derived from 'giants'/large apes in a transformation series over robust *Paranthropus* via *Australopithecus* (= primitive *Homo*) to more advanced and smaller *Homo* (today we know that the trend in size is the opposite).

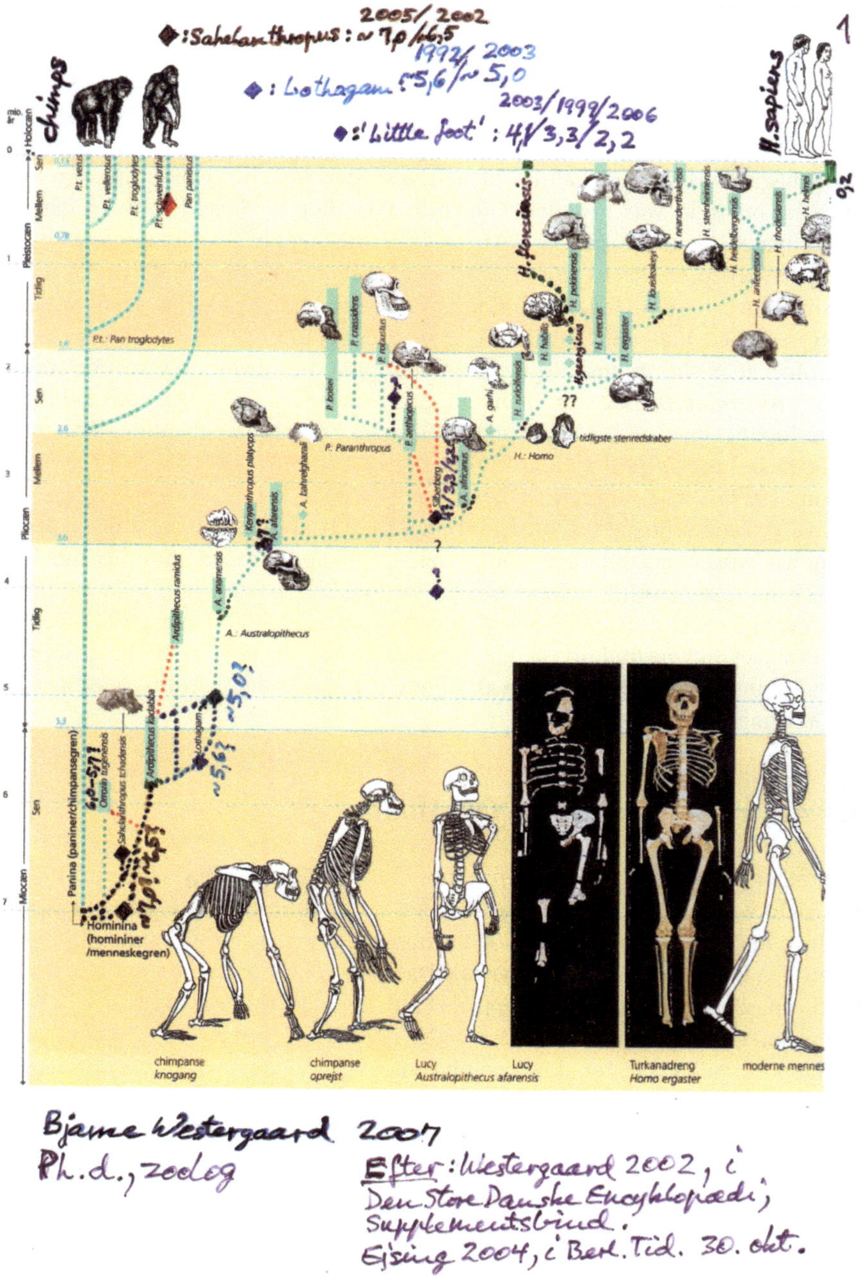

Fig. 9.1 Phylogenetic tree of chimps and hominids used by B. Westergaard in his revision (2002) of 'Human evolution' in 'Den Store Danske Encyclopedi' (Large Danish Encyclopedia) suppl. vol., and here with his further revisions and remarks added by hand. Used also by Bonde and Hoeg (2008) in their obituary on Bjarne W. in Yearbook of Danish Natural History Society after he had been murdered in Jan. 2008

Some have suggested that *africanus* could be common ancestor of both the robust and the *Homo* lineages (Day, 1965). The latter is refuted by finds of *Paranthropus* such as the 'Black skull' (WT 17000 (= West Turkana – Walker et al., 1986) – which should probably not be referred to the species *aethiopicus*, as the holotype jaw of this taxon, *Paraustralopithecus aethiopicus*, from Omo region, South Ethiopia (Arambourg & Coppens, 1968) has too small molar teeth (Groves, 1991); he refers it to an old *Homo* as *H. aethiopicus*). Its age is 2.5 m.y., that is, older than the youngest *africanus* (Thackeray et al., 2000) inclusive the holotype, 'Dart's baby' (Dart, 1925; Delson, 1988). The proper name for the 'black Skull' is *P. walkeri* by Ferguson (1989 – and see Groves (1991, pp. 254)).

This immediately makes a great difference for the age of *Homo*. In the first model, the oldest *Homo* is almost 2.5 m.y. (the so-called *H. habilis* and/or *H. rudolfensis* – see Leakey, 1994; Kimbel et al., 1996). In the second model, *Homo* must be much older, as some *africanus* from Makapansgat are probably a little over 3 m.y. old (Delson, 1988; Clarke. 1988).

9.5 The 'Robust' *Paranthropus*

This again implies that the sister group, *Paranthropus* (see Tobias, 1967; Grine, 1988; Groves, 1991), must have the same age within the clade Parhomo, named (by Bonde & Westergaard, 2004) for this combined *Homo-Paranthropus* group. But have 'robust' fossils that old ever been found? Usually, *Paranthropus* is also recorded as about 2.5 m.y. old, e.g. the age of the 'Black Skull' (WT 17000, so-called '*A. aethiopicus*', Walker et al., 1986; Kimbel et al., 1985 – disqualified above). But there is apparently an even more primitive *Paranthropus*, a fragmentary skull from Sterkfontein reconstructed by Clarke (1988) with concave face and the large molars, but no parietal crest, and even more primitive by being rather prognath and with much larger canines and incisors (not named – should, however, be critically compared with *A. garhi* of approximate same age (White et al., 1999; see below), which also has very large molars, and lacks a parietal crest). This is the same age as *A. africanus* from this locality, about 2½–3 m.y. But some records are almost always forgotten (at least not mentioned): In the lower part of the Omo section (Shungura Fm.), two very large deciduous molars have been found with an age about 2.9 m.y. (Coppens, 1975; Groves, 1991, p. 195), and it is quite likely they represent *Paranthropus*. And one should recall that at the beginning when hominids were found in the Afar region (Johanson & Taieb, 1976; Coppens, 1977; Taieb et al., 1976), the diversity of species was supposed to be two or three hominids, and some researchers later on thought that the so-called 'first family' at locality 333 represented early, robust australopithecines (e.g. Olson, 1985) – now known to have an age of 3.22 m.y. (Walter & Aronson, 1982). And it is sometimes indicated that there might be 'robust' types in Makapansgat (Tobias, 1989, but cf. Clarke, 1985 – originally called *A. prometheus* by Dart, 1948, now *A. africanus*), perhaps of a similar age over 3 m.y. In fact, the ideas of some people were that there was more than one taxon in most of the South African cave localities.

9.6 Our Stem-Mother 'Lucy'

Over the next few years after the find of the skeleton 'Lucy' (40% preserved, very little skull but the lower jaw) this very variable 'sample' over 3 m.y. old from the Afar region (Senut, 1978; Johanson & Taieb, 1976) was said to be more than one species (one *Homo*-like, one like *A. africanus* acc. to Johanson (1978), but a few month later considered the same taxon as known from jaw fragments from Laetoli in the Olduvai Gorge, Tanzania (White, 1977), half a million years older). This taxon was a new species called *Australopithecus afarensis*, but with the holotype jaw from Tanzania (Johanson, White, & Coppens, 1978). This will create eternal confusion, and this despite Coppens' (1977) demonstration that the lower dentition, especially the premolars are quite different in 'Lucy' (much more primitive, Coppens' gracile 'pre-*Australopithecus'* (e.g. 1994)), and the AL (= Afar Locality) 333 sample is more robust and advanced, *Homo*-like. These and other differences are said by White (1985; Kimbel, White, & Johanson, 1985; Johanson & White, 1979) to be bridged by intermediate forms in the samples, or they may be considered sexual differences. But this seems highly dubious, especially for the elbow and knee joints (Senut, 1978, 1980; Tardieu, 1979; Senut & Tardieu, 1985; both students of Coppens), as well as the heelbone and ankle (Stern & Susman, 1983; Susman et al., 1985). This 'taxon', as usual when White is involved, is claimed to be the ancestor of all younger hominids (White et al., 1981, 2009 – it is a remarkable incidence that one single person can, many times, find something so extremely rare as 'direct fossil ancestors' – and make other people believe it). It has become enormously popular, nicknamed as it is, from the Beatles' song (Johanson & Edey, 1981), and mostly palaeoanthropologists – surprisingly – seem to accept this taxon, of which even the name is not appropriate, as it should be *Praeanthropus africanus*, the name of the first hominid fragment found at Laetoli/Garusi in the 1930s (see Bonde & Westergaard, 2004; Strait et al., 1997). Below follows more on Lucy's bipedality and the controversy around her 'walking style', as well as problems with the famous foot prints in Laetoli, and the 'name-business'.

'Lucy', age 3.18 m.y. (Walter & Aronson, 1982) and some other Afar fossils constitute another, much more primitive species, dubbed *Homo antiquus* by Ferguson (1984) with the famous skeleton as holotype. But although it is far too primitive to be a member of *Homo*, it still has not received a proper generic name (in the Linnaean tradition), but is just called nov. gen. *antiquus* by Bonde & Westergaard (2004). To provoke a discussion of this important and primitive taxon (which due to its too young age is not ancestral to any other taxon/group, whether it has known autapomorphies or not, but must be a 'blind' side branch) this *new generic name will here be proposed as Afaranthropus* (Afar after this important region where Lucy's species so far seems to be endemic) with type species *A. antiquus* (Ferguson, 1984). The diagnosis of both genus and the single species can be taken as the character combination from Groves' analysis (1991, p. 223 fig. C and features favouring this cladogram C from the scheme p. 224) – despite the not so convincing autapomorphies for a proper cladistic characterization. The relatively long pubis might, however, be one more specialization creating a 'true' clade for 'Lucy' and her kin (Fig. 9.2). Discussion and synonomies by Groves (1991, pp. 249–251) as gen. indet.

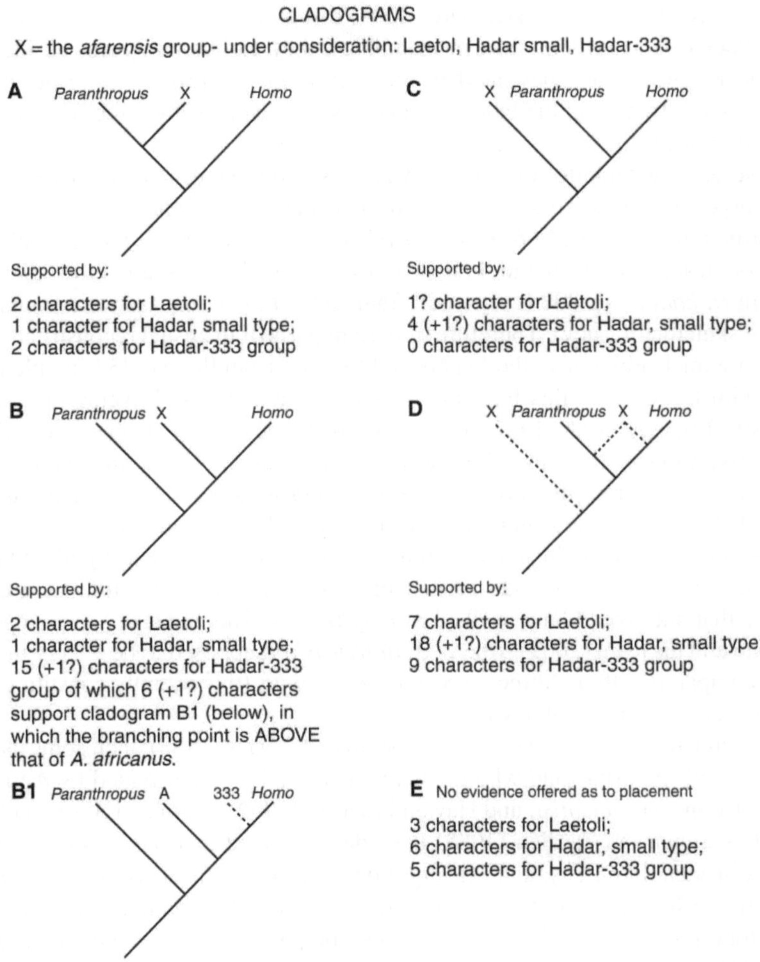

Fig. 9.2 Cladograms modified from Groves (1991) by addition of cladogram B1 showing in detail the position of *Homo hadar* Bonde and Westergaard (2004; = 333). Homo in the other diagrams comprises *Australopithecus africanus* (as a primitive member). x is that one of the three subdivisions of the polyphyletic *'Au.' afarensis*, which is tested in the resp. diagram. So B and B1 support inclusion of AL 333 fossils in *Homo*, while C indicates that 'Lucy', now *Afaranthropus antiquus* (Ferguson, 1984) is most primitive. For the Laetoli sample (incl. type of *'Au.' afarensis*) two charaters weakly favour each of the cladograms A and B, but overall similarity is with the primitive *Au. africanus* (acc. to Groves), so it can best be placed with some uncertainty as an early member of that lineage. This is very important, because this implies a minimum age ca. 3.7 m.y. for the split between *Homo* and *Paranthropus*, meaning that both of those lineages are nearly 4 m.y. old. If the Laetoli sample is part of *Australopithecus* (s. s.), then it can – as *Au. africanus* is occupied – retain the name *Au. afarensis* (unfortunately), but if not, then the name shifts back to *Praeanthropus africanus* with the Garusi maxilla as type specimen (and *afarensis* as synonym)

antiquus. We (Bonde & Westergaard, 2004) disagree with Groves concerning the much older Lothagam and Tabarin mandibles, they can certainly not be included in the same species as Lucy, as they are even more primitive, the former named *lothagamensis* by us and the latter *praegens* by Ferguson (1989 – see further below under *ramidus*, p. 167).

These are the 'gracile' or 'small' Afar fossils distinguished by Groves plus a rather large palate (showing that not all *antiquus* are small) which was used for the reconstruction of the skull of '*A. afarensis*' (Kimbel & White, 1980; Kimbel et al., 1984), often supposed to be the skull of 'Lucy', although almost all of the other parts are from *H. hadar*, AL 333 (only small fragments of the braincase were found with the type skeleton of Lucy). Therefore important parts of the usual reconstructions of 'Lucy' are misleading; also the fingers and toes are from the AL 333 sample (none found with Lucy). And this likewise counts for the claim by Lovejoy (1988, also Johanson, Lovejoy et al., 1982) and others, that Lucy had a striking gait, almost *exactly* like modern humans. This must be wrong, as the shape and 'flare' of the pelvis is different from the modern condition, and so is the relative length and the shape of the femoral neck and the pubis (critique also by Stern & Susman, 1983; Susman et al., 1985). So the muscles must have worked at least in a slightly different way, and furthermore her body proportions are unlike ours, making it extremely unlikely that she would have walked *exactly* like us. This mistake seems repeated by White and his team (2009), where '*A. afarensis*' is supposed to be a 'very modern biped' compared with its 'direct ancestor', *A. ramidus* (this actually constitutes two mistakes, or something unknown).

This controversy is reviewed in some detail by Cherfas (1983), incl. some aspects of the Laetoli footprints and who made them, a problem later treated by Anderson (1983) also in *New Scientist*, and Hay and Leakey (1992). Stern and Susman (1983) as well as Senut and Tardieu (1985) have demonstrated that the small specimens in the Afar area, like Lucy, and the larger ones like those from AF 333 are characteristically different concerning the postcranium, all with reference to mobility and ability for climbing and walking. In all cases, comparing elbows, knees, ankle joints and heelbones shows that the 'small' sample (Lucy) is much more ape-like, and the large specimens much more *Homo*-like. Lucy has retained many primitive features connected to climbing abilities, and she could well walk upright also, although not exactly a striding gait like modern humans, but probably with a slight bend in hips and knees. The large specimens were better adapted for walking on the ground. Lucy also had relatively long arms compared to the legs like in apes, while this measure is unknown for the larger specimens. Lucy has an unicuspid lower anterior premolar somewhat like apes, while the larger jaws tend to have a high inner cusp on that premolar like in man. Although male gorillas live more on the ground than the females which are more frequently climbing, it seems impossible to explain these differences just as sexual dimorphism; it is much more likely to be species differences between primitive *Afaranthropus antiquus* and the more *Homo*-like AL 333 – the same pattern which appeared from Groves' analysis of some different characters. And the two species are usually not found together on the same locality. The Lucy skeleton is from a higher member of the local formation with an age like 3.18 m.y., while the AL 333 sample is from the member immediately below with age 3.22 m.y. (Walter & Aronson, 1982). The Laetoli sample with the type of *afarensis* is a much older third species, its proper name being *Praeanthropus africanus* (above under Lucy).

9.7 Very Early Homo

The only profound cladistic analysis of this mess is that by Groves (1989, 1991 – and I have seen no valid arguments against it – those by White (e.g. 1985) seem so biased towards his original idea of one taxon that they are very difficult to use). Groves found to his own surprise, when the nearly 30 characters differing between the three localities/samples were analysed, that the 'robust first family' (AL 333) was quite advanced and *Homo*-like (Coppens, 1994; Senut & Tardieu, 1985 agree). This taxon even has several synapomorphies with *Homo* which are not found in the slightly younger *Australopithecus africanus* (Fig. 9.2). Recently, Westergaard and I (2004) have created a name, *Homo hadar* (or homo-hadar, non-Linnaean – see Groves, 1989, 1991, p. 260, where he almost implied this name in the headline) for this important taxon, the earliest, 3.22 m.y. old, representative for the *Homo* lineage (as distinct from *Paranthropus*) with the incomplete skull AL 333–45 as holotype (face lacking, but a juvenile one is known (see e.g. Olson, 1985; Johanson & Edgar, 1996), comparable to the type of *A. africanus* ('Dart's baby' of 1925)). We also indicated autapomorphies as diagnosis for *H. hadar* as a true clade using features listed by Groves. See a detailed discussion of this species ('unnamed') of *Homo* by Groves (1991, pp. 260–263). At loc. 333 (where now over 250 bones of the so-called 'first family' have been found) a recent find of a foot bone, metatarsal 4, has indicated – as expected from the heel and ankle – that the foot of this Hadar hominid is very *Homo*-like, and it indicates an arch in the foot like in modern humans (Ward et al., 2011), and therefore a human-like gait. This is clearly different from the condition indicated by the ankle of the more primitive skeleton of Lucy (*Af. antiquus* – see above under Lucy), although this is not the point made by the authors (incl. Johanson, Lucy's finder).

It appears that no-one has criticized and refuted the distribution of characters used by Groves (1991), so here his conclusion which now and then is confirmed, is as just shown, and many implications follow: AL 333 is a very early representative of the *Homo* lineage in East Africa – exactly like the Leakey family used to predict but now apparently will not admit, or even mention in a book on the origin and evolution of *Homo* (R. Leakey, 1994). So, unfortunately the traditional story told about these important fossils from Afar, also in the textbooks (e.g. Benton, 2005), is quite misleading, and it should clearly be revised and the taxonomy rectified.

These unexpected findings concerning AL 333 have several important implications:

(1) There must be representatives of the more primitive *Homo (Australopithecus) africanus* somewhere of at least the same age, probably somewhat older.
(2) Also, an early member of the robust *Paranthropus* lineage (sister group to *Homo*) must exist somewhere in Africa, over 3 m.y. old. Perhaps the so-called 'Little Foot' skeleton from Sterkfontein, S-Africa is of relevance here, because it clearly has a strong parietal crest and is rather prognath. Its age is said to be about 3.3 m.y. (Clarke, 2002) from an old 'pocket' in the limestone cave. Could it be the predicted old *Paranthropus*?
(3) Clearly, there is sufficient 'unknown' space in Central and Western Africa to accommodate such ancient species – almost nothing is known of fossiliferous layers of the right age in these regions.
(4) The large and slightly younger skull from Afar, AL 444-2, ca. 3 m.y. old – said to be the largest 'australopithecine' skull known, at least widest in the eye region (Kimbel et al., 1994 – nicknamed 'Boy of Lucy'), could rather be an advanced male of *H. hadar*, a species different from 'Lucy'. The former's relationship with the so-called 'advanced or primitive' (?) *'Australopithecus', A. garhi* (White et al., 1999) – suggested as more *'Homo-*like', but with big molars – has not been thoroughly investigated (see below under 'Stem-mothers').
(5) The Laetoli sample from Tanzania of age 3.6 m.y., incl. the holotype lower jaw LH 4 of *Australopithecus afarensis* (Johanson et al., 1978 – a most peculiar and unfortunate choice combining name and type, see Johanson & Edey, 1981; Tattersal, 1995, p. 152) is of more uncertain relationship. It was said by Groves (1991) to be 'most similar' to *A. africanus*. In that case, it could be the early *africanus* relative predicted above (under point 1), and one could argue that it should preserve its specific name, but as *Homo (Australopithecus) afarensis*. The nomenclature is more complicated, however: The Laetoli sample has always been considered just one single specific taxon (despite debate on who made the famous footprints). Therefore it comprises the first hominid fragment found in that locality in the 1930s, the so-called 'Garusi maxillary' (Kohl-Larsen & Reck, 1936), which was first named *Meganthropus africanus*. It was later considered unlikely to represent this Javanese 'genus', and was therefore referred to a new genus *Praeanthropus* (see the story in Strait et al., 1997). Accordingly, if the Laetoli sample is not believed to belong to *Australopithecus*, then its proper name is *Praeanthropus africanus* (syn. *A. afarensis*) as also used e.g. by Bonde & Westergaard (2004) – in our non-Linnaean nomenclature, however, it should be just one name, e.g. *praeanthropus,* for that terminal taxon/LITU.
(6) The relationship of that rather primitive species and others with ages around 3–4 m.y. like *'Australopithecus' anamensis* (Ward et al., 2001 – clearly not a *Homo (Australopithecus) sensu stricto*) and *Kenyanthropus* (Leakey et al., 2001) and *'A.' bahrelghazali*, a lower jaw from Tschad (Brunet et al., 1996), should obviously be cleared up. Our suggestion (Bonde & Westergaard, 2004) is shown in the tree and classification below, but note: without implying the reasonable splitting of *praeanthropus* (= *afarensis*) into three different taxa.

(7) An even more critical analysis would probably have a sequence of advancing sister groups: *lothagamensis* (Bonde & Wesytergaard, 2004), *anamensis, kenyapithecus, antiquus,* (which are more primitive than) *praeanthropus, bahrelghazali* below the split between *Paranthropus* and *Homo* (combined in the clade Parhomo by Bonde & Westergaard, 2004), but with the latter two 'species' possibly interchangeable with the Parhomo stem (indicated as *sedis mutandis*, of uncertain and interchangeable interrelationship). And as the faces and ear regions are unknown or very little known for all of the last three 'species', their relationship with the characteristic, flat-faced *kenyanthropus* skull may seem quite uncertain too. The AL 333 sample should be removed from the *praeanthropus/afarensis* group together with a few slightly older Afar specimens, they all clearly seem to belong in the *Homo*-lineage, though not as the most primitive members (see comments above).

(8) 'Lucy' (= *Af. antiquus*) is by no means the stem-mother/ancestor of all later hominids, like it is most often presented by the 'White-Johanson group' (e.g. Johanson & White, 1979 – now separated for good? – Kimbel et al., 1984, White et al., 2009) and by the popular press. She (if it is not a male – cf. Schmid, 1983) is quite primitive, but simply too late in the stratigraphy, and therefore must be an old, extinct side branch of the phylogenetic tree.

(9) This entire 'story' has great consequences for estimates of the (minimum) age of the *Homo* lineage – more than 3.2 m.y. or maybe more than 3.6 m.y., if *praeanthropus* is really closest to *H. [Australopithecus] africanus*, while the modern 'tradition' mostly claims about 2½ m.y. (e.g. Leakey, 1994; Kimbel et al., 1996), based on the earliest finds of *H. rudolfensis* and/or *H. habilis*, two taxa which Wood & Collard recently (1999a, 1999b) would even relegate to *'Australopithecus'* for entirely 'adaptationistic' reasons, making *Homo* less than 2 m.y. old. It will, however, be indicated below that quarrels about these ages and the boundaries of the 'genus *Homo*' are completely arbitrary. They have, as usually employed, no 'precise meaning' at all, but are consequences of sub-optimal classifications and often adaptationistic decisions about so-called 'key-characters'. In fact, the same counts for our own 'species', *Homo sapiens*!

(10) It should be absolutely obvious, that the traditional names and terms for *Australopithecus* and Australopithecinae (both incl. 'gracile' and 'robust' forms, and perhaps even older, more primitive taxa) are paraphyletic or polyphyletic concepts of no use in a consequent (natural) phylogenetic classification. Even in the vernacular form, 'australopithecines' can only be imprecise and spread confusion. This is entirely neglected by many prominent palaeoanthropologists like Berger et al. (2010) referring the clearly *Homo*-like new species *sediba* to a very 'broad' *Australopithecus* (incl. from (a broad) *afarensis* to *boisei*). This despite a phylogenetic analysis clearly showing *sediba* to be sister group of *Homo* (from *habilis* to (a broad) *erectus*) in the cladogram (fig. S 3 in Supp. Mat homepage), because all the derived features are held in common with *Homo* while similarities with *Australopithecus* clearly are in 'primitive' characters (cf. Berger's diagnosis of *sediba* and table 1). His analysis also supports *A. africanus* as the next sister group, that is, as a more

primitive *Homo*, as argued above.The same 'broad *Australopithecus*' also incl. *anamensis* is used explicitly by White et al. (2009 – in the thematic issue on *Ardipithecus* in *Science* – see p. 165), and in many of their texts they employ such 'grade groups' which are obviously paraphyletic and cannot be discussed in a precise way concerning relationships and phylogeny.

9.8 Stem-Mothers, *'Australopithecus'* and Australopithecines

So much for 'stem-mothers' like 'Lucy' or *'Australopithecus' afarensis*, or for her 'Boy' AL 444-2, or for 'ancestors' like the 'first family', *H. hadar* (AL 333), a neat little 'social group' (Radosevich & Retallack, 1988; Tattersal, 1989), or species like *'Australopithecus' anamensis*, none of which can convincingly be shown to be the closest relatives of *Australopithecus africanus*, the type species of this 'genus' (Dart, 1925). All of this makes the 'taxa' or groups *Australopithecus* (also with *A. garhi* from Ethiopia included [White et al., 1999]), and the subfamily Australopithecinae ('australopithecines') utterly paraphyletic. The 'genus' is almost polyphyletic as *Kenyanthropus* is excluded. This means nonsensical 'non-groups' of which one cannot speak precisely of neither phylogenetic relationships, their characteristic traits, nor literal extinction (see Bonde, 1975, 1977, 1981; Patterson, 1980). It appears extremely unfortunate that some modern anthropologists (Cela-Conde & Altbara, 2002) have tried to revive this counter-productive 'subfamily'. Perhaps even worse, Wood & Collard (1999a, 1999b) and Carbonell & Bermudez de Castro (2004) expand this 'group' by including some primitive species of *Homo* like *habilis* and *rudolfensis* in *'Australopithecus'* because of their presumed 'lack of certain key adaptations'. As Patterson argued (1980) it is difficult/impossible to characterize something by features it does *not* have – this creates 'non-groups' (paraphyletic).

There are so many myths and so much propaganda around this supposed 'taxon' (*'Au. afarensis'*), and the genus *Australopithecus* that it is very difficult for even professional (palaeo)anthropologists to disentangle.

Recently a still more unlikely, if not outright impossible, 'ancestor story' has been published by *Science* (Berger et al., 2010). Based on two new skeletons with skulls found in the Malapa Cave in Transvaal, S. Africa not far from the famous sites in the Sterkfontein Valley, a new species of *Australopithecus* is described and named *Au. sediba*, although it is pointed out that it has several derived *Homo*-like features not found in any other *Australopithecus* (see above point 10). Despite it is made clear in this way that this species is most closely related to *Homo*, it is still referred to *Australopithecus* anyway because of a number of primitive features they have in common (symplesiomorphies) – exactly the type of arguments which cannot create a classification with natural (monophyletic) groups, and therefore cannot contribute to the reconstruction of phylogeny or evolutionary history (presumbly the aim of such a paper).

9.9 Anthropologists, Ancestors, Adaptations and Annoyances

Now this has been known as a methodological fact for nearly fifty years since Hennig's book *Phylogenetic Systematics* (1966), so how come that some 'scientists' today can be completely ignorant of that, and how can that be accepted by reviewers of professional journals? Is it really impossible to convey such simple and logical biological methodology to anthropologists – or don't they care? Wood's & Collard's idea (1999) of moving some of the most primitive species of *Homo* over to *Australopithecus* would have the same effect, namely creating an even more useless paraphyletic taxon, '*Australopithecus*'. How come that they can propose such meaningless taxa without giving any reasonable 'explanation' but traditional, unfounded Simpsonian, adaptationalism? Lacking a supposed 'adaptation' (or belonging in a certain 'adaptive niche') is no character to be used in in taxonomy, systematics or phylogeny reconstruction – it is pure fantasy, 'just-so stories' of the most useless sort in an empirical science (cf. also under *ramidus*). Having supposed functional 'adaptations' is interesting, it just has nothing to do with classifications (Bonde, 1984a).

So what is the possible 'motivation' for such choice? In Berger's case, apparently, it has to do with being an (almost) 'ancestor', while everybody inclusive the authors themselves can see that this is not really the case – but this can be 'sold' to the journalists, media, and publishers of *Science* (same motivations for White, Johanson et al. concerning *'afarensis', ramidus* etc). Here, *'Au.' sediba* is obviously not an ancestor of *Homo*, because it is (even in the most favourable model) about 0.5 m.y. too young with its age of 1.8–1.9 m.y. There are early *Homo* (*rudolfensis* or *habilis*) about 2.5 m.y. old – in reality even much older *Homo* (see above), and Berger claims 'It is not possible to establish the precise phylogenetic position to the various species assigned to early *Homo*'.

If this were really what we know about *sediba*, then it should be classified as *incerta sedis* (or *sedis mutandis*), as *H. sediba inc. sed.*, at level with and sequensed with *H. rudolfensis, H. habilis* and the taxon of 'higher' *Homo* (*georgicus, ergaster, erectus, sapiens* etc.), and all of these would be *inc. sed.* So clearly, *sediba* is just another 'dead end', something Berger admits by stating that it 'represents a candidate ancestor of the genus [*Homo*], or a sistergroup to a close ancestor that persisted for some time after the first appearance of *Homo*' (precise meaning? *Science News*, 2010). If this is what Berger knows, then why not convey that information to the readers of the scientific paper and to the public, instead of relegating *sediba* to that mess of *'Australopithecus'* as traditionally (mis-)used? Then, on the other hand, *Au. africanus* is also just another primitive relative of *Homo* (as opposed to *Paranthropus*), and *H. hadar* from AL 333 is just a little more advanced (see above), so the real job would be putting *sediba* in a more precise phylogenetic relationship with all of these early members of the *Homo*-branch.

The interesting fact about *H. sediba* may well be that it could be near sympatric with *Homo* sp. (?*habilis*) from Sterkfontein, age between 1½ and 2 m.y., and/or the *Homo* sp. from Swartkrans lower horizon, Mb 1, originally called *Telanthropus capensis*, a mandible and a fragmentary part of a face. This is said by Groves (1991,

p. 239) to be 'indistinguishable from the 3733 taxon', which he classified as *Homo* sp. (unnamed) with the beautiful skull ER 3733 (ER = E of L. Rudolf = L. Turkana) implied as the type. This skull was later, by Zeitoun (2000), made the type specimen of *H. turkanensis* but it is by many people (incl. Bonde & Westergaard (2004)) referred to *H. ergaster* and considered an adault skull of age ca. 1.8 m.y. corresponding to the juvenile skeleton, 'Turkana boy' WT 15000, from the area W of L. Turkana which is roughly contemporaneous with the type jaw of *H. ergaster*, ER 992 (Groves & Mazak, 1975), age ca. 1.6 m.y.

Another *'Australopithecus'* which is in a similar way said to be '*Homo*-like' is *Au. garhi* (White et al., 1999) from Ethiopia, mentioned above, and also attempted by its authors to be 'sold' as ancestral to *Homo*. This is yet another 'taxon' which may be a mix of two different taxa from two localities: The skull appears 'primitive' and has very large cheek teeth, while the femur may be more advanced and *Homo*-like, but is from another nearby locality, so it could easily represent another taxon. Checking the list of characters described from the skull, most are very *Australopithecus africanus*-like, and very little, in fact, seems to point towards *Homo*. Its age is about 2½ m.y., possibly just a little older than the oldest skull fragments and tools traditionally referred to *Homo* (e.g. *H. rudolfensis* of 2.3 m.y.).

9.10 The *ramidus* Story – Our 'Roots'?

This everlasting story of 'famous ancestors' can easily be expanded with both recent and more ancient examples. From the front cover of *Nature*: the 'oldest ancestor' of everything human, the so-called *Australopithecus ramidus* (White et al., 1994 referring to 'our roots'), age ca. 4.5 m.y. from Middle Awash in Ethiopia. It was shown, however, already in the invited commentary to the paper by Wood (1994), that this species was too specialized in the skull base to be ancestral to all later 'hominids' (and thus implicitly should not have been referred to *Australopithecus*). In fact, it was not at all our successful 'ancestor', but rather the oldest human 'failure', a blind side branch. In our model of the phylogeny (Bonde & Westergaard, 2004 and below) even a long 'dead end', as the older Lothagam mandible (age about 5 m.y.) appears to have more advanced corpus and molars (we gave it a new LITU name, *lothagamensis*). Curiously, *Nature* allowed White et al. to publish the generic name *Ardipithecus* as a 'correction' to the original paper shortly after in 1995– well, somebody or several, including editors and reviewers, had made mistakes – presumably selling more issues that week with that front page and getting, in the long run, more quotations (like here).

Worse, even the specific name is dubious, as the new fossils from Ethiopia were not properly distinguished from the so-called 'Tabarin mandible' of about same age from North Kenya (Bonde & Westergaard, 2004) which Ferguson had already (1989) named as a subspecies, *Homo antiquus praegens* (in his opinion closely related to 'Lucy'), so *ramidus* is perhaps synonymous with *praegens* (for nomenclature it makes no difference that the latter was suggested as a subspecies – Johanson (1996) also vaguely indicates that *ramidus* and *praegens* may be the same species, and so does Tattersal (1995)). Perhaps this 'scientific story' is what Feyerabend (1975) means by 'anything goes' as a piece of propaganda for 'anarchistic science' against methods.

But seen from a phylogenetic perspective, the example *Ardipithecus* grew even more ridiculous as Haile-Selassie (2001) published some much older fossils from Ethiopia, with an age more than 5.5 m.y., as a 'subspecies' *A. ramidus kadabba*. As the new 'subspecies' (presumably in the traditional sense of chrono(-sub-)species, like in Simpson (1961) as modified by Bonde (1981, 2001)) clearly has a more primitive dentition than the original *A. ramidus ramidus*, this makes both the genus and species of *A. ramidus* paraphyletic, because the type-subspecies is more closely related to 'higher hominids' (see Fig. 9.3).

We simply classified this LITU as *kadabba,* noting that as described it lacks autapomorphies, and therefore cannot be distinguished from a 'true ancestor'. Because it is so fragmentary, this really does not mean much – and this counts to an even higher degree for the probably slightly younger LITU *lothagamensis,* a lower jaw fragment likewise without its own specializations.

As a curiosity, the single toe bone referred to this new 'subspecies' *kadabba* as the sole argument for its upright stance is about 0.5 m.y. younger than the type series of fossils – and obviously may quite likely belong to another taxon, as there are no points of comparisons. See also the nice and fanciful reconstructions in Lemonick and Dorfman (2001).

And 'curiosities' do not stop here, because White et al., Suwa et al., Lovejoy et al., etc. with an introduction by Gibbons (all 2009 in *Science*) have finally completed the long awaited and secretive task of describing the fragile partial skeleton referred to *A. r. ramidus* (found 1994, the year of description of the type material by White *et al.* in *Nature*). It is now claimed that this skeleton and its limbs and their proportions show that proper 'hominids' never went through a stage of knuckle walking like chimps, although this seemed indicated by the detailed anatomy of some hand-bones of both *'afarensis'* and the older *'A.' anamensis* (Strait *et al.,* 1997 – and the latter would imply a simple refutation of the supposed 'joint advanced feature' or synapomorphy, knuckle walking, in only chimps and gorillas). So we are not so closely related to chimps, it is argued in the 'popular press' and internet by its commentaries. How all that can be concluded from this remarkable skeleton (more complete than 'Lucy') beats me. But apparently, every published fossil hominid has to imply some sort of 'sensational news' (for reasons of publicity and funding, presumably, and ranking of publications from the 'group', department or institute or the journal – it is a competitive world, especially after Reagan and Thatcher). And it is getting worse with revised university legislation all over Europe, and enormous cuts in the state support to teaching and basic research, so we can expect more 'ancestors' and other 'sensations' in near future (see below).

Furthermore, the entire set of papers in *Science* are performed in an unusually adaptationistic style. In the conclusion by White et al. concerning 'paleobiology of early hominids', the word 'adaptive' is mentioned about ten times (adaptive shifts, plateaus, phases etc.). The entire narrative has the character of 'just-so stories' with the arguments not very convincing. The late Steve J. Gould would have made fun of this, presumably in one of his essays (see Gould & Lewontin, 1972). Why should such 'stories' be necessary, however interesting they may seem to some 'evolutionists'? I hoped they had died out more quickly after Simpson (e.g. 1961, 1976). Is it really informative with all these detailed scenarios where almost none of the arguments can be strictly tested? They are just fictitious, (science?) fictions, poems, if you wish. Lovejoy (2009) is even working within a framework of 'adaptive suites', said to be 'semiformal, largely inductive algorithms that causally interrelate fundamental characters that may have contributed to an organisms total adaptive pattern'. Does this mean that we now have to know, which characters are 'fundamental', and what is the 'total adaptive pattern'? Because one of the rationales behind this thematic issue of *Science* was the Darwin anniversary, and Darwin is quoted in almost every paper, I should like again to point to the quotation above (section on 'diversity') from the great hero about 'the immense amount of rubbish about the value of characters' – and to leave this as a warning.

9.11 Homo Ancestor Habilis?

So much for our 'oldest ancestor' *ramidus* – out on a limb – and we could go on with earlier and important examples: *Homo habilis* from Tanzania was considered an almost 'ideal ancestor' for the younger *Homo* spp. when it was described in 1964 (Leakey, Tobias, & Napier, 1964) and for about twenty-five years, when finally Groves' analyses (1989, 1991) showed that it has too narrow premolars, a specialization which prevents it from being an entirely convincing 'ancestor'. The proper, detailed description came in 1991 by Tobias (most of these Olduvai fossils found around 1960). And even here there has been, right from the beginning, some doubts whether the oldest specimens from Olduvai 'Lower Bed 1', like the type mandible, OH 7 (Olduvai Hominid 7, incl. parietal and postcranial fragments like the famous 'manipulative' hand supposed to be from the same young individual – the foot OH 8 is not from *habilis*, but more likely *P. boisei*), and the younger specimens (skulls, one with lower jaw) from the upper 'Bed 1' and bottom of 'Bed 2', are really the same taxon. And further, whether the fine skull ER 1813 from the East side of Lake Turkana, N. Kenya (ER = East Rudolf, the older German colonial name of the lake), found in the 1970s and of about the same age as the type (1.8–1.9 m.y.) also belongs in that taxon (see e.g. Rightmire 1993 contra Wood, 1992). ER 1813 has also been referred to *H. ergaster*, and the youngest *habilis* specimens were said to be more '*erectus*-like' (which they hardly are, but perhaps *ergaster* like).

This mess leaves us with a not so 'ideal ancestor'. The phylogenetic tree (Fig. 9.3) shows that in this time interval between 1½ m.y. and 2 m.y. the diversity of hominids was at its peak with about a dozen taxa worldwide, but only two outside of Africa, namely *Homo (Pithecanthropus) erectus* on Java and *H. georgicus* in SE-Europe. The remaining forms were living in Africa, and if Zeitoun's two new species (2000) from East Turkana are included, then about 7 of them lived more or less at the same time in the East African Rift valley – how to explain that in ecological and 'adaptationistic' terms?

9.12 H. *ergaster* – A 'True Ancestor'? – And the 'Flores Dwarf' or Hobbit

The next ancestral 'grade' on the ladder towards modern man is *H. ergaster*, mainly from Koobi Fora, East Turkana, age ca. 1.5–1.9 m.y. (the type mandible ER 992 youngest). And as can be seen on our tree, in our analysis we find it very difficult, even with the inclusion of the very excellent skull and skeleton of the 'Turkana Boy' (WT 15000; WT = West Turkana) from West of the lake (as by Westergaard & Bonde, 1986; Wood, 1992), to distinguish this taxon from an ideal ancestor of all the younger *Homo* spp. – so we classify it partly as an 'ancestor' by dividing it into two similar parts and repeat it twice in the classification. And this taxon should probably not be treated under the name *H. erectus* (or 'African *erectus*') as done by Leakey (1994 and his TV-series, Walker & Leakey, 1993; Bräuer, 1994; Rightmire, 1990, 1998), a name of an apparently characteristic taxon from East and SE-Asia which is somewhat more specialized (Bonde, 1976, 1977; Andrews, 1984; Groves, 1991; Wood, 1994; Anton, 1997; Anton et al., 2002). And it may turn out to be nearly just as old on Java (about 1.7 m.y. (Schwisser et al., 1994), but perhaps more likely only a little more than 1.5 m.y. (Larick et al., 2001)). Note, though, that Zeitoun's detailed cladistic analysis (2000) based on a large number of measurements from all the relevant skulls (each of them used as a 'terminal taxon' in the analysis), surprisingly placed WT 15000 in the middle of Asian *Pithecanthropus*, although it seems to lack the more evident specialisations of the 'proper' Asian *erectus* skulls (claimed by Walker & Leakey (1993) to be due to its young age, perhaps 9–10 y. in the midst of changing its milk-canines). Also analyses by Rightmire and Bräuer (cit. above) found no obvious differences between the African and Asian samples (see illustrations in Schwartz & Tattersall, 2003). Zeitoun also found that the adult skulls ER 3733 and 3883 are not adults of the same form as WT 15000, but represent two more primitive taxa, each characterized by almost a score of autapomorphies (specializations). Accordingly, he gave them two new separate species names, e.g. 3733 = *Homo turkanensis* (not used in Bonde & Westergaard, 2004), thereby indicating even more diversity between 1.5 and 2 m.y. Also Groves (1991) has 3733 as a separate taxon (unnamed).

The Caucasian finds from Dmanisi of an age about 1.7 m.y. are very '*ergaster*-like', perhaps even slightly more primitive, and the three good skulls seem quite variable, but they have all been dubbed *H. georgicus* (Gabounia et al., 2002; Vekua et al., 2002; Balter & Gibbons, 2000).

Anyway, the oldest part of *H. ergaster* (represented by a skull like ER 3733 in Bonde & Westergaard, 2004 and the classification below) could in principle, perhaps, be ancestral to all later *Homo* (but cf. again Zeitoun, 2000 placing it on a 'long sidebranch' as *H. turkanensis* and ER 3883 also on the next higher, long, specialised sidebranch). The younger part, incl. ER 3883, WT 15000 and the type jaw, ER 992, perhaps cannot be such common ancestors, because being of too young ages to be stem-forms of *Pithecanthropus*. However, they may be ancestral to the line towards neanderthals and sapients (the 'Euhomo' of Bonde & Westergaard, 2004).

And in this connection it should be mentioned that the controversial 'Hobbit' or Flores dwarf, *H. floresiensis* from a large cave on Flores, Indonesia (Brown et al., 2004; Morwood et al., 2005) including one buried skeleton about 18000 y. old and not much over 1 m tall, has a skull as primitive as this *'ergaster*-level'. This implies an age for this line of small *Homo* as more than 1½ m.y., and a very long 'ghost lineage'. This might indicate the earliest migration 'out of Africa'. Including it in the modern *H. sapiens* as abnormal or pathological dwarf seems out of the question as indicated by several studies of both skull and feet (Falk et al., 2005, 2007; Jungers & Baab, 2009), despite claims to the contrary (Martin et al., 2006; Obendorf et al. 2007; Oxnard, 2010 – first stated by the Indonesian palaeoanthropologist T. Jacob in newspapers and television, see Jacob et al., 2006).

The hope of finding ancient DNA in these very recent fossils from layers only about 12000 to ?28000 years old unfortunately seems to have faded away, because DNA cannot be preserved in such warm and humid conditions. It would have been very interesting to have a 'molecular clock' estimate of the approximate age of *ergaster, georgicus* or *erectus/Pithecanthropus* lineages. There are very old tools on Flores (Morwood et al., 1998) and the 'Hobbit' is sometimes 'associated' with these without any shred of evidence, and the tools seem too large anyway for that small human. They are about 0.8 m.y. old and the most interesting about them is clearly, that someone must have been crossing a reasonable stretch of water and the 'Wallace Line' to reach Flores even during the lowest water level during a glaciation – so that someone knew how to sail already at that date or before! (The world's oldest known boats are those from the Mesolithic Ertebölle culture in Denmark, only 7–8000 y. old). The passage to Australia, where the oldest fossil *Homo* are 50–60000 y. old (and not dwarfs) also runs through Flores and across the continental shelf known as 'Sahul Land' when dry during hard glaciations.

9.13 The '*erectus* Stage' – En Route Towards *Sapiens*?

The *'erectus* group' – by Bonde and Westergaard (2004) revived as clade Pithecanthropus (Dubois, 1892, and first used by Haeckel as a hypothetical *'P. alalus'*, the 'apeman with no language' (1866)) – is another classical 'mistake' as ancestor of *H. sapiens* (and Neanderthals). An almost universal claim some 40–60 years ago (e.g. Weidenreich, 1946; Day, 1965, textbooks like Romer, 1966; Carroll, 1987; Trinkaus & Shipman, 1992; Benton, 2005 wavering a little, and all popular literature, even today) – only old Louis Leakey strongly contradicted this (1963; Leakey & Goodall, 1969) and found his own *H. habilis* a better candidate, and therefore moved *erectus* out as a sidebranch. This placement seemed rather obvious when looking at the many characteristic specializations in the faces of both Java- and Peking-man (Bonde, 1976, 1989; Andrews, 1984; Wood, 1992, 1994 contra Rightmire, 1990 – Bonde & Westergaard revived the relevant older names *Javanthropus* and *Sinanthropus* for these two subclades).

Many seemed to have greatly favoured calling their pet-fossils by this popular 'ancestral name', such as Lumley (1982) for the 'Tautavel man' from Arago Cave – it seemed to be more interesting to have a 'true *erectus* ancestor' (perhaps for funding reasons?) rather than just another ancient Neanderthal, of which there are so many in France (the Arago skull and face and jaws have but one possible advanced feature in common with Pithecanthropus, all other similarities are 'primitive' symplesiomorphies). Likewise R. Leakey also 'hung his hat' on the well known 'ancestor *H. erectus*' for his pet skulls from Koobi Fora, ER 3733 and 3883. (Walker & Leakey, 1978; Leakey, 1994 – but see critique by Groves, 1991). And later *H. erectus* was used also for the even better find, the 'Turkana boy' from Nariokotome, WT 15000 (Walker & Leakey, 1993). But as mentioned above, it appears to lack the specializations of the face and skull and limp bone thickness of typical Asian *Pithecanthropus* (but cf. Zeitoun [2000] who referred those three specimens to as many different taxa!), and we prefer to classify them as a possible 'ancestor', *H. ergaster*, like most people do today (Westergaard & Bonde, 1987; Tattersal, 1995; Schwartz & Tattersal, 2003; Wood, 1992). Recall that Wood, in his huge monography on the Koobi Fora fossils (1991), was not allowed by the editor R. Leakey even to mention the name *ergaster*! The reason being that Leakey was furious over the publication of the name by Groves & Mazak (1975 – see comments and excuse by Groves (1991, p. 197)). By now it is fair to state that *H. ergaster* is a much more acceptable 'human ancestor' than *erectus*. The latter, on the other hand, with its youngest and most specialized representative, the Solo man, *H. (P.) soloensis*, now dated as less than 50000 y., has therefore become a late contemporaneous human with 'modern' *H. sapiens* in the far East, and presumably also with the 'primitive dwarf' *H. floresiensis*.

Therefore all the nonsense about an *erectus-sapiens* transition and what this implies of shifts in 'adaptations' etc. (Weidenreich, 1946; Campbell, 1963; Day, 1965; Romer, 1966; Rightmire, 1990; Wu & Poirier, 1995) is now utterly outdated – it never took place. The African *ergaster* and the equally old and enigmatic finds from Dmanisi in Caucasia of Georgia (Europe's oldest hominids, about 1.7 m.y.) called *H. georgicus*, with its great variation between the three skulls (Gabounia et al., 2000, Vekua et al., 2002; Balter & Gibbons, 2000) should rather be in focus of our 'ancestral interests'. They may hold the keys to the first (or second?) 'out of Africa', at least that migration which created Neanderthals and the line towards modern man (see also Stringer, 2002, 2003). There is also a partial skeleton associated with one skull and many other limb bones (Lordkipanidze et al., 2007). The skeleton is small like the skulls, which have cranial capacities between 600 and 780 cc., that is, in the range of *H. habilis*. Despite some variation between the three skulls, they are recognized as belonging to one species, and this is most similar to *H. ergaster* from E-Africa.The common 'stem lineage' of the neanderthals and the modern groups (dubbed Neanderthalia and Sapientia in Bonde & Westergaard, 2004) may well lie close to such fossils as *H. antecessor*, 0.8 m.y. from Atapuerca in N-Spain (Carbonell et al., 1995; Bermudes de Castro et al., 1997, 1999 – and recently a fragmentary lower jaw, provisionally referred to the same species, and stone tools have been found at a lower level, age 1.1–1.2 m.y. (Carbonell et al., 2008)) and *H. cepranensis*, 0.9 m.y. old from Italy (Mallegni et al., 2003). And these may represent the same 'species' on the other side of the Mediterranean, e.g. the Algerian Ternifine (Thigenif) skull and jaw, *Atlanthropus mauretanicus'* (Arambourg, 1963) and/or from Morocco the Salé skull, all claimed to be more than 0.5 m.y. old (Hublin, 1985 – if only one taxon, it should be named *H. mauretanicus*). The OH 28 skull from Olduvai may also be a candidate. The 1 m.y. old skull of Eritrea may also be of interest here (Abbate et al., 1998), as well as an Ethiopean skull of about same age (Manzi et al., 2003), but the interrelationships of those skulls and the older finds have never been satisfactorily analysed (cf. Rightmire, 1990, who referred the earlier African finds to *'erectus'*), although it should be obvious, that it is among those fossils from Europe and/or Africa we might find the ancestors of Neanderthals and ourselves. Great potential of the Mediterranean region. But in N-Europe there were people already 0.7 m.y. ago, as wittnessed by their flint tools from SE-England (Parfitt et al., 2005). In Denmark the oldest tools may be derived flints from sediments of the last interglacial in S-Jutland (Holm, 1996), but of a type (if they are tools at all) most similar to Tayacien and Clactonian in France and UK, earlier dated to the previous interglacial about 0.25 m.y., but now suppossed to be around 0.4 m.y. old. The northernmost occurrence of stone tools is probably 'The Cave of the Wolf' in Finland, said to be from the last interglacial, ca. 130000 y (Pettitt & Niskanen, 2005; Schultz, 2010). All of those tools must have been made by Neanderthalia.

9.14 Neanderthal and Sapient 'Ancestors'?

Can we ever expect to find such actual 'ancestors'? We believe it is not entirely impossible (Bonde & Westergaard, 2004), and that both the neanderthal and sapient lineages may show reasonable examples of such 'ancestral fossils' in an accretion model with about four stages (chronospecies) om each limb. Both lineages during the latest half a million years or more seem to have been separated and living in different geographic regions, some of them under quite different ecological regimes (Stringer & Gamble, 1993; Stringer & McKie, 1996; Trinkaus & Shipman, 1992; Trinkaus, 1983; Foley, 1987; Arsuaga, 1999; Schwartz & Tattersall, 2002). This age of the split seems to agree reasonably well with the ancient DNA evidence (Krings et al., 1997, 2000), and to be older than that indicated by traditional morphology often citing the neanderthal lineage as being about 0.3 m.y. (Tattersal's fossil record diagram 1995, partly also Stringer & McKie's diagram) and that of *Homo sapiens* as being about 0.13 m.y. (Stringer & Gamble, 1993, fig. 28), and Winkler and Schweikhardt (1982) take pre-neanderthals only back to ca 0.1 m.y, but pre-sapients (in Europe) ca. twice as long. But both of the latter ages are clearly misleading seen from a cladistic viewpoint (Bonde & Westergaard, 2004), as the socalled *Homo heidelbergensis* incl. its type specimen, the 'Heidelberg jaw' from Mauer has a few neanderthal specialisations in the dentition (e.g. size of dentition, weak taurodonty), the Petralona skull from N-Greece and the Arago face from S-France lack a 'fossa canina' like the neanderthals, and they likely represent the very early and primitive part of the neanderthal lineage, a clade we in 2004 dubbed 'Neandertalia' (see, Schwartz & Tattersal, 2001 about European fossils – earlier the *heidelbergensis* was seen as a possible 'ancestor' of both neanderthals and sapients – Bonde, 1976, 1977, 1981; Stringer, 1984, 1985; Groves, 1991). This lineage can now be 'followed through time' from the Heidelberg and perhaps the Petralona fossils being about 0.6–0.7 m.y. in an 'accretion model' of successive ancestors via *'steinheimensis'* (e.g. from Steinheim, Swanscombe, Reilingen (Dean et al., 1998) and Atapuerca SH (Arsuaga et al., 1997)). Not all the features of all the fossils will entirely follow the 'model', the Steinheim skull e.g. seem to have a fossa canina like sapients, and it has been used as an argument for two lineages in Europe. I once used this as an indication for possible gene flow between the two lineages when arguing, that they might be just 'subspecies' seen from a 'time-bio-species' viewpoint (Bonde, 1989) resulting in a 'hybrid' like the Steinheim skull with both fossa canina and the Neanderthal mark of a suprainiac fossa at the back of the skull (Hublin, 1988).

9.15 *Homo* and *H. sapiens*???

Within *Homo* and *H. sapiens* as generally used, there is a problem part of which to day is based on 'political correctness' (as expressed by Gould (1977) and Ghiselin (1997) and many others who do not want to be framed as 'racists'). This means acc. to the above two, that one ought not scientifically discuss the possible valid subdivisions of our own species and e.g. map their history, because such might have to do with discrimination – or rather some others might misuse it for purpose of racial discrimination. Knowing how much interest there is in the populations for following the history of their own family or 'ethnic group' or population, and the relevans of 'ethnic' medicine, this is a very peculiar constraint to impose on the scientists. Should scientists be censored because of a possibility of 'misuse' of their results, then physics and chemistry could shut down immediately. And biology, computer science, sociology and economics could follow soon.

Because of this legitimate interest in population history, your own 'roots', and the possibilities given by modern genetic techniques, it is worth pointing out that the pattern of relationships does not follow the boundaries of the traditional 'races' based on skin colour, shape of hair and other very superficial features. Many of these features can often be explained by selection pressure from the local environment and climate.

Such features and measurements of skulls (skull indices by the Swede Retzius in 1842), body shapes, height etc are classical parametres to characterise the human 'races' in the 19th century. These already had Linnaean names since his first small edition of 'Systema Naturae' (1735), namely classified under *Homo sapiens* ('know yourself') as *Europaeus* ('white'), *Americanus* (red), *Asiaticus* (yellowish) and *Africanus* (black). They were re-named shortly after 1800 by the Germann Blumenbach, based on his huge collection of skulls from all over the world. He added the Malaysian race, and he chose to call the white race ‘caucasicus’, because he had such a ‘perfectly’ beautiful female skull from Caucasia. From this type he thought the other races were developed by degeneration (today still believed by the ‘Watchtower’ sect in their propaganda material). This was a sort of ‘evolutionary theory’ also held by the great French naturalist Buffon (1749) concerning human races, and in 1766 he expanded that concept to all animals – but such changes are confined within a species. Here it is also worth noticing that his pupil Lamarck when describing the first proper evolutionary theory in 1809, exposed the idea that humans were derived from the most perfect ape, the chimpanzee (orang-utan was less perfect, and gorilla unknown), which as such was on its way to greater perfection, the general trend of changes (that is evolution) in all lines. Lamark’s general trend towards higher organisation has in modern times been taken up in theories about orthogenesis, often as evolution aiming at a ‘higher goal’ (e.g Teilhard de Chardin (1955)).

A refinement much later was mapping of such features as blood types, they did not follow the limits of the traditinal ‘races’, although Coon (1962) tried to apply the traditional race-concepts to a lot of different physical anthropology and compared with available fossils. His results were not very successful, and they were expressed in a ‘multiregional model’ (see below) of local, isolated, parallel development from fossil ‘*erectus*-like’ forms, which were not very obviously related to their postulated, modern *H. sapiens* ‘descendants’. And he had very few ‘negroid’ fossils to illustrate his claim, that the ‘negroes’ were the last ones to cross the ‘boundary’ between modern *H. sapiens* and their *erectus* forrunners although postulating that such took place all over the old world. Neither did the mapping of certain genes or other genetic markers (Bodmer & Cavalli-Sforza, 1976; Lewontin, 1982) have obvious success, and no consistent pattern appeared.

But the research on the diversity of human mitochondrial DNA (MtDNA) since 1987 (Cann et al.) indicated a picture (if a constant molecular clock is assumed) of origin of modern humans in Africa, where the largest population differences are found, and then spread to Eurasia. The last common ancestor of all living humans was calculated at roughly 200000 ± 100000 y., and the exodus from Afrika about half that old. This picture was quickly backed up by Stringer & Andrews (1988) based on fossil distribution through time. Our ‘stemmother’in Africa was dubbed the ‘mitochondrial Eve’ or ‘African Eve’ (Brown [1990] on the background and research history), and the early results and some controversies are reviewed by Johanson (1996), and Stringer and McKie (1996). Although reasonable criticisme of the original research and its methodology quickly came up, it turned out that most subsequent studies more or less confirm the general pattern, then called ‘out of Africa’ (Vigilant et al., 1991; Wilson & Cann, 1992; Horai et al., 1995), but the timing changed quite a bit. Search for ‘Adam’ based on Y-chromosomes gave a shorter time frame (Hammer, 1995), one study a very young common ancestor, less than 50000 y. And some studies of nuclear DNA also gave rather short ranges of time, about 100000. The age estimates have mostly been handled as if they gave the 'origin' as the age of the last common ancestor (LCA) of all living humans. But one should realise that all these calcultions probably rather give the age of a certain change in the genome, one or more mutations, and the 'African Eve' is more likely a position along the stem of our group somewhere before the split of the LCA. So these ages are maximum ages for the LCA, that could be much younger. And this is probably the case, and the agreement today lies somewhere between 100000 and 50000 y. for our LCA. This is apparently in agreement with the few relevant fossils of early, but entirely anatomical 'modern' people – none of these are over 40000 y. old (such as the Niah skull on Borneo), apart from earliest Australians perhaps 50–60000 y. The whole 'out of Africa' is treated in detail by Stringer and McKie (1996) in the book 'African Exodus'.

Certain rare mutations can characterize large groups of the living humans, and they have been mapped especially for sequences of haplotypes in the 'genographic project' by Wells (2002), so that by now the migrations both within and out of Africa can be followed in great detail because of a relatively small number of very rare and characteristic mutations. This is not in focus of this review, but can also be picked up in textbooks by Molnar (1998), Jobling et al. (2004), and in Denmark in excellent books by Jensen (2004, 2008) explaining the genetic background.

But how much do we know about the fossil record of even late Pleistocene hominids? Well, The 'Hobbit', *Homo floresiensis*, turned up in very young layers a few years ago, an entire skeleton of a completely unknown and very primitive branch of the phylogenetic tree (Brown et al., 2003). A neanderthal was recently identified from MtDNA in a bone fragment from southern Sibiria, and now another bone fragment, an isolated finger bone about 40000 y. old from the Denisova Cave in the Altai Mountains in S-Sibiria turns out to be a completely unknown type of hominid based upon its MtDNA, which show marked differences from both neanderthals and the *sapiens* lineage (Krause et al., 2010 – but clearly a neanderthal acc. to E. Willerslev, pers. comm. 2011). So big are the said difference that the new Denisova hominid should be the sister group of neanderthals plus sapients, and its origin can be calculated to about 1 m.y. ago, if the split between neanderthals and modern man was about 0.5 m.y. ago. With the new and very precise sequence methods, we are likely to encounter other surprises in the near future, and recalling the surprising find of the Hobbit such may come even from the palaeontological front.

9.16 Age and Limits of *H. sapiens*

What about our own species, presumably the most thoroughly surveyed species in the world? Is some agreement possible around this 'important species'? Apparently not – and from a cladistic viewpoint not at all. First, there are two competing models of the origin and evolution of our species, the 'multiregional model' and the 'out of Africa' or 'replacement model' mentioned above.

The multiregional model is vividly defended by only a few, like Wolpoff (1980, 1984), Thorne and Wolpoff (1992). This model derives living *H. sapiens* in parallel locally on all continents of the Old World from *H. erectus* stages, in some areas like Europe and the Middle East via a neanderthal stage, and with only little influence from other regions. Some hybridization is not excluded, but it is used as an explanation for the parallel development towards the more advanced *H. sapiens* in all regions, as 'well adapted' genes and traits are spread over the continents outcompeting the less favourable. From a biological species viewpoint where a species is a group of interfertile organisms isolated (concerning reproduction) from other species, then it is clear that one consequence of the model is that *H. sapiens* and *H. erectus* are actually the same species, just gradually changing through time (anagenesis or phyletic evolution), so there is no use for two names. And neanderthals form a subgroup of that species also. Wolpoff has also argued for this, and one could then easily subdivide this 'species' into chronosubspecies, one slowly replacing the other through time.

Unless one wants to follow such a scheme all the way back to the beginning of life (in a truely Lamarckian model of great numbers of parallel lines with the same evolutionary tendencies, eternal orthogenesis), then this will have to stop somewhere – but where? Well, earlier than about 2 m.y. there are only human fossils in Africa, so in some way it will presumably have to start there with a migration out of the contient after that date, and this corresponds well to the oldest fossils in Europe, *H. georgicus*, and the oldest Java populations, in both areas very 'erectus-like' and about 1.7 m.y. old., so this will also become the age of *H. sapiens* sensu lato. If this is not upheld, and the transition from erectus to sapiens-like occurs between species at different ages in different regions, then the question of the age of *H. sapiens* is troublesome. The downside of this model, apart from the age problem, is from a cladistic viewpoint the excessive parallel evolution, which is very un-parsimonious: the same shift occurring many times. In a cladistic model or reconstruction of the phylogeny, one would use the mutual derived/advanced features as potential synapomorphies showing that all *H sapiens* individuals are more closely related to each other than to any *H. erectus* lacking these traits, and this would imply that *erectus* became truly extinct (not just technically that the name stopped to apply).

In that way we are suddenly in the alternative 'model of replacement' and, as we saw, there has to be at least one out-of-Africa event ca. 2 ,m.y. ago. This model operates, however, with several migrations from Africa, and it is especially the last one or last few ones that are in focus when speaking of *H. sapiens*. As seen above, the genetic results seem to clearly confirm an 'out of Africa' model, and that modern type *H. sapiens* originated say 50000–200000 y. ago in (E-)Africa and from there expanded to the rest of the world replacing more archaic populations on the way. And from a genetic viewpoint, there seems not to be much evidence of mixing with the local, archaic populations (or there may be few characteristic genes that have been lost in our populations).

The replacement model avoids excessive parallel evolution and is therefore more simple and satisfactory from a cladistic methodological viewpoint, but this does not necessarily mean that it is the 'true' description of the history. It could also be better expressed in a system of natural groups of phylogenetic relationships (monophyletic or clades). In this model, the age of *H. sapiens* (or the subspecies *H. s. sapiens* depending on the reproductive relation to the neanderthals and the late *erectus*) is not so troublesome, if one can agree on how to define and recognise its origin. Is that when it finally split up into the first two subgroups still living today, or is it when our lineage split away from our sister(?sub-)species neanderthalensis (that is, a stem-definition)? Or is it somewhere in between?

The first definition from the point of diversification (the end of last common ancestor) would give a minimum age about 50–60000 y (the age of oldest representative from a subgroup, in this case australians which are said to be 50–60000 y. old). If the age corresponds to the split from neanderthals, then this is around, possibly over, 0.5 m.y. And if we pick one specific character or a functional character complex somewhere in between, then the age of the oldest fossil with that specific feature indicates a minimum age. Unfortunately, palaeontogists nearly always select the latter definition and point to a 'key character', in this case perhaps a chin and/or a vertical forehead, or if we could find a good indication, perhaps something like spoken language (perhaps of a certain complexity?) – and in this way we could quarrel from now to eternity without ever being able to agree upon the 'relevant' character. This is why a simple question about age can simply not be answered before we agree about some conventions.

In this way, both the inclusiveness and the age of *H. sapiens* is uncertain and completely arbitrary. We may give reasonably precise upper and lower limits, here minimum 60000 y and maximum ca. 0.5 m.y., but what you see in the textbooks and papers is something like ca. 0.2 m.y. (for the Kibish 1 fossil from Kenya), and before that it was ca 0.13 m.y. because that was the supposed age of those sediments found with the oldest so-called 'anatomically modern man'. And this is actually a misnomer, because it has too large eyebrow ridges (*tori supraorbitales*) for 'modern man', which can really not be traced further back than to 40–60000 y.

Exactly the same uncertainty counts for genus *Homo* or any other taxon in the system, it is completely arbitrary where the taxonomist chooses to 'cut off' the group, so all ages are completely arbitrary. This is why our system (Bonde & Westergaard, 2004) indicates only (minimum-)ages of splits in the tree, so that a group has the minimum age of the oldest fossil of its subgroups. And this age can also replace Linnaean rank, both relative and absolute at the same time. So here the age of *H. sapiens* is (40-) 60000 y., and 0.5 m.y. is the age of the (unnamed) group embracing *H. sapiens* (and Sapientia) and the sister group Neanderthalia, indicated by the oldest fossil, the Heidelberg jaw of the latter subgroup (ca 0.5–0.6 m.y.) – or from molecular biology, ancient DNA and the molecular clock (Krings et al., 1997 – here giving about the same result). We suggested that both ages be indexed: Sapientia + Neanderthalia: 0.5–0.6 m.y.; m: 0.5 m.y. (m: for molecular age).

How old then is *Homo*? Well here one consequent way of doing it would be all the way down to the split from chimps 6–7 m.y. ago – and actually some of the molecular biologists have recently suggested including chimps in *Homo* (Wildman et al., 2003). In this way, the split age within *Homo* would be 6–7 m.y. (molecular age probably a little less?). We have however, arbitrarily, decided to delimit *Homo* at the major extinct sister clade, the 'robust' *Paranthropus*, both lineages members of the clade Parhomo. The oldest fossil giving the minimum age in the phylogenetic model here is then *'afarensis'/praeanthropus* ca 3.7 m.y. on the *Homo*-line, and if this is not acceptable, then *H. hadar* ca. 3.4 m.y will indicate the minimum age (all fossils on the *Paranthropus* line are younger). We could have chosen another rather large sister group as 'cut off', namely *Pithecanthropus*, giving quite a different result (between 1½ and 2 m.y.). If so the traditional system would need a lot of new generic names for branches between *Pithecanthropus* and *Paranthropus*.

9.17 Evolution of 'Soft' Traits: Language

Only one example will be presented concerning 'non-fossilisable' features, and as this is about the 'Symbolic species', then it has to be about language. How can we judge the origin of spoken language, and is there a reasonable way of estimating its age? There must be a phylogenetic framework, and we here have one model based on fossils, but with some corners confirmed by molecular biology and clocks, such as the split-ages mentioned above. Here, an interplay between a well known phylogenetic model, that of the 'African Eve' theory, and a much less known model of all languages made by the old Polish historical linguist, R. Stopa in his hunt for the origin of the Indo-European language which also brought him to Africa.

The model by Vigilant et al. (1991) gave a tree of the interrelationships of modern populations based on MtDNA and with chimps as outgroup (the latter is important; the first model on 'Eve' by Cann et al. (1987) did not have an outgroup). The structure of Vigilant's model is approximately as follows: large difference between *Pan* and *Homo*, difference that can be translated to age spans, if the age of the split between the two groups can be estimated. (Had bonobo, *P. paniscus*, been incl. in the study, it should split away from the other chimps about halfway out the branch). All the sapient populations, of which many are from Africa (as opposed to Cann's study), are close together at the beginning of the *Homo* clade with only very small differences between them. The structure of their interrelationship is so that just about the dozen of branches first to split off are all from Africa, so these branches are long with relatively big differences between them, when compared to the closely packed branches at the top of the stem, whch are all from outside Sub-Saharan Afrika. Some of the very first groups to split off are some pygmees and !Kung-San people and several of the next branches are Bantu speaking people, then comes the rest of Africa and the rest of the world.

Stopa's studies in his books on the evolution of languages (1972, 1979) and his short review of his theory (1973) indicate that he finds 4–5 major levels in the early evolution of languages, and he translates many expressions/'sentences' between these levels of increasing complexity. In the 1972 book, Stopa analysed the African languages and compared them with Arabic and Indo-European on the one hand, and with 'Bushman' on the other side to see how sounds, clicks, phonemes and also body language (gesticulation) transforms from one language to the others.

Bushman (Khoisan) language appears to be the most 'primitive' (explained at length 1972, pp. 35 ff. and 1979, pp. 17–20), but also quite complicated and full of dental clicks and other sounds with the lips, as well af gesticulation when speaking. Words are very short, just one consonant and one vowel, can often mean several things, sentences very short, of one word or repeats of the same word, and there are no abstract words, only concrete ones. Counting goes one, two, three = many, and general terms are missing, such as 'to eat', there is only a word for eating something specific like meat, and no word for fruit, only words for concrete fruits.

But the most interesting aspect is his direct comparison (rather 'translations') of a number of expressions by chimps and Bushmen in different categories like 'calls connected with food', warning signals, calls for help, calls to partner of the other sex, cries of pain, laughter, and a sound when looking through the other chimps fur (for lice etc). These he describes (1972, pp. 34–35) as 23 calls ('words') of chimps and similar words of Bushmen, and in a large scheme pp. 50–57 comparing the 'chimp language' with six different Bushman languages.

These chimpanzee words he got exclusively from Yerkes' primate laboratory from the observations of Miss Blanche Learned (very suitable name) as published in a book by Yerkes and Learned in 1925. And there is indicated the number of observations for each utterance, most of them have been observed 100–300 times. I have checked this source, and admit that I had difficulties finding this exact 'list of words' or utterances. But never mind, in the 1979 book he expands the comparative schemes (pp. 100–103) as a 'dictionary' running from chimpanzee words to Bushman, then West Sudanic, and lastly Bantu, and in other parts of the books the 'translation' is continued to the level of other African languages + Hamitic + Indo-European.

The little summary of his linguistic studies (1973) and speculations about the origin of languages is interesting, because he tries to reconstruct the structure of a 'proto-language' between those of chimps and of Bushmen, the language of 'Homo fossilis' as he calls it. And finally, in the appendix, he characterises four levels of 'languages', the first being from animals, just sounds, no proper language, the second Bushmen, Hottentots, Zulu-Kafirs etc., the third Bantu, Hamites, some Caucasian and American (probably Na-Dene), and the fourth Europe, Asia, Australia, Oceania. He specifies the structure and how they are being spoken, how sounds are combined, and their cultural background like hunter-gatherers, fishermen, herdsmen, agriculture and trade. And further he specifies, for each group of sounds/languages, how they are produced in terms of form and function of the sounds. This is clearly a grade system going from 'primitive'/original to different levels of advancement.

If such a grade-scheme should be translated into a cladistic scheme of relationships, it would show the following pattern: animals (here of relevance Chimps) as sister group of the remaining groups combined. Within these Bushman-Hottentot sister group to the two advanced groups, Bantu + Hamites etc and the rest of the world. This pattern of relations is exactly included in the relationships based on MtDNA: Chimps – Bushmen – Bantu – rest of the world.

That these two completely independent and entirely different analyses should give the same result can hardly be coincident. It must be significant, and one would predict that the language of neanderthals would be like the reconstruction for the missing level, that of 'Homo fossilis'. Neanderthals do fit into the scheme based on DNA, but how to test the level of their actual language will presumably be difficult.

As to very early or very late origin of modern people's type of languages, I tend to believe it is fairly late, that is, an origin within the last million of years, but before we split away from neanderthals ca 0.5 m.y. ago, because the latter with their very large brains must have had a rather advanced language as well, but probably no more advanced than that of Bushmen. As the latter split off between 50000 and 100000 y. ago, the evolution of languages was probably very slow until humans rather quickly spread all over the world, also accellerating the pace of linguistic evolution. But language probably did not evolve very suddenly and full fledged as some linguists (e.g. Chomsky) seem to believe, with complicated grammar and everything from the beginning. Such does not exist in Bushman language and could well be something developed within the last 30–40000 y. in the Late Palaeolithic when almost all other sorts of people had gone extinct: erectus, almost the neanderthals, and rather soon the Flores dwarfs leaving us alone to chat.

That there has literally been a sort of 'proto language', a 'mother tounge'.or 'Ursprache' is quite obvious from the genetic research, because this points to some sort of rather recent bottleneck (less than 0.1 m.y.) with a population less than 10000 people as 'founders' for all living people. So whatever sort of languages there may have existed at that time, only that from one little group survived and evolved untill today. The mother tounge is something many linguists look upon with much scepticism, and some believe that the different recent language groups are so different that they must have originated separately. This is pure nonsense, even if languages way back originated several times independently, then there still is a common mother tounge for all the living languages – and it must be possible to reconstruct some aspects of it by cladistic methods.

One should remember that historical linguistics has actually employed cladistic methods or something very close to that for longer than the biologists (Platnick & Cameron, 1977), and that this has been based on a tradition from 'textual critics', the reconstruction of the history of old manuscripts and books, which have been copied from each other – a method several hundred years old. So surely the linguists have encountered all the same difficulties and controversies over 'ancestors', primitive features, classifications and the like. And there have been many attempts to reconstruct language evolution, not the least for Indo-European languages. Stopa's is a brave attempt to establish an overview of all languages and their common origin. Cavalli-Sforza et al. (1989) compared their tree based on genetic data for the world population with an evolutionary tree of all languages. They were satisfied with the match, but in reality the language tree was not sufficiently resolved to be very informative, as half of its groups were single branches of uncertain relationship, and therefore these groups could fit almost any other tree, so I believe the test failed, or was not very convincing. Stopa's language relations and their congruence with the MtDNA tree is much more significant, but should be specified in more detail.

How far back one can trace other features connected with language is much more uncertain. People have looked at brain endocasts for traces of symmetry and Broca's area and believe they can observe the latter 'bump' in *Homo rudolfensis*, ca. 2 m.y. ago, but whether this is really significant for location of a center for speach and has anything to do with an association center is very doubtful. Other 'soft' features which have no way of connecting directly or indirectly to the skeleton are clearly even worse to reconstruct from the fossils, and one will have to go to another abstraction level like supposed social relations or tool making to get a faint and very unsecure idea of language, conciousness and the like. But here is a 'free playground' for adaptationists and their 'just-so stories'.

9.18 Discussion

The proper phylogenetic framework establishing the 'simplest' relationship between the relevant groups based on analyses of their features (what else?) is essential for understanding the evolutionary history of man and his relatives inclusive the many fossils, be they 'complete', fragmentary or just scraps. Here the relationships, the phylogenetic tree, is based on cladistic methods, meaning that only shared, derived traits (synapomorphies; concerning morphology, molecules, behaviour etc.) count as evidence for close phylogenetic relationship. Another demand or ideal convention of this evolutionary philosophy here called 'phylocladistics' is that in phylogenetic systematics the 'natural' classification should portray these relationships (the 'tree of life') as precisely as possible and comprise taxa (named groups), be they mono-(holo-)phyletic, paraphyletic, metaphyla, ancestral, hybrids or of another biologically relevant sort, which may be expressed in a hierarchy.

Such classification and the corresponding evolutionary tree have to be the basis for speculations about the evolution of both the features on which the tree is based (most often morphology and molecules) and other features, e.g. 'soft' parts, development, genetics, 'adaptations', functional anatomy (e.g. upright stance and walking), behaviour, ecology, 'change' (evolutionary rates and 'kinds'), abilities for tool making, types of communication (language etc.) and conciousness, rituals, religions etc. That is, all these 'exotic' phenomena which most anthropologists seem to be mainly interested in, the origin of 'key adaptations', the 'real biological evolution', not just the dry bones or dull molecules (an exception to the 'dullness' seeming to be such genes from ancient DNA by which we can reconstruct features such as colours, earwax etc. of a 'fossil' organism, in case a 4000 years old inuit – see Nielsen et al., 2010).

So what are the relations between all these interesting features and phenomena and the phylogenetic tree and ditto hypotheses – and the 'data' upon which the latter is based? Here, it appears that closest to the 'data', the traits, features or characters is the simplest cladogram (or cladograms) as a symbol or a mapping of the distribution of features which employs as many of these as possible as synapomorphies (characterization of groups, clades, taxa), and therefore makes the groups maximally characterized (meaning that as many as possible generalisations can be made about these taxa; this is actually the reason why we use classifications at all – see e.g. Patterson, 1980; Nelson & Platnick, 1981; Bonde, 1984a). The 'data' here are put in quotes because they depend on hypotheses of homology (some sort of 'evolutionary sameness'), and are not just raw and 'objective' facts.

When a certain cladogram among several, perhaps many, possible ones is preferred, then that one can be transformed into a phylogenetic tree, implying the minimum number of changes of features on the tree, by adding some assumptions like time, and that the groups are 'blood related' in terms of 'nearness of common ancestry' that is one aspect, the cladistic one, of evolution (which some of us would consider a 'discovered fact' about nature), that features change along the branches of the tree, which is another aspect called 'phyletic evolution', and perhaps also that some taxa may be 'ancestral' to others (Bonde, 1977, 2001).

In general, many trees will be possible as corresponding to a certain cladogram, e.g. by just changing the timing of the splits a little (most of them are fossil based minimum ages anyway), or by changing position of sister groups (the sister clades of the tree may all be turned 180 degrees or interchanged), but these are insignificant banalities.

There are more principal differences between the possible trees when there are polytomies in the cladogram (expressing uncertainties) because for any trichotomy four different trees are 'equally good' (or possible), and this number increases alarmingly at multi-splits in the cladogram (Platnick, 1977; Patterson, 1980). From the tree (or rather from a characterstate tree) one or several precise classifications may be made which are 'isomorphic' with the tree, if certain conventions are obeyed (and disagreement can obviously exist concerning such conventions).

Should a certain phylogenetic tree ('stem tree'; or just a few of them) be preferable, then this may be seen as a model of the evolutionary history, and then further assumptions can be added about, say, economizing principles and functional anatomy implying certain behaviour and perhaps ecology. These phenomena can be 'mapped' onto the tree with implications for certain changes along the branches (or rather between the splits assumed to be 'speciations'). And even further assumptions can be added, if necessary, such as believing that these changes were 'adaptive', and stories about why some branches were more successful than others which died out, may be reconstructed. By asking and 'answering' these many 'why' questions (and not just 'how') by selecting preferred models among several possible, one presumably approaches the ideal of a complete 'evolutionary scenario'.

Now a characteristic feature of this 'layered' scientific process adding more and more assumptions about the world ('Nature') seem to be that, if controversies should rise about different models, the many extra assumptions make it very difficult to perform a precise test. Actually, it is only at the level of the cladogram that simple and precise tests based on 'data' can be carried out (more and/or 'better' characters by outgroup method = congruence among characters, which is really not a test against 'Nature'; or even closer to observable data by comparing with ontogeny [Patterson, 1983; Bonde, 1984b]).

Even at the next level of abstraction, that of the 'stem tree', so many extra assumptions have been added to make it less transparent what should be tested in case of evaluating two different models (trees), because a number (or all) of the assumptions might be wrong (presuming that the cladistic analyses at the lower level were all right). It is easy to see that the more assumptions, the further away from the 'data' and the more complicated the tests – if they are possible at all. Increasing the levels of abstraction evidently decreases the possibilities for precise tests. What may seem, instead, to increase when raising the level of abstraction is the tendency to use teleological explanations, as shown in my schematic diagram from 1984a. The more 'why' questions, the more 'purposefulness' – and probably more 'anthropocentrism'. This is the reason why S. J. Gould called the adaptive scenarios 'just-so

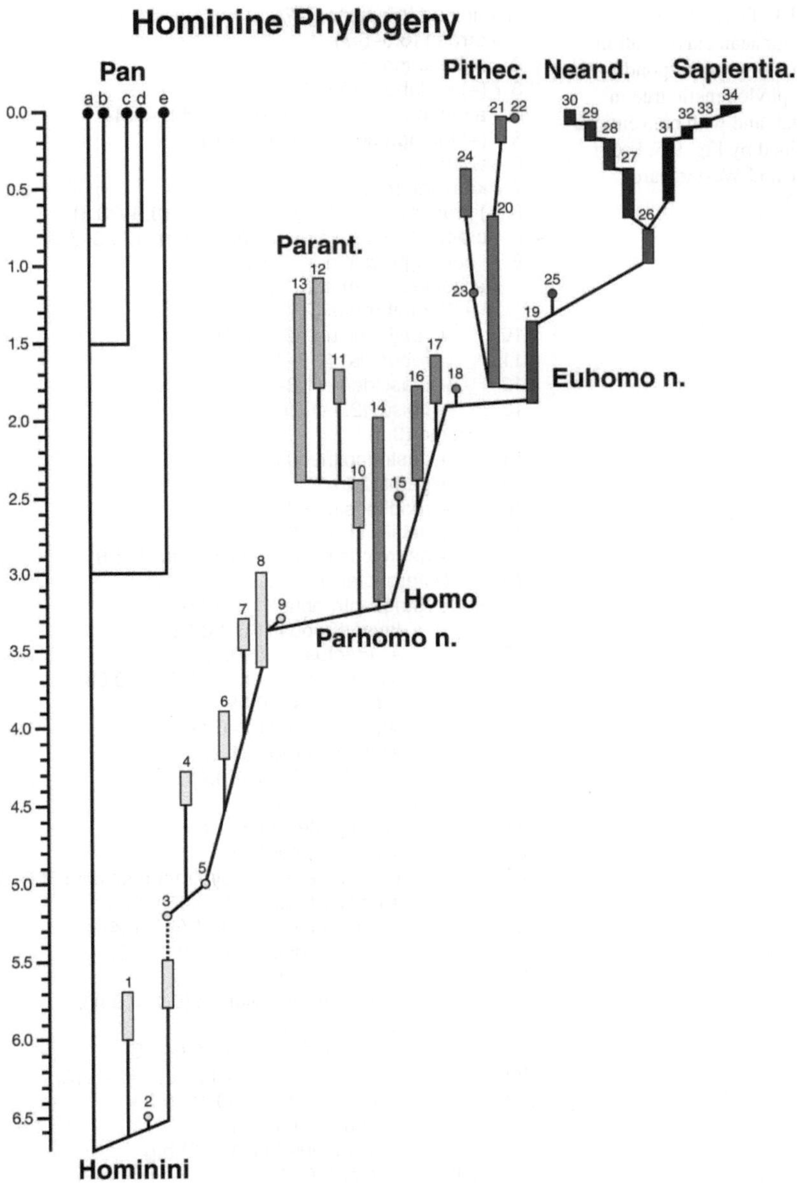

Fig. 9.3 Phylogenetic tree of hominins and Pan from Bonde and Westergaard (2004), but shaded version, based on characters and arguments in that paper, which is linked as pdf with the many references to the publisher's and N. Bonde's homepages – see p. 189. This tree is isomophic with the 'phylocladistic', non-Linnaean classification in Fig. 9.4

Fig. 9.4 Phylocladistic, non-Linnaean classification of Hominina corresponding to the phylogenetic tree in Fig. 9.3, and with conventions explained by Fig. 9.5. From Bonde and Westergaard (2004)

Hominina~~a~~ [/Sapiens~~a~~] (?6.5)
1 + orrorin (6.0–5.7)
2 + sahelanthropus (?6.5)
3 ? (+) kadabba* (5.8–?5.2)
4 + ardipithecus (?syn. praegens) (4.5–4.3)
5 ? (+) lothagamensis* nov. (?5.0)
6 + anamensis (4.2–3.9)
7 + kenyanthropus (3.5–3.3)
8 (+) praeanthropus* (syn. afarensis) (3.6–?3.3)
8 + praeanthropus* (syn. antiquua) mut. (?3.3–3.0)
9 ?+ bahrelghazali mut. (? 3.3)
Parhomo nov. mut. (3.2)
+ Paranthropus (2.7–1.1)
10 +) aethiopicus* (2.7–2.4)
11 + robustus* (1.9–1.7)
12 + crassidens (1.8–1.1)
13 + boisei (2.4–1.2)
Homo (3.2)
14 + australopithecus (3.2–2.0)
15 + garhi (2.5)
16 + rudolfensis (2.4–1.8)
17 + habilis (1.9–1.6)
18 + georgicus (?syn. wushanensis) (1.8)
19 (+) ergaster* (1.9–1.8)
+ Pithecanthropus (1.8–?0.08)
+ Javanthropus (1.8–?0.08)
20 +) erectus* (1.8–0.7)
21 +) erectur-soloensis* (?0.2–?0.08)
22 + soloensis (?0.08)
+ Sinanthropus (1.2–0.4)
23 +) lantianensis* (1.2)
24 + pekinensis (0.7–0.4)
Euhomo nov. (1.8)
19 (+) ergaster* (1.8–1.4)
25 + louisleakeyi (1.2)
26 (+) antecessor* (?syn.mauritanicus) (1.0–0.8)
+ Neandertalia (0.7–0.03;m0.6)
27 +) heidelbergensis* (0.7–0.4)
28 +) steinheimensis* (0.4–0.2)
29 +) aniensis* (0.2–0.1)
30 + neanderthalensis (0.1–0.03) b
Sapientia (0.6)
31 (+) rhodesiensis* (0.6–0.2)
32 (+) helmei* (syn. idaltu) (0.2–0.13)
33 (+) palestinus* (0.13–0.09)
34 sapiens (?0.07) b,g
[/Sapiens (m0.1–0.2) b,g
34.1 afer (?0.07)
34.2 australis (0.04)
34.3 europaeus (0.04)
34.4 asiaticus (?0.03)
34.5 americanus (0.01)]

LITU's from traditional monotypic genera are given the genus name to save the species name for any later subdivisions.

The classification of recent crowngroups and LITUs (Least Inclusive Taxonomic Units) is shown by subordination of their names in boldface, while totalgroups (marked: Nameø) with names of fossils (marked: +) mostly sequenced can easily be included or removed.

Paraphyletic groups are problematic and should be avoided, but can temporarily be classified by marking them with quotation marks. Significantly, a paraphyletic fossil group ("+ Name") may as a stemgroup not be extinct.

Informal ranks of recent crowngroups can be provided by their absolute age in million years in brackets: (m-number), as suggested by Westergaard (1989), and calculated by the molecular clock by Goodman *et alii* (1998, 1999).

Ages of fossils (here taken from Hertwig, 2002), giving minimum ages of taxa, are also put in brackets: (number), and informal ranks of extinct groups may be defined as their (minimum) time span (Farris' suggestion from 1976).

Further conventions are as follows; (?number): uncertain age; crowngroup (**Name**): last common ancestor and all its known descendants for a monophyletic group or clade of recent taxa; totalgroup (Nameø): crowngroup with addition of all known fossils from its paraphyletic stemgroup (usually not named), demarcated by recent diversification point and branch to recent sistergroup; ?+: fragmentary fossil of uncertain status; (+): possibly ancestral LITU (or part of LITU), where fossils are older than fossils from its recent descendant groups; +): same for extinct descendant groups; *; metaphyletic (plesiomorphic) LITU; syn.: synonymous name; inc.: *incerta sedis.* taxon with uncertain relationships; mut.: *sedis mutandis.* taxon being included in an unresolved polytomy; nov.: *nomen novum,* new name; b: taxon being a biological species ('biospecies'), isolated by internal reproductive isolation mechanisms; g: taxon being a gamogenetic species ('gamospecies'), integrated by gene flow (see Westergaard, 1989); a LITU usually corresponds to the smallest recognisable so-called phylogenetic species ('phylospecies'). With adequate conventions, it would also be possible to classify these different 'pluralistic' levels of species and hierarchies (Westergaard, 1989). Sistergroups are indented the same distance.

Fig. 9.5 Explanation of the conventions used for the classification Fig. 9.4. From Bonde and Westergaard (2004, pp. 46–47). Further in the pdf linked to the homepage, see p. 189

stories' (Gould & Lewontin, 1978). And philosopher I. Kant would presumably say (1795), that this is because our minds prefer such explanations, so we never see the world 'as it is'.

9.19 What Can Be Known About Hominid Evolution?

Clearly the evolution of the characters which were used in reconstructing the tree (Fig. 9.3) can be followed in detail: There are from 1 to 6 characters mentioned as synapomorphies for each group among the nearly 40 taxa, say three as a mean, giving about 120 features. And to these should be added a few autapomorphies for each terminal taxon, about two for each of these 35, giving another 70, that is in toto nearly 200 changes of characters. Of these, between 40 and 50 are on the line from the split from chimps to the last common ancestor of modern humans. So the change of these skeletal features can obviously be followed from split to split, but to discover the sequence of new traits between the splits among these up to 6 changes demands discovery of more 'intermediate' fossils with their specific combinations of features.

Concerning most of the 'soft' and other interesting features mentioned above in which we believe ourselves to be different from chimps, and which cannot directly be seen in fossils or be firmly correlated to skeletal traits, when did they occur? This is the issue of when did language originate, or 'hairlessness', or tool making or self-consciousness (Linnaei most important character for *H. sapiens*) or big female breasts or a big penis or association of certain centres in the brain.

The most precise that can be said is that it happened somewhere between the two splits just mentioned above, the origin of our lineage and the diversification of modern humans – that is, between about 7 and 0.1 m.y. ago, not a very precise result. And there is even a tendency to discover more and more of these 'human characteristics' in some chimps, meaning that their origin was most likely before our split from the chimps (e.g. tool making, self-conciousness – and may be even 'language', although not a 'spoken' one, but see Stopa, 1972, 1979, and above). Sophistication of tool making can only be followed in detail if we can make sure which taxon most likely made the tools from a certain deposit – and then, as always with fossil finds, they only give a minimum age of features and thereby have no direct association with tools. At the 'low level' within modern humans, the evolution of languages can be seen in a similar light and symbolized by evolutionary trees e.g. by the historical linguist Schleicher (1861, 1863, directly inspired by Darwin's 'tree of life', 1859) – and further inspiring his friend Haeckel (from 1866 the 'world champ' in phylogenetic trees). That the methods behind such trees in phylogenetic systematics and historical linguistics are almost identical was shown by Platnick and Cameron (1977).

9.20 Conclusion

The diversity of fossil hominid taxa is a lot larger than usually appreciated. Here is demonstrated a phylogenetic tree with 35–40 units/LITUs of fossil hominids and a corresponding 'phylocladistic' and non-Linnaean classification. The data and arguments for this recontruction is from Bonde and Westergaard (2004 – and can be seen as link from the homepage, also presenting the many references to the data background, which are not repeated here). Some few taxa are accepted as possible 'ancestors' and are placed on the branches not as terminal taxa (this seems to be in conflict with most other cladists). Therefore, it is a phylogenetic tree, elevated at least one abstraction level above a cladogram. Many other myths about 'direct ancestors' in the traditional phylogeny of hominids, such as Lucy, *Australopithecus afarensis, Ardipithecus ramidus/praegens, Australopithecus africanus, 'Au.' garhi, 'Au.' sediba, Australopithecus* in general, *Homo habilis,* and *H. erectus* are all shown not to be such 'ancestors' as claimed by their authors and many others.

Trying to trace the origin of the Indo-European languages, the Polish linguist R. Stopa has done a formidable study of the African languages, especially the Sub-Saharan ones and those of the Bushman tribes. In several books and large papers (1972, 1973, 1979) he has established that the Bushman languages are the most 'primitive' in the world characteristic with a large number of click and lip sounds, gesticulations and a very simple structure of the language, sentenses and words, lack of abstractions and no grammar. By comparing with other African languages, he has established a grade system of increasing complexity from Bushman languages over Bantu (which still has some clicks) to other African languages, Hamitic, and Indo-European languages. And interestingly, he claims he can directly translate from utterances by chimps to the equivalent expressions in the Bushman languages. The pattern of his hierarchical evolutionary model, chimp – Bushman – Bantu – other African languages plus the rest of the world, if interpreted in cladistic terms, can directly be compared with the model of relationships between the world populations based on MtDNA. Thus it becomes evident that the branching structure of the two models is exactly the same, which can hardly be a coincidence, so the two models confirm each other, and human language originated from ape utterances. This proves the utility of precise cladistic models of interrelationships, and the very complicated human family tree should be a warning against making very rash decisions about adaptationistic scenarios. The diversity of early humans in Africa was very high, especially in the interval between 1½ and 2 m.y. ago with about a dozen forms living at the same time worldwide, 10 of these in East and South Africa, 7 of them in the East African Rift valley at the same time.

Acknowledgments Thanks to the organisers for two very pleasant and in formative meetings in Copenhagen, and to the editors and publisher for their patience. Thanks also to the late Bjarne Westergaard, with whom I used to discuss lots of these problems with primate and hominid evolution and classifications, and cladistic methods and philosophy in general, he was a rich source of information – and thanks to his family who after his incomprehensible and unreasonable death in 2008 transferred his huge primate library to me and to the Natural History Museum (SNM). Further thanks to drs. Chris Stringer and Peter Andrews, Museum of Natural History, London, for discussions on hominids and phylogenetic systematics, and to Drs Ian Tattersal and Eric Delson with whom I enjoyed studying the many original fossil hominids from all over the world at the famous 'Ancestors' meeting in American Museum of Natural History in New York almost 30 year ago. I am grateful to my institute for workspace and fascilities as well as for support some years ago to journeys to symposia, where some of these ideas on hominids and cladistic classifications were presented e.g. in Oxford 2003.

References

A large number of references especially concerning the reconstruction of the phylogenetic tree from Bonde and Westergaard (2004) are not repeated here, but are available on the homepage of the publisher (http://extras.springer.com) as well as on NB's homepage (www.Institut.Geografi.Geologi.ku.dk/emeritus/Niels.Bonde) as link to a pdf. A long list with the remaining literature is also placed with a link on the homepage.

Anderson, I. (1983). Who made the Laetoli footprints? *New Scientist*, May 12, 1983, 373.

Berger, L. R., de Ruiter, D. J., et al. (2010). Australopithecus sediba: A new species of Homo-like australopith from South Africa. *Science, 328*, 195–204.

Bonde, N. (1977). Cladistic classification as applied to vertebrates. In M. K. Hecht, et al. (Eds.), *Major patterns invertebrate evolution* (pp. 741–804). New York: Plenum.

Bonde, N. (1984a). Functional anatomy and reconstruction of phylogeny. In E. Buffetaut, et al. (Eds.), *Actes du Symposium Paléontologique Georges Cuvier, Montbeliard, France 1982* (pp. 11–26). Imprim. Commune Montbéliard.

Bonde, N. (1989). Erectus and neanderthalensis as species or subspecies of Homo, with a model of speciations in hominiods. In G. Giacobini (Ed.), *Hominidae. Proceedings of the 2nd international congress of human paleontology* (pp. 205–208). Milan: Jaca Book.
Bonde, N. (2001). L'éspéce et la dimension du temps. *Biosystema*, vol. 19. *Systématique et Paléontologie –2001*, 29–62.
Bonde, N., & Hoeg, J. T. (2008). Bjarne Westergaard, 25.05.1948–29.01.2008. Da.Naturhist. Foren., Yearbook 2007/2008, pp. 83–88.
Bonde, N., & Westergaard, B. (2004). *Progress in hominid classification: Cladistic approaches.* Micselanea a E. Aguirre, lll, Paleoantropologia, pp. 36–57. Zona Arquelogica, Num. 4, (Museo Arquelogico Regional).
Brown, P., Sutikna, T., et al. (2004). A new small bodied hominin from the Late Pleistocene of Flores, Indonesia. *Nature, 431*, 1055–1061.
Carroll, R. (1987). *Vertebrate paleontology and evolution* (pp. 277–285). New York: W.H. Freeman and Co.
Cavalli-Sforza, L. L., et al. (1989). Reconstruction of human evolution: Bringing together genetic, archaeological and linguistic data. *PNAS USA, 85*, 602–606.
Cherfas, J. (1983). Trees made man upright. *New Scientist, 97*(1341), 172–178.
Coon, C. (1962). *The origin of races*. London: J. Cape.
Day, M. (1965). *Guide to fossil man* (4th ed., 1986). New York: Cassell.
Feyerabend, P. (1975). *Against methods*. London: Methuens.
Foley, R. (1987). *Another unique species*. New York: Wiley.
Gibbons, A. (2009). Ardipithecus ramidus: The view from Afar. *Science, 326*, 41–43.
Groves, C. P. (1991). *A theory of human and primate evolution*. New York: Oxford University Press.
Haile-Selassie, Y. (2001). Late Miocene hominids from the Middle Awash, Etiopia. *Nature, 412*, 178–181.
Johanson, D., & Edey, M. (1990). *Lucy, the beginnings of humankind*. London: Granada.
Johanson, D., & Edgar, B. (1996). *From Lucy to language*. London: Weidenfeld & Nicolson.
Jungers, W., & Baab, K. (2009). The geometri of hobbits: *Homo floresiensis* and human evolution. *Significance*. DOI: 10.1111/j.1740-9713.2009.00389.
Krause, J., Qiaomei, F., et al. (2010). The complete mitochondrial DNA genome of an unknown hominin from southern Sibiria. *Nature*. DOI: 10.1038/nature08976.
Leakey, L. B. S., Tobias, P., & Napier, J. (1964). A new species of the genus Homo from Olduvai Gorge. *Nature, 202*, 5–9.
Leakey, R., & Walker, A. (1976). *Australopithecus, Homo erectus* and the single species hypothesis. *Nature, 261*, 572–574.
Pettitt, P., & Niskanen, M. (2005). Neanderetals in Susiluola Cave, Finland, during the last interglacial period. *Fennoscandia Archaeologica, XXII*, 79–87.
Romer, A. S. (1966). *Vertebrate paleontology*. Chicago: University of Chicago Press.
Schmid, P. (1983). A reconstruction of the skeleton of A.L. 288-1 (Hadar) and its consequences. *Folia Primatol, 40*, 283–306
Senut, B., & Tardieu, C. (1985). Functional aspects of Plio-Pleistocene hominid limb bones: Implications for taxonomy and phylogeny. In E. Delson (Ed.), *Ancestors: The hard evidence* (pp. 193–201). New York: A.R. Liss.
Stern, J. T., & Susman, R. L. (1983). The locomotor anatomy of Australopithecus afarensis, *American Journal of Physical Anthropology, 60*, 279–317.
Stopa, R. (1973). Hominization. *Journal of Human Evolution, 2*, 371–378.
Stopa, R. (1979). *Clicks, their form, function and their transformtion or how our ancestors were gesticulating, clicking and crying*. Acta Scient. & Litterar. 561, Universitas Iagellonica., Krakow.
Stringer, C., & Gamble, C. (1993). *In search of the Neanderthals*. London: Thames & Hudson.
Stringer, C., & McKie, R. (1996). *African exodus. The origin of modern humanity*. London: Cape.
Tattersal, I. (1995). *The fossil trail*. New York: Oxford University Press.

Tobias, P. V. (1967). *The cranium and maxillary dentition of Australopithecus (Zinjanthropus) boisei. Olduvai Gorge 2*. Cambridge: Cambridge University Press.
Vekua, A., Lordkipanidze, D., et al. (2002). A new skull of early Homo from Dmanisi, Georgia. *Science, 297*, 85–89.
Vigilant, L., Stoneking, M., et al. (1991). African populations and the evolution of human mitochondrial DNA. *Science, 253*, 1503–1507.
Wells, S. (2002). *The journey of a man. A genetic odyssey*. London: Penguin.
White, T., Asfaw, B., et al. (2009). Ardipithecus ramidus and paleobiology of early hominids. *Science, 326*, 75–86.
White, T. D., Suwa, G., & Asfaw, B. (1994). Australopithecus ramidus, a new species of early hominid from Aramia, Ethiopia. *Nature, 371*, 306–312.
White, T. D., Suwa, G., & Asfaw, B. (1995). Australopithecus ramidus, a new species of early hominid from Aramia, Ethiopia. *Nature, 375*, 88.
Wolpoff, M. H. (1980). *Paleoanthropology*. New York: Knopf.

Chapter 10
The Tripod Effect: Co-evolution of Cooperation, Cognition and Communication

Peter Gärdenfors, Ingar Brinck, and Mathias Osvath

Abstract This article concerns the co-evolution of hominin cooperation, communication and cognition. Certain hominin ecologies seem to have relied on cognitive foresight. The capacity of planning for future needs, combined with more developed cooperative skills, opened up the cognitive niche of cooperation towards future goals. Such cooperation requires complex intersubjectivity (theory of mind). We analyze five domains of intersubjectivity: emotion, desire, attention, intention, and belief; and argue that cooperation towards future goals requires, among other things, joint intentions (we-intentions). We scrutinize the cognitive and communicative conditions for reciprocal altruism, found in some species; and indirect reciprocity, a form of cooperation typical in the hominin line.

Sharing intentions and beliefs about the future requires communication about what is not present in the current environment. Symbols are efficient tools for this kind of communication, and we argue that the benefits of cooperation for the future selected for the evolution of symbolic communication. In line with recent models describing how indirect reciprocity might develop into an evolutionarily stable strategy, we emphasize the need for yet more complex intersubjectivity and symbolic communication, including a minimal syntax.

Our argumentation triangulates hominin cognition, cooperation, and communication, showing how these interdependent factors mutually reinforce each other over the course of evolution. The new take in this article is the combined analyses of cooperation and cognitive mechanisms. Finally, our theses are linked to archaeological evidence.

10.1 Program

The focus of this article is the co-evolution with communication and cognition of typically human forms of cooperation. The basic assumption is that certain ecological settings, present in hominin evolution, harbored selective pressures that fostered

P. Gärdenfors (✉)
Department of Philosophy, Lund University, Lund, Sweden
e-mail: Peter.Gardenfors@lucs.lu.se

T. Schilhab et al. (eds.), *The Symbolic Species Evolved*, Biosemiotics 6,
DOI 10.1007/978-94-007-2336-8_10, © Springer Science+Business Media B.V. 2012

cooperation. The new forms of cooperation presume cognitive foresight and complex forms of intersubjectivity that probably were less developed in the hominin predecessors. Our central thesis is that the presence of these cognitive mechanisms enabled the cooperation towards future goals to co-evolve with symbolic communication.[1]

Our argument triangulates cognition, cooperation, and communication. We submit that each of these domains contains several interdependent sub-components, and that these sub-components appeared gradually, influencing the development of other cognitive domains. The new take in this article is the emphasis on the evolution of hominin cooperation coupled with analyses of the relevant cognitive and communicative mechanisms.

Cooperation, symbolic communication, and cognition are intertwined in convoluted co-evolutionary processes. Our triangulation begins with two mechanisms typical of human cognition: foresight and intersubjectivity. The second part of the triangulation concerns the role of cooperation in hominin societies as a selective pressure in the evolution of human cognition and communication. We examine those aspects of the hominin ecology that promoted advanced forms of collaboration and an improved understanding of other minds. We discuss the role of two kinds of cooperation typical among humans: cooperation towards future goals and indirect reciprocity. Reciprocal altruism ("you scratch my back; I'll scratch yours") is found in several animal species. Indirect reciprocity is a more extreme form of altruism: "I help you and somebody else will help me." The conditions for this to evolve as an evolutionarily stable strategy have recently been modeled (e.g., by Nowak & Sigmund, 2005). The crucial aspect for our argument is that these conditions presume complex forms of intersubjectivity and communication.

As a third part of the triangulation, we take a further step towards an ecologically grounded explanation of the function of human language (see also Bickerton, 2002; Gärdenfors & Osvath, 2010). To explain uniquely human adaptations such as language, it is not sufficient to present an evolutionary scenario that shows how useful linguistic skills might have been for human ancestors. What needs explaining is not just why those ancestors evolved language, but also why chimpanzees and other apes *did not*, even though they share a fairly recent common ancestor. It must be shown that language was useful and adaptive specifically for the human line and not for the chimpanzee line (Brinck, 2001; Bickerton, 2002; Gärdenfors, 2004; Johansson, Zlatev, & Gärdenfors, 2006). Many scenarios concerning the origin of language fail this test. The answer to why linguistic capacities started to evolve in human ancestors lies in the ecological conditions under which they lived.

Our analysis starts from a fairly traditional distinction between signaling systems and symbolic communication. As regards symbols, we focus on their referential, functional, and combinatorial properties. Our main conclusions are that cooperation towards future goals requires symbolic communication; and that indirect reciprocity

[1] Our strategy is thus similar to that of Whiten (1999). However, we put more emphasis on analyzing the sub-components in the three domains.

requires more complex symbolic communication, including names and markers for roles (and thus a minimal syntax). At the end of the article, these claims will be linked to potential supporting archaeological evidence.

Connecting the three parts, we argue that the cooperative behavior of agents must be judged in relation to their cognitive and communicative adaptations. These factors do not seem to have been sufficiently considered in game-theoretical analyses. For example, in Lehmann and Keller's (2006) classification of models of the evolution of cooperation and altruism, only two parameters related to cognition and communication are included: a one-period "memory" parameter, defined as the probability that "an individual knows the investment into helping of its partner at the previous round"; and a "reputation" parameter, defined as the probability that "an individual knows the image score of its partner" (Lehmann & Keller, 2006, p. 1367). We shall argue that these parameters are insufficient.

To summarize, our analysis of cooperation combines a game-theoretical account with an ecological/cognitive analysis. Our main purpose is to show how communication, based in intersubjectivity and cognitive foresight, opens up new possibilities for cooperation in hominin societies. Tomasello (1999) writes about the "ratchet effect" of human culture. We would rather propose the *tripod effect* of cognition, cooperation, and communication, lifting human culture by successively lengthening of each of the legs. This process has allowed hominins to raise their culture in a way that resulted in a qualitatively different form from that of other species.

10.2 Characteristics of Human Cognition

There have been many proposals as to which cognitive features are uniquely human: tool use, symbol use, mental time travel, etc. One by one, it has been shown that these features can be found, at least to some extent, in other animal species. The upshot may be that human cognition does not have any truly unique elements (Brinck, 2008b). Rather, there might be unique combinations of elements or more developed mechanisms in some domains (or both). Here we focus on three mechanisms that appear to be particularly advanced in humans: detached representations, cognitive foresight, and intersubjectivity. In Section 10.3, we will argue that they are necessary for human forms of cooperation.

10.2.1 Detached Representations and Inner Worlds

To give intelligible descriptions of many phenomena in animal and human cognition, it is useful to distinguish between cued and detached mental representations (Gärdenfors, 1996, 2003). A *cued* representation refers to something in the current external situation of the subject or something that triggered it in a recent situation. Delayed responses, in the behaviorist's sense, are cued representations resulting from perceptual categorization. When, for example, a particular object is categorized as food, the subject will act differently than if it had been categorized as a potential mate.

By contrast, *detached* representations stand for objects or events that are not present in the subject's current or recent external context and so could not directly trigger the representation.[2] A memory that can be evoked independently of the context in which it was created is an example of a detached representation. The behavior of a chimpanzee who walks away from a termite mound, peels the leaves off a twig to make a stick, returns to the termite mound, and uses the stick to fish for termites is difficult to explain without ascribing to the chimp a detached representation of a stick and its function. Detached representations make it possible to refer to entities that occur only in the imagination. For example, unicorns, elves, and trolls inhabit our imaginative worlds and invade our language, but do not exist in the external world.

Having a detached representation requires that present sensations are suppressed; otherwise they will come into conflict with the representation (Glenberg, 1997). This places new demands on mental capacities. The suppression of information appears to be managed by the frontal lobes of the brain, which have expanded most rapidly during hominin evolution. The frontal lobes are crucial for skills such as planning, fantasizing, and self-control.

The "inner world" of an individual is a collection of detached representations and their interrelations. Donald (1999, p. 143) argues that something like an inner world is a precondition for rehearsing skills: "hominins had to *gain access to the contents of their own memories*. You cannot rehearse what you cannot recall." There are strong indications that humans have more complex inner worlds than other animals, in particular, due to human capacities for creating cultural artifacts and detachment of time, and the conscious reflection (mentioned by Donald) made possible by these capacities (Gärdenfors, 2003; Sterelny, 2003; Arp 2006; Brinck, 2008b). Gomez (2004, p. 20) argues that the prolonged immaturity in the children of apes and, especially, humans results in a greater capacity to form representations, which in turn leads to greater cognitive and behavioral flexibility.

10.2.2 Cognitive Foresight

Many animal species rely on forecasts of the future. Those which mainly depend on predicting stable and recurring scenarios solve the forecasting problem with comparably rigid mechanisms. Innate releasing mechanisms are one way of solving problems that relate to future scenarios. The urge for food hoarding in the autumn is a prime example of such mechanisms. Nest building, migration, and hibernation can also be governed by innate mechanisms. Ecological niches with a higher variability of possible futures require a corresponding flexibility in the organisms' systems of anticipation. Associative learning, coupling current stimuli with future outcomes, constitutes a system that is advantageous in a changing environment. This kind of

---

[2] Detached representations are called secondary representations in Suddendorf and Whiten (2001) and decoupled representations in Sterelny (2003).

learning requires a sufficient stability in the environment to enable reinforcement learning of relationships during one individual's life.

Associative learning is too rigid for flexible foresight, which might involve novel elements or scenarios. Humans rely on their episodic system when making such foresights. Episodic foresight appears to make us pre-experience potential future events by sensorial simulations of such events (simply put: by using "the inner eye"). This ability is closely connected to the capacity for episodic memory, first described by Tulving (1972). The defining characteristic of an episodic system is the first-person experience of re- or pre-experienced events. Episodic memories enable recall – as opposed to mere factual knowledge, which is characteristic of semantic memory. Projection into possible futures is most likely based on the same cognitive system as episodic memory. Episodic foresight constructs events rather than reconstructing them, as is the case with episodic memories (see e.g. Suddendorf & Corballis, 1997, 2007; Atance and O'Neill, 2005; Suddendorf & Busby, 2005; Tulving, 2005). Accumulated neurological data confirm that episodic foresight and memory – and perhaps even intersubjectivity – rely on the same core brain network (for a review see Buckner & Carroll, 2007). Episodic retrospection and prospection are often referred to together as *mental time travel* (e.g., Suddendorf & Corballis, 2007).

The foresight system is used in deliberative planning for the future: a voluntary act aimed towards a forecasted future. Such planning is often divided into two broad categories: planning for an immediate need and planning for a future need. The categories pinpoint the mechanisms that instigate planning. In the former case, a current drive state, which *precedes* any foresight, is present in the individual. In the latter case, the drive state that triggers the planning *follows from* foresight. This distinction is used as a behavioural criterion in animal studies on episodic-like planning. An inhibitory ability is also needed for planning for a future drive state, in order that the current state can be disregarded in favor of a possible future one. Exactly how foresight itself is triggered is still subject to research.[3]

To explain self-projection into the future, we focus on two kinds of self-awareness (Brinck, 1997; Brinck & Gärdenfors, 1999). *Indexical self-awareness* occurs when the agent develops a dynamic representation of egocentric space, in which items in the surroundings are located relative to the agent's vantage point. There is no explicit self-representation, but the self is implicitly situated in egocentric space. The agent represents the surrounding objects as units that are causally connected over time, with which she can therefore interact causally. Indexical, self-relative representations are necessary to move the agent to action; they occur in at least all mammals and birds.

Detached self-awareness emerges when the agent has self-representations that concern her future or past conditions, and can imagine herself as having different

[3] Nevertheless, it is fairly safe to conclude that some sort of cuing takes place in the present situation: a cue in the physical or "mental" environment; if not, it would be the result of chance, just popping up in the brain.

properties in different domains. This paves the way for the self-ascription of properties that the agent does not in fact have, and thus for counterfactual reasoning about the self. Together, indexical and detached self-awareness allow an agent to think of herself now, relative to the objects at hand and the environment she is located in, while projecting herself into future situations where the things around her, her needs, and her possibilities for action have changed.

It has been argued that the ability to plan for future needs is exclusive to humans (e.g., Atance & O'Neill, 2005; Gilbert & Wilson, 2007; Gulz, 1991; Köhler, 1921; Noble & Davidson, 1996; Roberts, 2002, 2006; Suddendorf & Busby, 2005; Suddendorf & Corballis, 1997, 2007; Tulving, 2005). The Bischof-Köhler hypothesis asserts that non-humans are unable to use flexible cognitive prospection beyond their present needs (Suddendorf & Corballis, 1997). However, recent studies suggest episodic foresight in great apes, with a similar ability in corvids (Mulcahy & Call, 2006; Naqshbandi & Roberts, 2006; Correia, Dickinson, & Clayton, 2007; Raby et al., 2007; Osvath & Osvath, 2008; Osvath, 2009).

The earliest clear indications in hominins of planning with foresight are found among the remains of the Oldowan stone tool culture, which was founded at least 2.6 Mya (Toth, 1985; Osvath & Gärdenfors, 2005).[4] Hominin life on the savannah resulted in selective pressures to change diet from a predominantly vegetarian one to a more protein- and fat-based one. As a consequence, the day ranges of the early hominins must have been longer than those of extant apes. There is clear evidence that, already more than two million years ago, transport of artifacts (at least the stone tools) was an important trait of the Oldowan culture (Toth, 1985). Most probably, there were selective advantages to transporting tools and carcasses. Given the positive results of the great ape studies and the phylogenetic proximity of these hominins to *H. sapiens*, it is very likely that they used episodic foresight. As a consequence, hominin life on the savannah during the Oldowan era opened up many new forms of cooperation towards future goals (Sterelny, 2003; Plummer, 2004).

10.2.3 Intersubjectivity

Whereas cognitive foresight requires some form of self-awareness, many forms of cooperation additionally demands other-awareness and *intersubjectivity*. Indexical self-awareness becomes intersubjective when the agent's egocentric space contains other agents and not just objects, and when the agent can react to those agents' experiences. Detached self-awareness becomes intersubjective when, from an allocentric point of view, the agent perceives herself as one agent among others exchanging information about themselves.

In developmental psychology, intersubjectivity is frequently described as a deliberate sharing of experiences about objects and events (cf. Trevarthen & Hubley,

---

[4] Which species used these stone tools is not fully established. It might have been several hominin species: for example, A. gahri, H. habilis, and later perhaps H. ergaster.

1978; Stern, 1985). In philosophical and psychological debate, it is commonly known as having a "theory of mind", with an emphasis on belief. Premack and Woodruff (1978) originally distinguished between having a theory of others' intentions and motives, on the one hand, and having a theory of their knowledge and beliefs, on the other. They claimed that apes have a theory of mind in the first sense, and suggested testing them for the second.

We interpret intersubjectivity as the capacity to react, or respond, to others' manifest states of mind, including their emotions, attentions, intentions, desires, and beliefs. Updating the analyses in Gärdenfors (2003) and Brinck (2008a), we argue that intersubjectivity is multi-faceted and decomposable into at least five capacities[5]:

(1) Reacting to the *emotions* of others.
(2) Reacting to the *desires* of others.
(3) Reacting to the *attention* of others.
(4) Reacting to the *intentions* of others.
(5) Reacting to the *beliefs* and *knowledge* of others.

As we will argue in Section 10.2.4, this ordering of intersubjective capacities is supported by the phylogenetic and ontogenetic evidence from the hominin lineage.

10.2.3.1 Reacting to Emotions

The ability to share others' emotions is often called *empathy*. Bodily and vocal expressions of emotions communicate the agent's negative or positive experiences; they are most obvious among social animals. Preston and de Waal (2003, p. 4) define empathy as "any process where the attended perception of the object's state generates a state in the subject that is more applicable to the object's state or situation than to the subject's own prior state or situation." The phrase "more applicable" indicates that the subject, at least so some extent, mimics the emotional state of the object. Preston and de Waal argue that, as a mechanism linking perception and action, empathy in a basic form is available to most, if not all, mammal species. This suggests that empathy has comparatively early evolutionary roots. They see a clear

---

[5] We have not found any other researcher who grades what is involved in intersubjectivity in this way. Tomasello (1999, p. 179) distinguishes between three levels of a child's development: seeing others as agents, seeing others as intentional agents, and seeing others as mental agents. His second and third levels correspond roughly to our capacities 4 and 5. Baron-Cohen (1994) proposes that "mind reading" has four components: an intentionality detector (ID), an eye-direction detector (EDD), a shared-attention mechanism (SAM), and a theory-of-mind mechanism (ToMM). Wellman and Liu (2004) write about a "scaling of theory-of-mind tasks" when they address the "sequence of understandings" in children's developing intersubjectivity. They do a meta-analysis of studies of children's understandings of different forms of intentions, emotions, desires, and knowledge. There is a considerable overlap with the capacities presented here. Stern (1985) and Brinck (2008a) distinguish between interaffectivity, interattentionality, and interintentionality without emphasizing the distinct roles of desires and beliefs.

evolutionary motivation for its emergence in the ability to recognize and understand the behavior of conspecifics.[6]

10.2.3.2 Reacting to Desires

Whereas emotions express how an individual feels, desires express an individual's positive and negative *attitudes* towards objects, events, and other agents. Because desires (in this sense) are relational (or referential), reacting to desires is a more complex capacity than reacting to emotions.[7] Infants react to the desires of others, as expressed for example by pointing or facial expressions, by ten months of age. A slightly more advanced form of this capacity is understanding that others may not have similar desires as oneself. This capacity develops before reacting to beliefs (cf. Flavell, Flavell, Green, & Moses, 1990; Wellman & Liu, 2004) – as the following study shows. In a non-verbal test on 14- and 18-month-old infants (Repacholi & Gopnik, 1997), the infants were presented with a plate of broccoli and a plate of crackers. All of the infants preferred the crackers. The experimenter indicated her preference for one of the foods by producing emotional expressions of either disgust or pleasure. Then she reached out her hand and asked the infant to give her food. The older infants gave her the food for which she had indicated her preference, even if it was different from their own. The younger ones always gave the experimenter the crackers.

The capacity for reacting to desires has been little investigated in animals. One positive example involves the way that begging gestures in primates are sometimes rewarded. Experiments with chimpanzees about so-called other-regarding preferences indicate that they are not willing to deliver food (even at a very low cost to themselves) to unrelated group members (Silk et al. 2005); whereas Burkart, Fehr, Efferson, and van Schaik (2007) show that marmoset monkeys provide food to others altruistically. Burkart et al. speculate that the relevant difference between the species is that marmosets are cooperative breeders (like humans), while chimpanzees are not.

10.2.3.3 Reacting to Attention

Besides primates, the capacity to follow another's gaze towards a target is well attested in canines, goats, dolphins, and corvids (Povinelli & Eddy, 1996; Emery,

---

[6] It is characteristic of facial signals, such as the play-face expressions of chimpanzees and gorillas, that they carry emotional rather than referential meaning.

[7] Understanding the desires of others does not require representing their beliefs: thinking "A wants X" does not require thinking "A believes that X is most desirable.". Thus, desires need not be "value beliefs" as Flavell et al. (1990) maintain. Similarly, ascribing a desire to somebody does not entail an ascription of explicit intention: thinking "A wants X" does not require thinking "A intends to choose X." As long as the desired object is present or a physical representation of it exists in the shared context, the agents' behaviorally manifest attentional states reveal their intentions to act (Brinck, 2001, 2004).

2000; Bugnyar, Stöwe, & Heinrich, 2004; Kaminski et al., 2005). *Gaze following* can occur by a simple contagion mechanism. *Shared attention* results when two agents notice that they are simultaneously attending to the same target; it is well documented among great apes (Hare, Call, Agnetta, & Tomasello, 2000). It goes a long way towards the construction of a "consensual reality" that can be communicated about, but it does not quite reach it (Brinck, 2004). *Joint attention* results when the agents have attention (or eye) contact while sharing attention to a target; the prolonged eye contact signals mutual awareness and promotes communication about the target. The ability to engage in joint attention has not, so far, been established conclusively in non-human primates (though see Gomez, 2007; Leavens, Hopkins, & Bard, 2005, 2008 for a different view).

10.2.3.4 Reacting to Intentions

Because the human cognitive system takes self-motion as a cue for goal-directedness, goal-directedness and *intentions to act* can be grasped directly from observation of behavior. Gergely and Csibra (2003) argue that infants primarily interpret instrumental actions not by their causality but by their efficiency: a function of the visible goal of the action and the physical constraints of the agent's situation (visual conditions, the presence of obstacles, etc.). Csibra (2003) distinguishes this capacity for a *teleological understanding of action* from the intersubjective capacity to perceive *communicative contexts of action* from others' gaze and contingent reactivity. Understanding *referential intentions* requires paying attention to others' intentions to interact (Brinck, 2008a); it includes joint attention.

Sharing goals in terms of sharing intentions for instrumental action is sometimes mistaken for sharing intentions for interaction, which involves communication. For instance, Tomasello et al. (2005, p. 675) seem to neglect this distinction when they trace the human "ability to participate with others in collaborative activities with shared goals and intentions: shared intentionality" to the infant's capacity for understanding goal-directed physical action by twelve months of age. As we will argue below, shared intentionality involves having joint intentions and presupposes other cognitive and communicative capacities than does sharing goals in physical contexts of action.

The capacities for sharing referential intentions and intentions to interact are essential to flexible cooperation. We hold that *joint intention* requires that the agents (i) share an intention to interact, (ii) react to each other's individual intentions to act, and (iii) coordinate their respective future-directed intentions. For example, in building a tower of blocks, a child may understand that the adult holding the tower steady makes it possible for the child to place new blocks; the child may then be able to reverse roles with the adult. Complementary role taking develops around eighteen to twenty-four months of age (cf. Brownell & Carriger, 1990). Very likely, the capacity for so-called *we-intentions* (cf. Searle, 1995) arises at this stage; these intentions concern the agents as a unit and their shared relation to a global goal; they typically strengthen individual agents' motivation for working towards the common goal.

10.2.3.5 Reacting to Beliefs and Knowledge

The most advanced test of intersubjectivity is to investigate a subject's understanding of what others *believe* or *know*. Tomasello and Call (2006) review the experimental evidence whether chimpanzees know what others have seen. Interesting results come from a series of experiments by Hare et al. (2000; Hare, Call, & Tomasello, 2001), where a subordinate and a dominant ape were placed in two rooms on opposite sides of a third room. Both could see food placed in the open or behind barriers in the middle room. The problem for the subordinate ape was that the dominant ape would take all the food she could see. Sometimes the subordinate ape could see food on the side of the barrier that the dominant ape could not see. It was found that the subordinate ape went for the food that only it could see much more often than the food that both could see. Tomasello and Call (2006, p. 375) conclude that the chimpanzees "know not only what others can and cannot see at the moment, but also what others have seen in the immediate past."

It is easier to test human children, since one can communicate verbally with them from a fairly early age. The most common experimental method uses the so-called *false belief test* (see e.g. Perner et al., 1987; Gopnik & Astington, 1988). Nonverbal versions of the test show that infants have an implicit understanding of false beliefs by the age of 2;11 as revealed by instances of anticipatory looking (Clements & Perner, 1994; Garnham & Ruffman, 2001; Carpenter, Call, & Tomasello, 2002). In one nonverbal version of the test (Call & Tomasello, 1999), human infants performed as well as in the verbal version, while the apes that were tested failed.

Human adults can have higher-order knowledge and beliefs, such as "of course I care about how you imagined that I thought you perceived I wanted you to feel." This capacity forms the basis of *joint beliefs*, which are involved in so-called *common knowledge.* Joint beliefs open up typically human forms of collaboration – for example, conventions and contracts – as we shall argue in Section 10.3.5.

10.2.4 Phylogenesis and Ontogenesis of Intersubjectivity

From the analysis in the previous section of the five forms of intersubjectivity, it should be clear that humans exhibit forms of intersubjectivity that are not readily found in other animals. In particular, humans have a well-developed competence for reacting to beliefs, and also excel at forming joint intentions and joint beliefs.

The five forms of intersubjectivity are not independent: reacting to emotions and reacting to desires both concern other agents' feelings; but while reacting to emotions is dyadic, reacting to desires is triadic, revealing other agents' evaluations. The capacities for reacting to attention, intentions, and beliefs all reveal something about other agents' current interests, thereby permitting inferences about what those agents will do next. Reacting to attention can be seen as a special case of reacting to intentions, the target of perceptual attention corresponding to the intended goal.

Phylogenetically speaking, when looking down our list one can detect a decreasing number of species that exhibit the different forms of intersubjectivity, even if the data is, in many cases, incomplete. There is evidence that at least the great apes represent others' goals and intentions to act; but when it comes to reacting to intentions to interact, the situation is unclear. As far as we know, there is a lack of evidence for joint intentions in non-human species (Tomasello et al., 2005). As we shall see, this capacity is crucial for human cooperation. Chimpanzees, and possibly other apes, understand that "seeing is knowing" but react differently from adult humans to others' beliefs and knowledge, and there is no evidence that they form joint beliefs.

From an ontogenetic perspective, human infants react to others' emotions from very early on. They develop the capacity to react to others' desires – as expressed by pointing, facial expression, vocalization, and reaching – by the age of ten months. They react to attention to the self by the age of two months (Reddy, 2003), start following gaze to a target within their visual field around the age of six months (D'Entremont, 2000), and follow gaze in any context by the age of eighteen months. Note that gaze will provoke different reactions if initially oriented towards the infant's attention than if directly oriented towards a target of action. Four-month-old infants follow gaze if preceded by eye contact (Farroni, Johnson, & Csibra, 2004); by seven months of age they follow pointing accompanied by head turn (Striano & Bertin, 2005). Infants react to others' intentions to act by between the age of ten and fourteen months (Csibra, 2003); they react to communicative, referential intentions by around eighteen months of age (Franco, 2005).

In brief, phylogenetic evidence, and ontogenetic evidence from human children, roughly support a similar categorization and ordering of the different forms of intersubjectivity, while the data on the ontogeny of apes is, so far, too scarce to have any significance here.

10.3 Types of Cooperation and Their Cognitive Requirements

We have claimed that capacities for detached representations, cognitive foresight, and intersubjectivity – in the forms of joint attention and intention as well as reacting to beliefs and knowledge – have played a crucial role in the evolution of human cognition. Next, we shall present five types of cooperation and use the material from previous sections to explain which cognitive capacities each type presupposes. In doing so, we shall take an evolutionary perspective and use a game-theoretical approach. The various types of cooperation are presented in roughly their order of complexity – which, arguably, matches a decreasing number of species that exhibit them. The communicative skills required for the different types of cooperation will be discussed in Section 10.4.

Broadly defined, to cooperate is to work together for a joint benefit. Cooperation can occur by the mere coordination of behavior, but can equally presuppose a mutual sharing of representations of means and goals. Few kinds of group behavior could be called cooperation in the literal sense of the word. Among those that cannot, we find the visceral coordination of behavior: for example, termites building termite mounds. Some of the types of cooperation we describe in this section depend on elaborate long-term planning and negotiation; therefore they require cognitive foresight, as well as complex kinds of intersubjectivity.

In true cooperation, all participants are actively involved. For instance, when a leader manipulates other subjects in order to reach a personal goal, we do not say that the other subjects are cooperating with the leader. Cooperation demands that all agents are in control of their own behavior relative to the goal. A narrower definition of cooperation concerns situations in which joint action presents a dilemma, the paradigmatic examples being the prisoners' dilemma (PD) game and iterated versions of it. Here, the focus will be on the narrower definition.

Cooperative games have been studied within two frameworks. In general, traditional game theory has assumed that cooperation takes place between individuals of the species *Homo oeconomicus*, who are ideally rational and have a perfect "theory of mind." Alternatively, within an evolutionary framework (Axelrod & Hamilton, 1981; Maynard Smith, 1982), the players are the genes of different animal species, assumed to have no rationality or cognitive capacities at all. However, their strategies can slowly adapt, over a number of generations, via the mechanisms of natural selection. A key concept is that of an *evolutionarily stable strategy* (Maynard Smith, 1982). This is a strategy that, if adopted by a population of players, cannot be improved upon by any alternative strategy that initially is rare in the population. Within the evolutionary framework, intersubjectivity is considered irrelevant. The differences of the two frameworks in their assumptions about the players' cognitive and communicative skills are seldom discussed. We shall argue that, in order to understand the evolution of cooperation, cooperative behavior must be judged in relation to the agents' cognitive and communicative capacities.

10.3.1 In-Group Versus Out-Group

A simple way to determine cooperation in an iterated PD or similar game is to divide individuals into an *in-group* and an *out-group*. The basic strategy is then to cooperate with everybody in the in-group and defect against everybody in the out-group. Many species solve this problem via olfaction: for example, bees from a different hive smell differently and are treated with aggression. For another kind of example, Dunbar (1996) speculates that dialects have evolved to serve as markers of the in-group among humans. In general though, in-group cooperation does not require intentional communication and cognitively only presumes a perceptual mechanism that separates members of the in-group from the rest. In general, the in-group requires *some* form of marker that helps distinguish the in-group from the out-group. Evolutionary biologists (e.g. Zahavi, 1975) stress that such markers should be hard to fake, in order to exclude free riders from the in-group.

The evolutionary origin of in-group formation lies in kin selection. In a kin group, the outcomes of a PD game must be re-evaluated due to the genetic relatedness of the individuals. In many cases, this makes cooperation the only evolutionarily stable strategy. In other words, a game that is a PD game on the individual level may be perfectly cooperative, when the genes are considered to be the players. Kin groups may then form kernels from which larger cooperative groups develop (Lindgren, 1997, p. 351).

10.3.2 Reciprocal Altruism

In an iterated PD game, one player can *retaliate* against another's defection. Trivers (1971) argues that this possibility can make cooperation more attractive and lead to what he calls *reciprocal altruism.* Such cooperation can be formulated as a slogan: "you scratch my back; I'll scratch yours." Field studies of vampire bats (Wilkinson, 1984), fish, and primates have reported the presence of reciprocal altruism. Still, the evidence for reciprocal altruism in non-human species is debated; what reciprocal altruism there is seems to be largely restricted to short-term reciprocity.

Some laboratory experiments show negative results (Stephens, McLinn, & Stevens, 2002; Hauser, Chen, Chen, & Chuang, 2003). In line with this, Stevens and Hauser (2004) argue that the cognitive demands of reciprocal altruism have been underestimated. They claim that reciprocal altruism will be evolutionarily stable in a species only if the temporal discounting of future rewards is not too steep, the discrimination of the value of the rewards sufficient to judge that what an altruist receives back is comparable to what it has given itself, and memory capacity sufficient to keep track of interactions with several individuals.

Although the methodology of the experiments varies, studies of temporal discounting reveal that rates differ drastically between the few species that have been investigated (Kacelnik, 2003). Humans have a comparably low discount rate: a prerequisite of the cognitive foresight that was presented in Section 10.2.2. Studies on great apes reveal that, in some food contexts, they perform at level with humans (e.g. Rosati, Stevens, Hare, & Hauser, 2007; Osvath & Osvath, 2008). What factors – ecological, cognitive, neurological, etc. – explain the discounting rate of a particular species is an interesting problem that should receive more attention within evolutionary game theory.

Countering the arguments of Stevens and Hauser (2004), de Waal and Brosnan (2006, pp. 103–104) have proposed distinguishing three types of reciprocity. (1) *Individual reciprocity* is based on symmetrical features of dyadic relationships: kinship, or similarities in age or sex. This is the cognitively least demanding form of reciprocity: "it requires no scorekeeping since reciprocation is based on pre-existing features of the relationship. . . . All that is required is an aversion to major, lasting imbalance in incoming and outgoing benefits" (de Waal & Brosnan, 2006, p. 104). (2) In *attitudinal reciprocity*, individual willingness to cooperate depends on the attitude that one's partner has recently shown. It involves few cognitive capacities: only that individuals are recognized over time and the "attitude" of the partner in various interactions remembered (or learned by some conditioning mechanism). As de Waal and Brosnan write, this kind of reciprocity is "contingent upon the partner's immediate preceding behavior." If it can be described as a tit-for-tat strategy, the cooperation need not involve any intersubjectivity except for reacting to emotions. No intentional communication is required for any form of reciprocal altruism;

but the interaction is enhanced when desires are expressed: e.g., by begging. (3) In *calculated reciprocity*, "individuals reciprocate on a behavioral one-on-one basis with a significant time interval." Calculated reciprocity requires more advanced memory of the interactions, reacting to the desires of others and some form of score-keeping. This type of reciprocity is the most difficult to achieve, since it "requires memory of previous events, some degree of score-keeping, partner-specific contingency between favors given and received" (de Waal & Brosnan, 2006, p. 104). The reciprocity discussed by Stevens and Hauser (2004) is closest to this type of reciprocity; therefore it is not surprising that it would be the most difficult type to achieve. Van Schaik and Kappeler (2006, p. 11) write: "Chimpanzees are capable of the 'calculated reciprocity'..., but this mechanism may be rare among other species, if it occurs at all."

Reacting to the partner's attitude is a general mechanism for improving cooperation. Charness and Dufwenberg (2006, p. 1580) introduce the concept of *guilt aversion* to explain why individuals behave more cooperatively than game theory predicts. According to them, "decision makers experience guilt if they believe they let others down." That is, considering the desires of one's partners and having an emotional mechanism that makes one avoid letting them down changes the payoff matrix of, for example, a PD game: the player ends up being more cooperative than traditional game theory predicts. The situation is analogous to that of kin selection, where a PD game is also turned into a cooperative game, when the inclusive fitness of the kin is considered.

Charness and Dufwenberg (2006) claim that guilt aversion involves having beliefs about others' beliefs. However, this claim about the intersubjectivity of decision makers seems unnecessarily strong for the mechanism they are investigating. It is sufficient that the decision maker can react to others' desires by feeling guilty. In spite of this reservation, the notion of guilt aversion seems central to analyzing many forms of human cooperative behavior and merits further theoretical and experimental investigations.

The extent to which guilt aversion exists in other species is an interesting question.[8] The so-called *ultimatum game* could be a benchmark for testing guilt aversion and thereby, indirectly, the capacity to represent the desires of others. In this game, a proposer is offered a sum of money, which she is free to divide in any proportion with a responder. The crucial feature is that the responder can accept or reject the offer. If she accepts, both players receive the proposed division; but if she rejects, both receive nothing. With human subjects, proposers typically offer responders 40–50% of the amount given, and responders generally reject offers under 20%. These results indicate that the responders want the proposers to take their desires into account and will punish a proposer who is too unfair.

By contrast, Jensen Call, and Tomasello (2007) showed that, when chimpanzees were used as subjects and proposers given a choice between an 8-2 and a 5-5 offer,

[8] This question relates to the studies concerning other-regarding preferences presented in Section 10.2.3.

75% chose the 8-2 offer, which was rejected by only 5% of the responders. Even more drastically: when the proposers could choose between a 10-0 and an 8-2 offer, 46% choose the 10-0 offer, which was then rejected by 44% of the responders. These results indicate that, in food-related tasks, chimpanzees do not exhibit guilt aversion as strongly as humans. This might be a result of certain rules that chimpanzees adhere to when it comes to food possession; but it could also suggest a more general lack of guilt aversion. Bräuer, Call, and Tomasello (2009) found differences between chimpanzees and bonobos suggesting that bonobos generally expect equity and might perceive violation of it more acutely.

10.3.3 Cooperation Towards Future Goals

Many types of animal cooperation do not seem to require detached representations (Brinck & Gärdenfors, 2003). If the goal is present in the immediate environment – say, food to be eaten or an enemy to be fought – the collaborators do not need to coordinate their representations of it before acting. If, on the contrary, the goal lies in the future, a shared representation of that goal must be produced before planning can begin or joint action be taken.

We have suggested that the hominin lifestyle opened up new forms of cooperation about future states involving coordinated action towards a common goal – as well as, probably, complementary role taking and role reversal. Thus, Plummer (2004, p. 139) writes that "competition with large carnivores may have favored cohesive groups and coordinated group movements in *Homo habilis*, cooperative behavior including group defense, diurnal foraging (as many large predators preferentially hunt at night) with both hunting and scavenging being practiced as the opportunities arose, and the ability (using stone tools) to rapidly dismember large carcasses so as to minimize time spent at death sites." If this hunting was organized in the same way as in extant hunter-gatherer societies, it would be an example of *division of labor* and require both complementary role taking and joint intentions. We will return to the cognitive and communicative demands of labor division in Section 10.4.2.

Another important aspect of cooperation towards future goals is that the value of the goal cannot be determined in the current situation (Brinck & Gärdenfors, 2003). In contrast to a goal that is already present on the scene – say, a certain kind of prey – the value of a future goal must either be estimated individually by each agent or calculated jointly. The former strategy requires the capacity for cognitive foresight; the latter presupposes a rather elaborate communication system.

To sum up the cognitive requirements: cooperation towards future goals involves not only cognitive foresight, but also the recognition of individuals, the inhibition of current desires, and the capacity to react to the emotions, desires, attention, and intentions of others. Thus, our theory generates the testable prediction that, among extant species, only humans will exhibit cooperation towards future goals. The central question concerning what kind of communication is required for this joint-intention-involving collaboration will be addressed in Section 10.4.2.

10.3.4 Indirect Reciprocity

In social species, individuals often face a decision whether or not to cooperate. In the analyses in standard game theory of prisoners' dilemma and similar games, it is taken for granted who the potential collaborators are. In practice, however, the most important question is: how do you know *whom* to cooperate with? (See e.g. Dessalles, 2007, p. 360.) When choosing one's partner, *trust* may be one factor, the *competencies* of one's collaborator another.

As we have noted, reciprocal altruism is found in some other animal species and is common among humans. That said, in humans one often finds more extreme forms of altruism: "I help you and somebody else will help me." So far, there is no evidence that this form of cooperation – called *indirect reciprocity* – exists in nonhuman animals.[9]

Nowak and Sigmund (2005, p. 1291) show that – under certain conditions – indirect reciprocity can function as an evolutionarily stable strategy. However, as we shall see in Section 10.4.3, their explanation depends on strong assumptions about the interactors' communicative capacities. The key concept in their evolutionary model is that of individual *reputation* (see also e.g. Leimar & Hammerstein, 2001; Panchanathan & Boyd, 2004; Fehr, 2004). Typically, the reputation of an individual *i* is built up as members of the society observe *i*'s behavior towards third parties and convey what they have observed to other members of the society. This form of communication requires detached representations. The level of *i*'s reputation can then be used by any other individual when deciding whether to help *i* in a situation of need.

Nowak and Sigmund (2005) term a strategy first order if the assessment of an individual *i* in a certain group depends only on *I*'s actions. More sophisticated strategies distinguish between *justified* and *unjustified* defections. justified in defecting as a form of punishment. However, the donor then runs the risk that his own reputation drops. To prevent this, the donor should communicate that the reason she defects is the receiver's bad reputation, making it possible for other individuals to understand her intentions.

Nowak and Sigmund (2005) categorize a strategy as second order if it depends in part on the reputation of the receiver and third order if it additionally depends on the reputation of the donor. They then show that only eight of the possible second- and third-order strategies are evolutionarily stable, and that all of these depend on the distinction between justified and unjustified defection. The intersubjectivity required by such comparisons includes empathy and reacting to desires and attention, but also, more importantly, reacting to the *intentions* of the donor.

[9] A possible exception is revealed in a study by Warneken and Tomasello (2006). They let three human-raised juvenile chimpanzees watch while a human attempted, but failed, to achieve a succession of goals. Sometimes the chimpanzees helped the humans, mainly in situations where the human reached for objects but failed to grasp them. By contrast, Warneken and Tomasello (2006) showed that 18-month-old human infants did help an adult experimenter in a large majority of the cases involving various scenarios. See also Warneken, Hare, Melis, Hanus, and Tomasello (2007).

Nowak and Sigmund (2005, p. 1291) also claim, without spelling out the details, that indirect reciprocity seems to require some form of "theory of mind." An individual watching a second individual (a donor) not helping a third in need (a receiver) must judge whether or not the donor is non-cooperative as a punishment for a receiver of bad reputation, or is simply being non-cooperative. Such judgment involves understanding the intentions of the donor. Thus, indirect reciprocity at least demands the ability to (a) recognize other players over time; (b) remember and update their reputation scores (and one's own); (c) represent others' emotions, desires, attention, and intentions in such a way that one can judge whether a particular donor action is "good" or "bad"; and (d) inhibit current desires in order to help somebody in need.[10] Again our theory predicts that only humans will exhibit indirect reciprocity. We will return to the communicative requirements on how reputation is spread in Section 10.4.3.

It should be stressed that reputation is not something visible to others – unlike status markers, such as a raised tail among wolves. Rather, each individual in a society must either keep a private account in memory of the reputation of others, or be able to communicate with others about the reputation of particular individuals. Kuhn and Stiner (2008) write that body decorations and emblems may have been used as status markers in hominin societies. Semmann, Krambeck, and Milinski (2005) have demonstrated experimentally that building a reputation through cooperation is valuable for future social interactions; not only within but also *outside* one's own social group.

The trust that is built up in reciprocal altruism is dyadic: it is a relation between two individuals. By contrast, reputation is an emergent *social* phenomenon, involving most, if not all, members in the group. Nowak and Sigmund (2005, p. 1296) write: "Indirect reciprocity is situated somewhere between direct reciprocity and public goods. On the one hand it is a game between two players, but it has to be played within a larger group." Socially shared information about the reputation of different individuals is a characteristic feature of the game.

There may be other factors than those considered by Nowak and Sigmund influencing the reputation of an individual. In many situations, people are not only willing to cooperate but also ready to punish free riders. Punishing behavior is difficult to explain, both because it is costly and because the cooperative non-punishers may benefit (Fehr & Gächter, 2002). Sterelny (2003, p. 131) notes that, for cooperation to become important for hominin societies, punishment of defection must be cheap. Barclay (2005) presents evidence that the reputation of a punisher increases over time and that people are more willing to cooperate with punishers. Such a mechanism entails that punishing behavior that is costly in the short run will eventually become beneficial for the punisher: i.e., punishing behavior is rewarded

[10] Van Schaik, and Kappeler (2006, p. 15) write: "the three basic conditions for reputation are individual recognition, variation in personality traits, and curiosity about the outcome of interactions involving third parties." In our opinion, these criteria are too weak, since they do not include that reputation is communicated and therefore requires flexible, referential communicative skills.

in the long run (see also Sigmund, Hauert, & Nowak, 2001); furthermore, it can stabilize cooperation in iterated PD-style interactions (Lindgren, 1997). Symbolic communication enhances this tendency, since it can turn punishment into common knowledge. In contrast to human conduct, there are no observations of altruistic punishment in natural animal behavior (van Schaik & Kappeler, 2006, p. 15) – although de Waal and Brosnan (2006) report a case of experimentally induced refusal to cooperate that is costly for the non-cooperative individual.

10.3.5 Conventions, Commitments and Contracts

In human societies, many forms of cooperation are based on *conventions* (Lewis, 1969). Conventions may be established without explicit communication; but, of course, communication makes the presence of a convention clearer. The central cognitive requirement for conventions is that they presuppose enduring *joint beliefs* or common knowledge. For example, if two cars meet on a gravel road in Uganda, both drivers know that this coordination problem has been solved in this society numerous times before by driving on the left-hand side; both know that both know this; both know that both know that both know this, and so on. They both then shift to the left without any hesitation.

In general, joint beliefs form the basis for much of human culture. Commitments and contracts are special cases of joint-belief-involving cooperation about the future. When you commit yourself to someone to do some action, you intend to perform the action in the future; the other person wants you to do it and intends to check that you do it; and you and they have joint beliefs concerning these intentions and desires (Dunin-Kepliz & Verbrugge, 2001). Commitments cannot arise unless the agents have the capacity for cognitive foresight and can form joint beliefs.

Contracts represent a uniquely human activity. The reason again is that a contract, among other things, presumes that the partners have joint beliefs: if we agree, for example, that I shall deliver a duck next time I am successful in my duck hunting, in exchange for the axe you have given me now, then I believe that you believe that I will deliver the duck, and you believe that I believe that our agreement will then be fulfilled – and so on. A contract also depends on the possibility of future sanctions: if I don't deliver the duck, you or the society may punish me for breaching the agreement. Like commitments, contracts presume cognitive foresight. Furthermore, the reputation mechanism that supports indirect reciprocity will benefit the formation of contracts.

10.4 The Evolution of Symbolic Communication

This section will bring together the concepts and material from the preceding sections in an account of the communicative demands on two of the types of human cooperation discussed in Sections 10.3.3 and 10.3.4: cooperation towards future goals, and indirect reciprocity. We will argue that a major force behind the evolution of human language is that it enhances these forms of cooperation. In addition to describing the characteristics of any particular system of communication, it is desirable to give an independent evolutionary – ecological or sociocultural – explanation of why this particular kind of system has arisen. In the following sections we shall present our account of the ecological and social forces that have generated symbolic communication as a uniquely human system.

We will start from a distinction between *signals*, *icons*, and *symbols*. On the basis of these notions, we will analyze the role of different forms of communication and how they support cooperation.

10.4.1 From Signaling to Symbolic Communication

A decisive difference between language and signaling is that signals merely indicate what is present in the environment. Bees only dance directly upon returning to the hive when they have found nectar. Vervet monkeys only seem to signal when danger is immediate. However, *icons* and *symbols* make it possible to communicate about things that may not even exist. Both are tools of communication; both can be expressed in various ways, such as by sounds, pictures, or gestures. Icons and symbols – and also acts of pointing (by proxies or stand-ins) – retain meaning in the absence of the referent. Iconic and symbolic communication can be used to offload the demands of intersubjective cognition: with icons and symbols, one can communicate one's emotions, desires, intentions, and beliefs, so that one's addressee need not rely only on mind reading. An additional property of symbols, not in general shared with icons, is that their meanings can be related in systematic ways. Deacon (1997) emphasizes logical relations, while Gärdenfors (2000) discusses similarity relations.

The signals used in communication are more or less identical in all members of an animal species, barring sexual and age differences. A symbol, on the other hand – and, in many cases, an icon, too – is a convention that one must *learn* if one is to use it as a communicative tool. Symbols vary widely between different groups.

In all probability, the transition from a signaling system to an iconic or symbolic language has not been made in one step. Bickerton (1990) among others has proposed a stage in the evolution of language when a *protolanguage* – containing only the semantic components of language but not the syntactic markers – was used. Like language, protolanguage is *compositional*: complex signs are composed out of individual, meaningful signs that relate systematically to each other and to other, similar, signs. It is also *conventional*: a sign is a part of mutual knowledge and must be learned. However, protolanguage is not grammatical. According to Bickerton (1990), *Homo ergaster/erectus* mastered a protolanguage, but it was not until *Homo sapiens* that a language with a grammatical structure emerged.[11] For purposes of

[11] Partly following Donald (1991), Zlatev, Persson, and Gärdenfors (2005) argue that, in addition to protolanguage and language with syntax, one can distinguish communication systems based on

this article, we need not subscribe to Bickerton's (1990) theory of protolanguage. The only assumption we need is that there is a stage in the evolution of language where compositional and conventional symbols emerged.[12]

10.4.2 *Cooperation About Future Goals Requires Symbolic Communication*

In recent work (Brinck & Gärdenfors, 2003; Gärdenfors, 2003, 2004; Osvath & Gärdenfors, 2005; Gärdenfors & Osvath, 2010), we have proposed a strong connection between a lifestyle dependent on cognitive foresight on the one hand, and the evolution of symbolic communication on the other. In brief, our argument is that symbolic communication makes it possible to cooperate towards future goals *that are not mutually manifest to the agents*: i.e., neither known to be repeated at a certain interval, nor present in the shared environment. Goals that so far exist only in an agent's imagination – along with the means to attain them – can be shared socially by using the manifest symbols that represent them.

Clearly, communicating about non-present entities presupposes capacities for detached representations and cognitive foresight. It additionally requires a means for making one's representations accessible to others and thus for referring to absent objects and agents. In this case, signaling will not do. One cannot point or use an alarm call to refer to a deer that one saw down by the river yesterday; but one can use a symbol to do so. The symbol takes the place of the deer in the exchange of ideas.

An additional requirement is that the means for communication should make it possible to describe *non-actual* scenes. Theories about the evolution of symbolic communication abound. Most of them miss that the act of describing a scene is most valuable when the scene is not present. Describing something that happened in the past or an activity that lies in the future requires symbols that can have different *functions*, corresponding to actions and objects and agents. It requires that symbols can be put together in a variety of ways, making possible combinations such as <Go, Mountain>, <Mountain, Deer, John>, <Boy, Man, Go, Woman>. Such combinations are ambiguous, but because the communication is either verbal or gestural – and consequently the communication between speaker and listener is direct – the problem is minor. Besides, the physical context in

dyadic and triadic mimesis. They argue that untrained apes can communicate at least via dyadic mimesis: e.g., by gestures (see also Pika et al., 2003). However, the natural gestural communication of apes does not constitute a protolanguage because it is not conventionalized (and thus not symbolic) and has only limited compositionality. On the other hand, the communications of Kanzi, Koko, and other language-trained apes satisfy the criteria for protolanguage. Jackendoff (1999) likewise distinguishes several stages in the evolution of syntactic structures.

[12] Partly conventionalized compositional icons may be sufficient for many situations, although fully developed symbols are more efficient.

which the communication takes place will help disambiguate the exchange, even without any explicit efforts by the speakers.

The sharing of imagined scenes is a *sine qua non* of cooperation towards future goals.[13] The goal is not present, so the actions leading to it must be imagined and their consequences evaluated. For example, by uttering (or signing) "meet mountain well sunset", one individual can convey to another an image of a scene – a mountain – that constitutes a plan in respect to a future goal, one that is mutually beneficial to both individuals. If the scene is something new, there is no way a signaling system can communicate these imagined elements.

A particular form of future-oriented cooperation is division of labor, which, as a cultural phenomenon, seems to be unique to the human species. Simple commands can be expressed with the aid of combinatorial icons or symbols, in accordance with the various ways of dividing the tasks within a group. As regards the pragmatics of communication, commands and questions are more central modes of communication for setting up cooperation than are declarative statements.

The long ranging lifestyle of the Oldowan hominins resulted in selective pressures favoring a communicative system that made it possible for the members of a group to share detached representations of future collaborative scenes. The evolution of symbolic communication could have lead to a solution that was, in evolutionary terms, locally optimal for a society built around cooperation towards future goals.[14] Iconic communication, in the form of miming (Donald, 1991), can achieve much towards the coordination of scenes; but arbitrary symbols are more efficient for the task.

Showing the evolutionary importance of cooperation towards future goals, Deacon (1997, pp. 385–401) suggests that the first form of symbolic communication was marriage agreements: that is, deliberate commitments to pair bonding. He argues that there was strong evolutionary pressure in hominin societies to establish relationships of exclusive sexual access. He writes (Deacon, 1997, p. 399) that such an exclusive sexual bond "is a prescription for future behaviors." Even though we do not know of any evidence that marriage agreement was the first form of symbolic communication, still, we find this claim interesting in the discussion of early cognitive foresight.

In brief, our thesis is that cooperation towards future goals co-evolved with symbolic communication, presumably with gestural communication as an intermediary stage. However, without the presence of well-developed cognitive foresight and intersubjectivity, the potential for cooperation, and the ensuing selective pressures that resulted in symbols, would not have emerged. Unlike most other theories of the evolution of language, the ideas presented here explain why humans are alone in having symbolic language. Being able to cooperate towards future goals requires

---

[13] Arp (2006) argues, albeit from a different perspective, that scene visualization is unique to the Homo line.

[14] This kind of local optimality is one of the four conditions that Dessalles (2007, pp. 173-174) presents for protolanguage.

planning with cognitive foresight, and complex forms of intersubjectivity. Both these cognitive capacities are far more developed in humans than they are in other species.

10.4.3 Indirect Reciprocity Requires Syntactic Markers

One could argue that indirect reciprocity is a way of preparing for future cooperation, since it is, in effect, a behavior indicating whom can be trusted to be cooperative. If we follow the model presented by Nowak and Sigmund (2005), the success of indirect reciprocity depends heavily on the mechanism of reputation. However, they do not analyze what kind of communication is required for strategies based on reputation. First of all, the communication system of the group must be capable of referring to individual members *in their absence*: e.g., by names or proxies. No known animal signaling system has this capacity. Second, to capture the second-order strategies required for evolutionary stability in Nowak and Sigmund's model, the communication system must be able to express relations such as "*x* was good to *y*" and "*y* was bad to *x*" and distinguish these expressions from "*y* was good to *x*" and "*x* was bad to *y*".[15] Such expressions involve assigning two individuals in a particular interaction different *roles*: the rudiments of agent and patient. Such roles are difficult to convey unambiguously without using syntactic markers.

The communication required by indirect reciprocity concerns different aspects of whom you can trust. The information often concerns absent individuals and hence can be characterized as gossip. Gossip normally contains expressions of the form "who did what to whom." It involves identifying roles that require some form of syntax. Thus, according to the theory presented here, gossip plays a role in the evolution of language; however, it does not function as a replacement for grooming as Dunbar (1996) suggests.

Indirect reciprocity may not be the only causal factor behind syntax: there may be other ecological pressures that have led to more complex language. One candidate is planning for division of labor – a special case of cooperating towards future goals – which also involves assigning roles for future actions to different individuals. If the individuals are present, perhaps such assignment can be achieved with the aid of pointing or other forms of indexicals. However, if the individuals are not present, the availability of syntax will facilitate the coordination of future actions.

The need for communication depends on the size of the group facing the situations in which a potential for cooperation exists. In a tightly knit group where everybody sees everybody else most of the time, a reputation mechanism is not needed. One can compare this with how the ranking within such a group is established. If you observe that *x* dominates *y* and you know that *y* dominates you, no

[15] In this context, it should be noted that Nowak and Sigmund's (2005) model considers only two kinds of interaction: "good" and "bad." It goes without saying that, in real groups, communication about reputation builds on finer nuances.

(aggressive) interaction or communication is needed to conclude that x dominates you. On the other hand, in large and loosely connected groups, these mechanisms are inadequate; and so some form of communication about non-present individuals is required.

A consequence of this argument is that the evolutionary mechanisms favoring indirect reciprocity will be stronger when the population increases and contacts between people become increasingly distant (Gamble & Steele, 1999; Leonard & Robertson, 2000). The conditions for reciprocal altruism seem similar to those for the cultural innovations analyzed by Shennan (2001). However, since hominin populations have probably mostly been small, and there have been bottlenecks (Ambrose, 1998; Richerson, Boyd, & Heinrich, 2003), the selective significance of indirect reciprocity may have varied considerably over time.

10.5 Connections to Archaeology

Our arguments have built on a theoretical analysis of the cognitive and communicative prerequisites for the specifically human forms of cooperation. As directions for future research, we want to draw attention to the further empirical support that may be obtained from new archeological data or new analyses of existing such data. One finds evidence for cognitive foresight in the Oldowan culture (Toth, 1985). Of course, evidence of cooperation towards future goals would help to determine when protolanguage evolved. Three areas seem to be particularly relevant to this problem.[16] First, signs of cultural division of labor (e.g. Hawkes, O'Connell, Blurton, Alvarez, & Charnov, 1998; O'Connell, Hawkes, & Blurton Jones, 1999; Bird, 1999) would be evidence of communication about future cooperation. Second, big game hunting seems to require such communication. The spears found at Schöningen (Thieme, 1997) suggest that such hunting was practiced by *Homo heidelbergensis* 400 Kya.[17] The findings from this site await detailed analysis; their cognitive and communicative implications are, as yet, far from clear (but see Haidle, 2009). Third, large constructions – dwellings or other kinds of built structure – would also be indicative of communication about future cooperation. Although disputed, the Terra Amata construction (Villa, 1983) – which, it has been estimated, was built 400 Kya – is a possible candidate.

If our arguments are correct, it is sufficient to establish the presence of symbolic communication if there is clear evidence in any one of these three areas. All in all, we believe the archaeological evidence available at present supports the view that cooperation towards future goals – and thus the use of combinatorial symbols – was

---

16 We wish to thank Miriam Haidle, Steven Mithen, and April Nowell for helpful discussions on these matters.

17 Thieme (1999) writes: "equipped with outstanding technical talents for processing wood, they were already in this early period qualified to plan a future large game hunt with special weapons, to organize it, to coordinate it and to successfully accomplish it" (our translation).

established at least 400 Kya, well before the evolution of *Homo sapiens*. Although our argument is different, this conclusion accords with Bickerton's (1990) proposal that protolanguage emerged with *Homo erectus*.

It would be interesting to discover during which time period evidence for indirect reciprocity can be established in the archeological data: for example, evidence of individual reputation being of social importance. If our analysis in Section 10.4.3 is correct, such a datum would provide an upper bound to when the first syntactic markers evolved among hominins.

The role of ornaments and body decoration as reputation "markers" could be an interesting target for archaeological analysis. For example, Kuhn and Stiner (2008, p. 43) argue that ornaments have, as their main communication target, strangers at a "middle distance" of social relations: that is, individuals who share the same cultural background but do not know the wearer personally.[18] This kind of information is clearly important for deciding whether or not to cooperate.

10.6 Conclusion

Our main argument has been that cognitive foresight and increasingly complex forms of intersubjectivity opened up new forms of cooperation involving future goals and indirect reciprocity. Historically, such cooperation has secured selective advantages for the members of the most cooperative hominin groups. New forms of cooperation created a need for a communication system for sharing detached representations about future goals and the reputation of other individuals. We have argued that the required system must be symbolic. Our two central theses are, first, that cooperation towards future goals requires combinatorial symbolic communication; and, second, that indirect reciprocity additionally requires items such as names and role markers, and thus a minimal syntax.

Table 10.1 summarizes the cognitive and communicative requirements of the different forms of cooperation as they have been analyzed in Sections 10.2, 10.3, and 10.4. The list is not intended to be complete: in particular, the forms of cooperation could be given a more fine-grained analysis (see e.g. Gärdenfors, 2008). We have only included the cognitive and communicative features that are characteristic of each particular form of cooperation. As one goes down the list, one observes a general increase in cognitive and communicative demands. Cooperation towards future goals, indirect reciprocity, conventions, commitments, and contracts are forms of cooperation that are, so far, found only in humans.

In this article, we have presented a theory that connects different forms of cognition and communication with different forms of cooperation, within an evolutionary framework. A future extension of the research program might consider *individual*

[18] In agreement with our argument, Kuhn and Stiner (2008, p. 48) conclude: "if ... beads and other forms of body ornamentation represent a new way of communicating, then it follows humans must already have been using symbols to communicate when ornaments first appeared in Paleolithic sites."

Table 10.1 The cognitive and communicative requirements on different forms of cooperation

Type of cooperation	Cognitive demands	Communicative demands
In-group out-group	Recognition of group member	None
Reciprocal altruism (attitudinal reciprocity)	Recognition of individuals, (minimal) memory of interactions	None, but enhanced by expressing emotions and desires
Reciprocal altruism (calculated reciprocity)	Recognition of individuals, memory of interactions, inhibition of current desires, reacting to the desires of others	Expressing emotions and desires
Cooperation towards future goals	Recognition of individuals, cognitive foresight, reacting to emotions and desires, joint intentions	Symbolic (or iconic) communication
Indirect reciprocity	Recognition of individuals; inhibition of current desires; reacting to emotions, desires, attention and intentions of others	Symbolic communication in a language with names and syntax for roles
Cooperation based on convention	Joint beliefs	None, but enhanced by symbolic communication
Commitment and contract	Recognition of individuals, cognitive foresight, joint beliefs	Symbolic communication

differences in intersubjective, foresight and communicative capacities, and aim at connecting these to variations in cooperative behavior.[19] The analysis presented in Table 10.1 could then be used to generate empirically testable connections between these factors.

Acknowledgments Most of the work for this article was done as part of the EU project *Stages in the Evolution and Development of Sign Use* (SEDSU). We want to thank the participants of the project for helpful comments and ideas. Peter Gärdenfors gratefully acknowledges support as a Senior Individual Researcher from the Swedish Research Council as well as support from the Linnaeus project *Thinking in Time: Cognition, Communication and Learning* during the final stages of the work. The research of Ingar Brinck was financed by the Bank of Sweden Tercentenary Foundation through the Centre for Cognitive Semiotics at Lund University.

References

Ambrose, S. (1998). Late Pleistocene human population bottlenecks, volcanic winter, and differentiation of modern humans. *Journal of Human Evolution, 34*, 623–651.

Arp, R. (2006). The environment of out Hominin ancestors, tool-usage, and scenario visualization. *Biology and Philosophy, 21*, 95–117.

[19] We wish to thank Ernst Fehr for this suggestion.

Atance, C. M., & O'Neill, D. K. (2005). The emergence of episodic future thinking in humans. *Learning and Motivation, 36*, 126–144.

Axelrod, R., & Hamilton, W. D. (1981). The evolution of cooperation. *Science, 214,* 1390–1396.

Barclay, P. (2005). Reputational benefits for altruistic punishment. *Evolution and Human Behaviour, 27*, 325–344.

Baron-Cohen, S. (1994). How to build a baby that can read minds: Cognitive mechanisms in mindreading. *Cahiers de Psychologie Cognitive, 13*(5), 513–552.

Bickerton, D. (1990). *Language and species*. Chicago: The University of Chicago Press.

Bickerton, D. (2002). Foraging versus social intelligence in the evolution of protolanguage. In A. Wray (Ed.), *The transition to language* (pp. 207–225). Oxford: Oxford University Press.

Bird, R. (1999). Cooperation and conflict: The behavioral ecology of the sexual division of labor. *Evolutionary Anthropology, 8*(2), 65–75.

Bräuer, J., Call, J., & Tomasello, M. (2009). Are apes inequity averse? New data on the token-exchange paradigm. *American Journal of Primatology, 71*, 175–181.

Brinck, I. (1997). *The indexical 'I'. The first person in thought and language*. Dordrecht: Kluwer.

Brinck, I. (2001). Attention and the evolution of intentional communication. *Pragmatics and Cognition, 9*(2), 255–272.

Brinck, I. (2004). Joint attention, triangulation and radical interpretation: A problem and its solution. *Dialectica, 58*(2), 179–205.

Brinck, I. (2008a). The role of intersubjectivity for the development of intentional communication. In J. Zlatev, T. Racine. C. Sinha, & E. Itkonen (Eds.), *The shared mind: Perspectives on intersubjectivity* (pp. 115–140). Amsterdam: John Benjamins.

Brinck, I. (2008b). From similarity to uniqueness: Method and theory in comparative psychology. In L. S. Röska-Hardy & E. M. Neumann-Held (Eds.), *Learning from animals? Examining the nature of human uniqueness* (pp. 155–170). London: Psychology Press.

Brinck, I., & Gärdenfors, P. (1999). Representation and self-awareness in intentional agents. *Synthese,* 118, 89–104.

Brinck, I., & Gärdenfors, P. (2003). Co-operation and communication in apes and humans, *Mind and Language, 18*(5), 484–501.

Brownell, C. A., & Carriger, M. S. (1990). Changes in cooperation and self-other differentiation during the second year. *Child Development, 61,* 1164–1174.

Buckner R. L., & Carroll, D. C. (2007). Self-projection and the brain. *Trends in Cognitive Sciences, 11*, 49–57.

Bugnyar, T., Stöwe, M., & Heinrich, B. (2004). Ravens, *Corvus corax* follow gaze direction of humans around obstacles. *Proceedings of the Royal Society of London, B, 271*, 1331–1336.

Burkart, J. M., Fehr, E., Efferson, C., & van Schaik, C. P. (2007). Other-regarding preferences in a non-human primate: Common marmosets provision food altruistically. *PNAS, 104*, 19762–19766.

Call, J. & Tomasello, M. (1999). A nonverbal false belief task: The performance of children and great apes. *Child Development, 70*(2), 381–395.

Carpenter, M., Call, J., & Tomasello, M. (2002). A new false belief test for 36-month-olds. *British Journal of Developmental Psychology, 20*, 393–420.

Charness, G., & Dufwenberg, M. (2006). Promises and partnership. *Econometrica, 74*, 1579–1601.

Clements, W. A., & Perner, J. (1994). Implicit understanding of belief. *Cognitive Development, 9*, 377–397.

Correia, S. P. C., Dickinson, A., & Clayton, N. S. (2007). Western scrub-jays anticipate future needs independently of their current motivational state. *Current Biology, 17*, 856–861.

Csibra, G. (2003). Teleological and referential understanding of action in infancy. *Philosophical Transactions of the Royal Society, B 29*, 447–458.

Deacon, T. (1997). *The symbolic species*. London: Penguin Books.

D'Entremont, B. (2000). A perceptual-attentional explanation of gaze following in 3- to 6-months-olds." *Developmental Science, 3*, 302–311.

Dessalles, J.-L. (2007). *Why we talk*. Oxford: Oxford University Press.

De Waal, F. B. M., & Brosnan, S. F. (2006). Simple and complex reciprocity in primates. In P. M. Kappeler & C. P. van Schaik (Eds.), *Cooperation in primates and humans* (pp. 85–105). Berlin: Springer.

Donald, M. (1991). *Origins of the modern mind*. Cambridge, MA: Harvard University Press.

Donald, M. (1999). Preconditions for the evolution of protolanguages. In M. C. Corballis, & S. E. G. Lea (Eds.), *The descent of mind* (pp. 355–65). Oxford: Oxford University Press.

Dunbar, R. (1996). *Grooming, gossip and the evolution of language*. London: Faber and Faber.

Dunin-Kepliz, B., & Verbrugge, R. (2001). A tuning machine for cooperative problem solving. *Fundamenta Informatica, 21*, 1001–1025.

Emery, N. J. (2000). The eyes have it: the neuroethology, function and evolution of social gaze. *Neuroscience and Biobehavioral Reviews, 24*, 581–604.

Farroni, T., Johnson, M. H., &Csibra, G. (2004). Mechanisms of eye gaze perception during infancy. *Journal of Cognitive Neuroscience, 16*(8), 1320–1326.

Fehr, E. (2004). Don't lose your reputation. *Nature, 432*, 449–450.

Fehr, E., & Gächter, S. (2002). Altruistic punishment in humans. *Nature, 415*, 137–140.

Flavell, J. H., Flavell, E. R., Green, F. L., & Moses, J. L. (1990). Young children's understanding of fact beliefs versus value beliefs. *Child Development, 61*(4), 915–928.

Franco, F. (2005). Infant pointing. In N. Eilan, C. Hoerl, T. McCormack, & J. Roessler (Eds.), *Joint attention, communication, and other minds* (pp. 129–164). Oxford: Oxford University Press.

Gamble, C., & Steele, J. (1999). Hominid ranging patterns and dietary strategies. In H. Ullrich (Ed.), *Hominid Evolution: Lifestyles and Survival Strategies* (pp. 396–409). Leipzig: Edition Archaea.

Gärdenfors, P. (1996). Cued and detached representations in animal cognition, *Behavioural Processes, 36*, 263–273.

Gärdenfors, P. (2000). *Conceptual spaces: The geometry of thought*. Cambridge, MA: MIT Press.

Gärdenfors, P. (2003). *How homo became sapiens: On the evolution of thinking*. Oxford: Oxford University Press.

Gärdenfors, P. (2004). Cooperation and the evolution of symbolic communication. In K. Oller & U. Griebel, (Eds.), *The evolution of communication systems* (pp. 237–256). Cambridge, MA: MIT Press.

Gärdenfors, P. (2008). The role of intersubjectivity in animal and human cooperation. *Biological Theory, 3*(1), 1–12.

Gärdenfors, P., & Osvath M (2010). The evolution of anticipatory cognition as a precursor to symbolic communication. In R. K. Larson, V. Déprez, & H. Yamokido (Eds.), *The evolution of human language: Biolinguistic perspectives* (pp. 103–114). Cambridge: Cambridge University Press.

Garnham, W. A., & Ruffman, T. (2001). Doesn't see, doesn't know: Is anticipatory looking really related to understanding or belief. *Developmental Science, 4*, 94–100.

Gergely, G., & Csibra, G. (2003). Teleological reasoning in infancy: The naive theory of rational action. *Trends in Cognitive Sciences, 7*, 287–292.

Gilbert D. T., & Wilson, T. D. (2007). Prospection: Experiencing the future. *Science, 317*, 1351–1354.

Glenberg, A. (1997). What memory is for. *Behavioral and Brain Sciences, 20*(1), 1–55.

Gomez, J. C. (2004). *Apes, monkeys, children, and the growth of mind*. Cambridge, MA: Harvard University Press.

Gomez, J. C. (2007). Pointing behaviors in apes and human infants: A balanced interpretation. *Child Development, 78*(3), 729–734.

Gopnik, A., & Astington, J. W. (1988). Children's understanding of representational change, and its relation to the understanding of false belief and the appearance-reality distinction. *Child Development, 59*, 26–37.

Gulz, A. (1991). *The planning of action as a cognitive and biological phenomenon*. Lund: Lund University Cognitive Studies 2.

Haidle, M. (2009). How to think a simple spear? In S. de Beaune, F. Coolidge, & T. Wynn (Eds.), *Cognitive archaeology and human evolution* (pp. 57–75). Cambridge: Cambridge University Press.

Hare, B., Call, J., Agnetta, B., & Tomasello, M. (2000). Chimpanzees know what conspecifics do and do not see, *Animal Behaviour, 59*, 771–85.

Hare, B., Call, J., & Tomasello, M. (2001). Do chimpanzees know what conspecifics know? *Animal Behaviour, 61*, 139–151.

Hauser, M. D., Chen, M. K., Chen, F., & Chuang, E. (2003). Give unto others: Genetically unrelated cotton-top tamarin monkeys preferentially give food to those who altruistically give food back. *Proceedings of the Royal Society of London, B 270*, 2363–2370.

Hawkes, K. J., O'Connell, F., Blurton, N. G., Alvarez, J. H., & Charnov E. L. (1998). Grandmothering, menopause, and the evolution of human life histories. *PNAS, 95*, 1336–1339.

Jackendoff, R. (1999). Possible stages in the evolution of the language capacity. *Trends in Cognitive Sciences, 3*(7), 272–279.

Jensen, K., Call, J., & Tomasello, M. (2007). Chimpanzees are rational maximizers in an ultimatum game. *Science, 318*, 107–109.

Johansson, S., Zlatev, J., & Gärdenfors, P. (2006). Why don't chimps talk and humans sing like canaries? *Behavioral and Brain Sciences, 29*(3), 287–288.

Kacelnik, A. (2003). The evolutioin of patience. In G. Lowenstein, D. Read, & R. F. Baumeister (Eds.), *Time and decsioin: Economic and psychological perspectives on intertemporal choice* (pp. 115–138). New York: Russell Sage Foundation.

Kaminski, J., Riedel, J. Call, J., & Tomasello, M. (2005). Domestic goats, *Capra hircus*, follow gaze direction and use social cues in an object task. *Animal Behaviour, 69*, 11–18.

Köhler, W. (1921). Zur psychologie des schimpansen. *Psychologische Forschung, 1*, 2–46.

Kuhn, S. L., & Stiner, M. C. (2008). Paleolithic ornaments: Implications for cognition, demography and identity. *Diogenes, 214*, 40–48.

Leavens, D. A., Hopkins, W. D., & Bard, K. A. (2005). Understanding the point of chimpanzee pointing: Epigenesis and ecological validity. *Current Directions in Psychological Science, 14*, 185–189.

Leavens, D. A., Hopkins, W. D., & Bard, K. A. (2008). The heterochronic origins of explicit reference. In J. Zlatev, T. Racine, C. Sinha, & E. Itkonen (Eds.), *The shared mind: Perspectives on intersubjectivity* (pp. 187–214). Amsterdam: John Benjamins.

Lehmann, L., & Keller, L. (2006). The evolution of cooperation and altruism – A general framework and a classification of models. *Journal of Evolutionary Biology, 19*, 1365–1376.

Leimar, O., & Hammerstein, P. (2001). Evolution of cooperation through indirect reciprocation. *Proceedings of the Royal Society London, B 268*, 745–753.

Leonard W., & Robertson, M. L. (2000). Ecological correlates for home range variation in primates: Implications for human evolution. In S. Boinski & P. A. Garber (Eds.), *On the move: How and why animals travel in groups* (pp. 628–648). Chicago, IL: University of Chicago Press.

Lewis, D. K. (1969). *Convention*. Boston: Harvard University Press.

Lindgren, K. (1997). Evolutionary dynamics in game-theoretic models. In B. Arthur, S. Durlauf, & D. Lane (Eds.), *The economy as an evolving complex system II* (pp. 337–367). Reading, MA: Addison-Wesley.

Maynard Smith, J. (1982). *Evolution and the theory of games*. Cambridge: Cambridge University Press.

Mulcahy, N. J., & Call, J. (2006). Apes save tools for future use. *Science, 312*, 1038–1040.

Naqshbandi, M., & Roberts W. A. (2006). Anticipation of future events in squirrel monkeys (*Saimiri sciureus*) and rats (*Rattus norvegicus*): Test of the Bischof-Kohler Hypothesis. *Journal of Comparative Psychology, 120*, 345–357.

Noble, W., & Davidson, I. (1996). *Human evolution, language and mind: A psychological and archaeological inquiry*. Melbourne: Cambridge University Press.

Nowak, M. A., & Sigmund, K. (2005). Evolution of indirect reciprocity. *Nature, 437*, 1291–1298.

O'Connell, J. F., Hawkes, K., & Blurton Jones, N. G. (1999). Grandmothering and the evolution of *Homo erectus*. *Journal of Human Evolution, 36*, 461–485.

Osvath, M. (2009). Spontaneous planning for future stone throwing by a male chimpanzee. *Current Biology, 19*, 190–191

Osvath, M., & Gärdenfors, P. (2005). *Oldowan culture and the evolution of prospective cognition.* Lund: Lund University Cognitive Studies, 122.

Osvath, M., & Osvath, H. (2008). Chimpanzee (*Pan troglodytes*) and orangutan (*Pongo abelii*) forethought: self-control and pre-experience in the face of future tool use. *Animal Cognition, 11*, 661–674.

Panchanathan, K., & Boyd, R. (2004). Indirect reciprocity can stabilize cooperation without the second-order free rider problem. *Nature, 432*, 499–502.

Perner, J., Leekam, S., & Wimmer, H. (1987). Three-year-old's difficulty with false belief: The case for a conceptual deficit. *British Journal of Developmental Psychology, 5*, 125–137.

Pika, S., Liebal, K., & Tomasello, M. (2003). Gestural communication in young gorillas (*Gorilla gorilla*): Gestural repertoire and use. *American Journal of Primatology, 60*(3), 95–111.

Plummer, T. (2004). Flaked stones and old bones: Biological and cultural evolution at the dawn of technology. *Yearbook of Physical Anthropology, 47*, 118–164.

Povinelli, D. J., & Eddy, T. J. (1996). What young chimpanzees know about seeing. *Monographs of the Society for Research in Child Development, 61*(2), Serial No. 247.

Premack, D., & Woodruff, G. (1978). Does the chimpanzee have a theory of mind? *Behavioral and Brain Sciences, 4*, 515–526.

Preston, S. D., & de Waal, F. (2003). Empathy: Its ultimate and proximal bases. *Behavioral and Brain Sciences, 25*, 1–72.

Raby, C. R., Alexis, D. M., Dickinson, A., & Clayton, N. S. (2007). Planning for the future by western scrub-jays. *Nature, 445*, 919–921.

Reddy, V. (2003). On being the object of attention: Implications for self-other consciousness. *Trends in Cognitive Sciences. 7*(9), 397–402.

Repacholi, B., & Gopnik, A. (1997). Early understanding of desires: Evidence from 14 and 18 month olds. *Developmental Psychology, 33*, 12–21.

Richerson, P. J., Boyd, R. T., & Heinrich, J. (2003). Cultural evolution of human cooperation. In P. Hammerstein (Ed.), *The Genetic and Cultural Origins of Cooperation* (pp. 357–388). Cambridge, MA: MIT Press.

Roberts, W. A. (2002). Are animals stuck in time? *Psychological Bulletin, 128*, 473–489.

Roberts, W. A. (2006). The questions of temporal and spatial displacement in animal cognition. In E. A. Wasserman & T. R. Zentall (Eds.), *Comparative cognition: Experimental explorations of animal intelligence* (pp. 145–163). New York: Oxford University Press.

Rosati, A. G., Stevens, J. R., Hare, B., & Hauser, M. D. (2007). The evolutionary origins of human patience: Temporal preferences in chimpanzees, bonobos, and human adults. *Current Biology, 17*, 1663–1668.

Searle, J. (1995). *The construction of social reality*. New York: Free Press.

Semmann, D., Krambeck, H.-J., & Milinski, M. (2005). Reputation is valuable within and outside one's own social group. *Behavioral Ecology and Sociobiology, 57*, 611–616.

Shennan, S. (2001). Demography and cultural innovation: A model and its implications for the emergence of modern human culture. *Cambridge Archaeological Journal, 11*(1), 5–16.

Sigmund, K., Hauert, C., & Nowak, M. (2001). Reward and punishment. *PNAS, 98*, 10757–10762.

Silk, J. B., Brosnan, S. F., Vonk, J., Henrich, J., Povinelli, D. J., Richardson, S. J., et al. (2005). Chimpanzees are indifferent to the welfare of unrelated group members. *Nature, 437*, 1357–1359.

Sterelny, K. (2003). *Thought in a hostile world.* Oxford: Blackwell.

Stern, D. N. (1985). *The interpersonal world of the infant: A view from psychoanalysis and developmental psychology*. New York: Basic Books.

Stephens, W., McLinn, C. M., & Stevens, J. R. (2002). Discounting and reciprocity in an iterated prisoner's dilemma. *Science, 298*, 2216–2218.

Stevens, J. R., & Hauser, M. (2004). Why be nice? Psychological constraints on the evolution of cooperation. *Trends in Cognitive Science, 8*, 60–65.

Striano, T., & Bertin, E. (2005). Social-cognitive skills between 5 and 10 months of age. *British Journal of Developmental Psychology, 23*, 1–11.

Suddendorf, T., & Busby, J. (2005). Making decisions with the future in mind. *Learning and Motivation, 36*, 110–125.

Suddendorf, T., & Corballis, M. C. (1997). Mental time travel and the evolution of human mind. *Genetic, Social and General Psychology Monographs, 123*, 133–167.

Suddendorf, T., & Corballis, M. C. (2007). The evolution of foresight: What is mental time travel and is it unique to humans? *Behavioral and Brain Sciences, 30*, 299–351.

Suddendorf, T., & Whiten, A. (2001). Mental evolution an development: Evidence for secondary representation in children, great apes, and other animals. *Psychological Bulletin, 127*(5), 629–650.

Thieme, H. (1997). Lower Paleolithic hunting spears from Germany. *Nature, 385*, 807–810.

Thieme, H. (1999) Altpaläolithische Holzgeräte aus Schöningen, Lkr. Helmstadt. *Germania, 77*, 451–487.

Tomasello, M. (1999). *The cultural origins of human cognition*. Cambridge, MA: Harvard University Press.

Tomasello, M., & Call, J. (2006). Do chimpanzees know what others see – or only what they are looking at? In S. Hurley & M. Nudds (Eds.), *Rational animals* (pp. 371–384). Oxford: Oxford University Press.

Tomasello, M., Carpenter, M., Call, J., Behne, T., & Moll, H. (2005). Understanding and sharing intentions: The origins of cultural cognition. *Behavioral and Brain Sciences, 28*, 675–735.

Toth, N. (1985). The Oldowan reassessed: A close look at early stone artifacts. *Journal of Archeological Science, 12*, 101–120.

Trevarthen, C., & Hubley, P. (1978). Secondary intersubjectivity: Confidence, confiding, and acts of meaning in the first year. In A. Lock (Ed.), *Action, gesture, and symbol: The emergence of language* (pp. 183–229). New York: Academic.

Trivers, R. L. (1971). The evolution of reciprocal altruism. *Quarterly Review of Biology, 46*, 35–57.

Tulving, E. (1972). Episodic and semantic memory. In E. Tulving & W. Donaldson (Eds.), *Organization of memory* (pp. 381–403). New York: New York Academic Press.

Tulving, E. (2005). Episodic memory and autonoesis: Uniquely human? In H. Terrace & J. Metcalfe (Eds.), *The missing link in cognition: Evolution of self-knowing consciousness* (pp. 3–56). New York: Oxford University Press.

Van Schaik. C. P., & Kappeler, P. M. (2006). Cooperation in primates and humans: closing the gap. In C. P. van Schaik & P. M. Kappeler (Eds.), *Cooperation in primates and humans* (pp. 85–105). Berlin: Springer.

Villa, P. (1983). *Terra amata and the middle pleistocene archaeological record of Southern France*. Berkeley, CA: University of California Press.

Warneken, F., Hare, B., Melis, A. P., Hanus, D., & Tomasello, M. (2007). Spontaneous altruism by chimpanzees and young children. *PloS Biology, 5*, 1414–1420.

Warneken, F., & Tomasello, M. (2006). Altruistic helping in human infants and young chimpanzees. *Science, 311*, 1301–1303.

Wellman, H. M., & Liu, D (2004). Scaling of theory-of-mind tasks. *Child Development, 75*, 523–541.

Whiten, A. (1999). The evolution of deep social mind in humans. In M. C. Corballis & S. E. G. Lea (Eds.), *The descent of mind: Psychological perspectives on hominid evolution* (pp. 173–193). New York: Oxford University Press.

Wilkinson, G. S. (1984). Reciprocal food sharing in the vampire bat. *Nature, 308*, 181–184.

Zahavi, A. (1975). Mate selection – A selection for a handicap. *Journal of Theoretical Biology, 53*, 205–214.

Zlatev, J., Persson, T., & Gärdenfors, P. (2005). *Bodily mimesis as the "missing link" in human cognitive evolution*. Lund: Lund University Cognitive Studies, 121.

Part III
The Cognitive and Anthropological Connection

Chapter 11
Language as a Repository of Tacit Knowledge

Harry Collins

Abstract The relationship between language and practice has been badly and profoundly misunderstood. There is an intimate relationship at the collective level and the content of language is formed by the practices of a community. It is not the case, however, that an individual has to engage in all the practices of a community in order to acquire the language and the practical understanding that goes with it. Were this not the case there could be no societies: societies depend on the division of labour and any profound division of labour depends on practical understanding by those who do not themselves practice what they have to understand. Were it not the case that individuals could learn language without practicing it then the speech of the congenitally wheelchair-bound, or blind, would be noticeably limited. The chapter works through this position, contrasting it with the position developed by philosophers of practice such as Hubert Dreyfus, describes experiments which support the position, and explores the question of the extent to which this means experience can be captured by language.

11.1 Introduction: The Main Argument

Computer enthusiasts, of the group sometimes referred to as the 'artificial intelligentsia,' tried to persuade us that human abilities including fluent language speaking could be captured in programs if only those programs were complicated enough. Some of those who thought hard about the matter, such as philosopher Hubert Dreyfus (1972), argued that they could not be so captured. They argued that human knowledge was not confined to symbols but involved physical involvement with the world. In this they were right – human knowledge does *involve* physical involvement with the world. I am going to argue below that the nature of that 'involvement' was misunderstood, but let it go for the time being. For the moment we can agree that Wittgenstein was right when he said, 'if a lion could speak we would not understand him.'[1] And he was right in saying that we would not understand him because a

[1] Ludwig Wittgenstein (1953, p. 223).

H. Collins (✉)
SOCSI, Cardiff School of Social Sciences, Cardiff University, Cardiff, Wales CF10 3WT, UK
e-mail: CollinsHM@cf.ac.uk

T. Schilhab et al. (eds.), *The Symbolic Species Evolved*, Biosemiotics 6,
DOI 10.1007/978-94-007-2336-8_11, © Springer Science+Business Media B.V. 2012

lion's physical engagement with the world is so different to ours. Since our language was not built up out of experiences pertaining to the physical world of lions – such as using your teeth and claws to rip raw flesh from zebras that have just been pulled down and killed – it does not contain the knowledge of lions. A talking lion might be able to express what doing lion-like things was like in the symbols of Lionese, but Lionese would not mean anything to us that corresponded with what it means to talking lions. Hence, Hubert Dreyfus entitled one of his earliest papers on the topic 'Why Computers Must Have Bodies in Order to Be Intelligent.'[2] He called it that because he could see, quite correctly, that our knowledge had to do with the way we used our bodies. The same applies in a slightly different way to groups of people who live in different social settings: our language cannot capture what it means to divine a witch using the Azande poison oracle and the language of the Azande cannot capture what it means to take out a mortgage because, though we have similar bodies, we use them in very different ways.[3]

Unfortunately, the inference was inexact. Nobody noticed the problem nor, at the time, was there any reason to notice the problem, because Dreyfus was fighting a good war with enormous courage in spite of the contumely being heaped upon his head and it was vital that he win it. Nevertheless, 'war-damage' was being inflicted and now that the war is over we can see that it was pretty serious. The damage is that language and symbols have been equated; in throwing out the strings of digital symbols used by computers as a potential repository of human knowledge, language, which, on the face of it, is composed of symbols rather than physical activity, was also thrown out. The result is that the possibility that language could be a repository of human knowledge has been excluded from consideration in a doctrinaire way. The argument is that if understanding Lionese involves zebra-ripping then anyone or anything who/which understands the language must have ripped zebras; the language, it was said, cannot itself capture what it is to rip zebras, only the language plus the practice can contain it.

The fallacy is this: It is true that there could be no Lionese as we know it without zebra-ripping among lions at the collective level. This is what I have elsewhere called the 'social embodiment thesis' – the collectivity of lions develops its specific language as a result of the typical form of lions' bodies and the typical things they do with them. Not every individual who speaks Lionese with fluency has, however, to be a zebra-ripper. Lionese is the collective property of lions and their 'form-of-life' but that does not stop non-lions (or individual non-zebra-ripping lions if such there are), from acquiring it if they put in enough effort. The 'minimal embodiment thesis' claims that not every member of the collectivity has to have a lion-like body or engage in lion like activities to become fluent in the language of the collectivity

[2] Dreyfus (1965).

[3] Another incarnation of the argument is to do with the incommensurability of Kuhnian 'paradigms' in science. (Kuhn, 1962).

and certain individuals can, therefore, get by with a minimal body.[4] This is possible because a language, like Lionese, itself contains some or all of the knowledge and understanding of the practical activities that went to build it up in the collectivity of active lions, so that to acquire the language with fluency is to acquire some or all of that knowledge and understanding. To put this another way, the language of Lionese, when fluently exercised, does contain much or all of the tacit knowledge of what it is to be a zebra-ripper.

To acquire a language with the degree of fluency needed to capture the meaning that goes with it is not a trivial accomplishment. Such fluency is, as a matter of fact, nearly always acquired along with the carrying out of the practices pertaining to the language and that is another reason why it has not been noticed that it can be acquired without engaging in the practices – it is a rare thing. It is also why it remains the case that if lions could speak most of us would not understand them. Nevertheless, one or two of us, who had put in the effort that it takes to become fluent in Lionese from immersion in the spoken discourse alone, could understand lions in the absence of claws and ripping-teeth. It may be the case that strings of digital symbols cannot capture zebra-ripping but it is not the case that *language* cannot capture zebra-ripping because *language is more than strings of digital symbols*.[5]

What follows is that, *contra* Dreyfus, it has not been proved that a computer must have a body to be intelligent. The argument so far, if it is correct, shows that there is logical space for a computer to learn a practice-based language without practice because there is logical space for humans to learn a practice-based language without practice. This does not prove that computers *can* acquire such a language – I believe that they cannot while I will also argue that humans can. The reason I think computers (as we know them) cannot acquire language, is that they merely manipulate digital symbols not that they do not have bodies. It could be argued that the reason they merely manipulate symbols rather than engage in linguistic discourse is because they don't have bodies but if individual humans can acquire language without bodies, or without much in the way of bodies, then it seems odd to impose the need for bodies on computers.

That a human does not need a body, or much of a body, to speak fluently, and therefore capture the 'tacit knowledge' of a community whose physical practices have not been shared, should be obvious. Everyone in a wheel chair can do it![6] The reason it has not been obvious, and the reason it has been such hard work to establish the point, was, I believe, and as I have intimated, to do with the role of symbols in the artificial intelligence debate. It was felt that to allow fluency relating to bodily experience to entities without proper bodies would allow fluency, and hence human knowledge, to computers. Thus Dreyfus, to be consistent, also has to argue that

[4] See Collins (2004a) for the 'social embodiment thesis' and its counterpart, the 'minimal embodiment thesis'.

[5] For the difference between symbol manipulation and language see the 'The Transformation-Translation Distinction' in Collins (2010, p. 25).

[6] See, for example, Collins (2004b).

sports coaches or commentators who have not played a game to a high level cannot coach or comment properly and that non-surgeons cannot talk sensibly about the kind of cuts a surgeon makes:

> You may have mastered the way surgeons talk to each other but you don't understand surgery unless you can tell thousands of different cuts from each other and judge which is appropriate. In the domain of surgery no matter how well we can pass the word along we are just dumb. So is the sportscaster who can't tell a strike from a ball until the umpire has announced it.[7]

Arguers like Dreyfus have to say such things because they have to say that even humans who have not been engaged in the physical activities they try to talk about cannot talk about them fluently.

But the barrier was being put in the wrong place. To win the good war against the 'intelligence' of computers one had to argue, not that they must have bodies, but that digital symbols located in an entity that was not embedded in social life were not sufficient for it to become and remain fluent in language.[8] The crucial point is that even a fully embodied human being that is isolated from society cannot become or remain fluent in the language. The example of feral children exhibits the first part of the prohibition, the growing apart of physically separated cultures exhibits the second part.[9]

The logic of the two positions is set out in cartoon form in Fig. 11.1 which has language speakers in the left half and entities that fail to speak language in the right. The Dreyfus position (the top boxes) concentrates on individuals: it claims that only those who practice (represented by hammer and anvil) can speak 'Hammer-and-Anvil Language' (HAL) fluently – case 'A'. Anyone who is isolated from hammer and anvil – case 'B' – (dotted boxes indicate isolation) cannot be fluent in HAL and is bound to speak a different language at best. Since computers – case 'C' – cannot engage in any physical activity they cannot be fluent in any language.

The position being argued for here is that a language like HAL can only develop if there is a community of hammerers on anvils but that the language, once developed, has, to some extent, 'a life of its own.'[10] An individual who immerses themselves in that language – case 'D' – can acquire it and speak it fluently even though they do not do any hammering themselves. On the other hand, those isolated from the HAL community – case 'E' – (dotted boxes indicate isolation once more) will not be able to speak HAL even if they do use hammers and anvils. Furthermore,

---

[7] Selinger, Dreyfus, and Collins (2007 at p. 737).

[8] It was unfortunate that one of the most salient arguments against the potency of computers, John Searle's 'Chinese Room' argument, starts by positing the existence of a completely fluent computer. Actually, the Chinese Room would not work without it being embedded in society through the medium of a human being (Collins, 1990, 2010).

[9] See the quotation from H G Well's 'Country of the Blind' below.

[10] I am sure Dreyfus too would believe that a community of hammerers is necessary to develop the HAL language but he does not work out the consequences, concentrating on the practices of individuals.

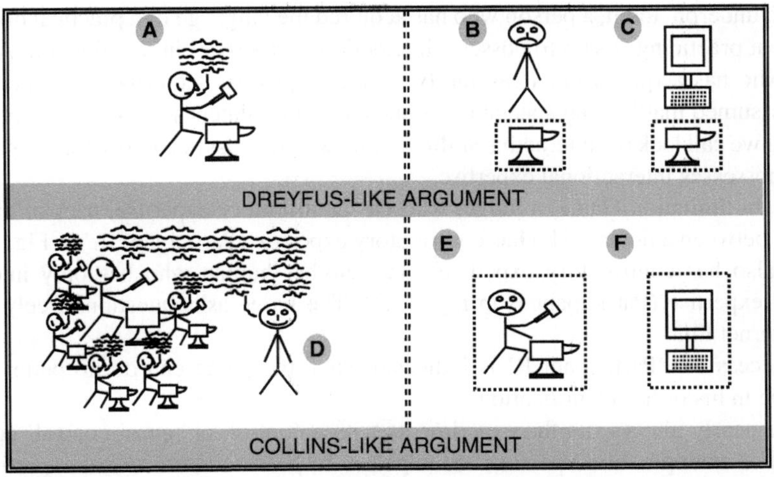

Fig. 11.1 The language barrier: Dreyfus-like (*upper*) and as argued here (*lower*)

humans physically isolated from all societies cannot speak any language with fluency and computers – case 'F' – always being isolated from society even when physically located within it, cannot speak any language with fluency either. But the problem is the isolation, not the inability to practice.[11]

In sum, a language is formed in intimate relationship with the practices of the speaking community – whether zebra-ripping or hammering – but that does not mean that each individual who speaks the language has to engage in those practices; they can learn the language from immersion in the discourse alone as in case 'D'. At the same time, engaging in the practice without immersion in the discourse, as in case 'E', does not enable the discourse. The rest of this chapter is a commentary on this argument and on the nature of language [12].

11.2 Experiments on Interactional Expertise

Lest the example of the disabled is insufficient on its own, we have conducted a series of experiments designed to show that it is possible to become fluent in the language of a practice-based domain without practicing (e.g., Collins, Evans, Ribeiro, & Hall, 2006). These experiments involve the 'Imitation Game.' The Imitation Game is similar to the 'Turing Test' but the entity being tested is a human not a computer. In these experiments, and the theoretical developments

[11] I am sure Dreyfus would agree that isolated individuals and isolated computers could not speak. Again, however, it is a matter of what you concentrate on. He concentrates on their activity rather than their isolation.

[12] See also Collins (2011) for development of the point.

which underpin them, a person who has acquired the language of a practical domain without practicing is said to possess 'interactional expertise' in that domain. A person who has acquired practical ability is said to possess 'contributory expertise.' It is assumed that barring special cases such as inarticulateness or social isolation – which we call lack of 'interactive ability' – one who possesses contributory expertise also possesses interactional expertise.

In the Imitation Game, a judge, who has contributory expertise, tries to distinguish between a human who has contributory expertise (Participant 'A' in Fig. 11.2, who also has contributory expertise), and another human who has only interactional expertise (Participant B in Fig. 11.2). The judge asks questions freely over an internet link.

Success for 'Participant B' is indicated when the judge can do no better than chance in his or her identifications.

We nearly always run the experiments with a control, or 'quasi-control' condition. For example, blind persons are expected to possess interactional expertise in sighted-persons' discourse because they have been immersed in the language of the sighted all their lives. Running the experiment with a blind person pretending to be a sighted person is called the 'chance condition' because we expect the judge will do not much better than chance when he/she tries to identify the participants. The quasi-control or, 'identify condition,' has the reverse polarity. In the case of the blind we ask sighted persons to try to imitate the blind using a blind judge. Here we expect the judge to be able to identify the participants because the sighted have not been immersed in the discourse related to the practices of the blind.

The experiments that we did with the blind showed that the balance of correct guesses over failures to identify in the chance condition was 13% whereas in the identify condition it was 86%. Exact chance in the chance condition should not be expected because there are many ways to catch the pretending person out: they have to lie convincingly as well as demonstrate their possession of sighted discourse.

The experiment has also been run successfully with both polarities on the colour-blind, those with perfect pitch, a sociologist long immersed in the language of gravitational-wave physics pretending to be a gravitational-wave physicist, 'gays' pretending to be 'straights' and active Christians pretended to be non-Christians.[13]

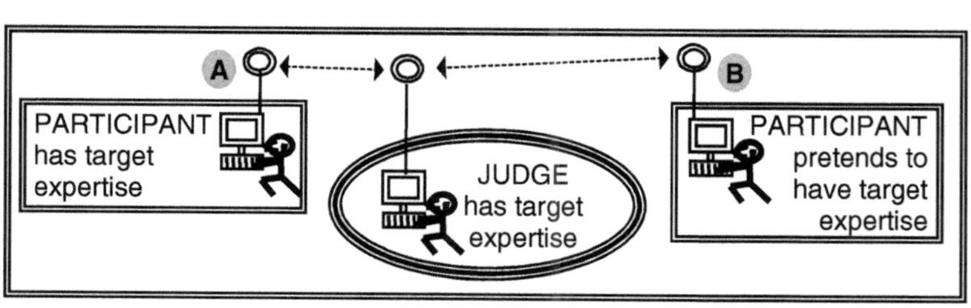

Fig. 11.2 The imitation game

[13] Latest results can be found at www.cf.ac.uk/socsi/expertise

Readers should think about which they expect to be the chance condition and which the identify condition in each case. From here on I will take it that the existence of interactional expertise – that which is represented on the left hand side of the lower box in Fig. 11.1, has been established.[14] It now remains to explore the notion in more detail.

11.3 Domain Specific Languages and Language in General

The idea of interactional expertise contradicts the idea that each individual must have a body, that is capable of doing the practical activities corresponding to the concepts of the language, if linguistic fluency is to be attained. This claim has been disputed.[15] The confrontation between Dreyfus and Collins on this matter is clear in the case of domain specific languages. Thus Dreyfus insists that one cannot speak fluently about the cuts used in surgery unless one is a surgeon, one cannot commentate expertly on a sport unless one has played it, one cannot comment on chess unless one is an expert chess-player, and so forth. Collins, on the other hand, insists that all this can be accomplished in the absence of the corresponding practical experience so long as the immersion in the corresponding discourse is sufficiently profound; this explains why, according to Collins and *contra*-Dreyfus, non sports players can be excellent coaches and commentators and why the disabled can be fluent in the discourse of those who are not disabled. Furthermore, the only way smooth division of labour in complex areas, such as large technological projects, can be understood is via the mediation of interactional expertise: in such circumstances we have to be able to understand each other's jobs if we are to work smoothly in a team even if we cannot execute those jobs. Collins believes that the arguments and experiments strongly favour his viewpoint on this disagreement.[16]

When it comes to the ability to acquire language as a whole, as opposed to domain specific languages, the argument is more nuanced. Collins has to accept that languages cannot develop in the absence of certain bodily features – lungs, larynx, ears, certain brain developments, and so forth. That there are no speaking lions almost certainly has to do with the fact that they do not have these bodily features. Thus, human-like bodies are essential to the development of languages – and that is why only humans possess them in the first place – the social embodiment thesis. To save any possible confusion, this does indeed mean that no collectivity of computers could acquire a language of its own – 'Computerise', as it were. And it would not be able to do this for the reasons that Dreyfus provides – computers are not suitably

[14] For a 'philosophical' discussion of the notion see Collins et al. (2006) or Collins and Evans (2007).

[15] For example, Selinger et al. (2007), Collins (2008).

[16] For a recent analysis of the relationship between 'contributory expertise' and 'interactional expertise' see Collins (2011).

embodied. It does not mean, however, that, simply in virtue of the absence of a body, a single computer could not acquire language from the surrounding human society.

In humans, the requirement for a body also applies only at the species (collective) level and not to individuals. Thus, the congenitally profoundly deaf, who do not normally acquire fluency in the native language of the hearing, can acquire it if given intense and special help from an early age – such as learning to recognise sound through vibrations transmitted through a balloon. It seems reasonable to suppose that those born without the ability to form words could also learn to do so if given prostheses and specialised help. The most natural hypothesis appears to be that the body required by an *individual* who is to learn a native language is minimal: on the basis of this argument, it looks as though something close to a 'brain in a vat' could learn to speak if given the right prostheses to allow deep immersion in the spoken discourse.[17] Others have argued that it would be impossible to learn human-like languages without the senses of back and front, forward and backward, possession and loss, and so forth that come with a body.[18] As soon, however, as one begins to strip away the need for practice, and the corresponding body parts that correspond with understanding in every domain, there seems no obvious point at which the stripping-away has to stop. Therefore, my claim would be that though a community of brains-in-vats, would, like lions, and computers, not have the right kind of bodies to acquire any language as a species, individual brains-in-vats could acquire the ordinary human language of a fully embodied human linguistic community. It seems to me that the onus is now on those who want to argue for the need for a body in every individual who/which is to acquire language to show where and why the stripping away of bodily features has to stop.

11.4 Can Interactional Expertise Give Rise to a Self-Supporting Language

It seems probable that 'interactional' languages are parasitical. Let us define the language pertaining to a domain of practice as an 'interactional language.' So, the language of gravitational wave physics can be called 'Interactional Gravitational Wave language' or 'IGW.' For the sake of exposition we will also define CGW, which is the language spoken by those with contributory expertise in gravitational wave physics. According to the interactional expertise hypothesis, IGW and CGW are one and the same thing, so long as those who speak IGW are in continuing contact with those who speak CGW! Now imagine that enough people to support a

[17] A 'brain-in-a-vat' is here assumed to be different to a computer. A computer, (here by definition) is a digital symbol manipulator or *transformer*. It is not immersed in language, only in symbols, irrespective of its physical connections to the rest of society. The brain-in-the-vat, we assume, has some mysterious properties that allow it to be immersed in meaningful language, as opposed to symbols, if the right prostheses are added. It should be thought of as a human from whom more and more bodily parts have been stripped away.

[18] Selinger et al. (2007).

living natural language, each of whom was a speaker of IGW/CGW, went to live on a distant island where there was no contact with gravitational wave physics. Once they landed on the Island they would have to cease doing gravitational wave physics so we will refer to them as speakers of IGW. It seems probable that the IGW they spoke would soon cease to bear much resemblance to the language of those doing, or interacting with gravitational wave physicists who were left behind. This is for two reasons: the 'host language' – CGW – would be changing all the time in response to changes in the technology whereas IGW as spoken on the island would not be so affected – it would cease to be IGW and become 'Island-IGW'; at the same time Island-IGW would change as it responded to the new demands of Island society and practices. Island-IGW would come to look like the Pacific Islanders' cargo-cult language – including talk of 'cargo' and its correlates continuing long after the planes had left but bearing less and less resemblance to the living contributory language associated with the domain of air-freight. Island-IGW and IGW/CGW would pull apart. This does not mean that Island-IGW would cease to be a language *sui-generis* it would just mean that it would no longer be IGW because it would have less and less to do with gravitational waves – it would just be a new language. For IGW and CGW to remain identical, those who only speak but do not practice have to remain in contact with the lived activity of the community of practitioners.

H.G. Wells makes the point in his short story: 'The Country of the Blind.' He imagined a people who were completely blind and were isolated in a hidden valley.

> For fourteen generations these people had been blind and cut off from all the seeing world; the names for all the things of sight had faded and changed; the story of the outer world was faded and changed to a child's story ...[19]

11.5 Does Interactional Expertise Carry Experiences?

What we call the 'Strong Interactional Hypothesis' is, to use the terminology of the last section, that IGW and CGW, and their equivalents in every other domain of practice, are identical. The hypothesis is that those with maximal interactional expertise are indistinguishable from those with contributory expertise in any test involving language alone.

This implies that the judgements made by such persons would also be identical. This is the implicit, but generally unnoticed, assumption behind all assessment committees where large issues are at stake and the committee members cannot have the skills associated with each and every facet of what has to be judged: it is the assumption behind what probably counts as good management.[20] In the

[19] H. G. Wells (1911) 'The Country of the Blind.' The quote can be found on page 474 of the Odhams collected edition of Wells's works. One might illustrate the point further with the metaphor of the immune system: however well a child is prepared for the biological environment via the antibodies in its mother's milk, isolate it from dirt and its immune system will start to fail – it will no longer be ready for interaction with the changing world of infective agents.

[20] See Collins and Evans (2007), and Collins and Sanders (2007), for more on this.

Imitation Game judges will spend a lot of time asking participants to make technical judgements as the ability to make sensible judgement is a good test of expertise.

Even if the Strong Interactional Hypothesis is true, having interactional expertise is not necessarily going to carry contributory expertise with it. It seems likely that no amount of talking to bike-riders is going to enable one to get on a bike and ride if one has never ridden before. So we know that what happens in the body of an interactional expert is not the same as what happens in the body of a contributory expert (and this is the essential point behind the experiments on the blind and similar). But *how does it feel* to be an interactional expert as opposed to a contributory expert?

There is some evidence from sports psychology that 'visualising' an activity can help in the performance. This might be explained by the fact that some of the neurons that 'fire' when an activity is visualised – the mirror neurons – are the same as some of those that fire when an activity is practised.[21] It is tempting to say, then, that the experience is the same.

It is hard to know how one might test for identity of experience given that, if the Strong Interactional Hypothesis is true, systematically different way of experiencing a domain will not be revealed in differences in the words that are spoken about it or judgements made within it. For example, we would not expect that the invented accounts of First World War experiences written by Sebastian Faulks, who did not serve in the trenches, would be obviously different to the accounts of Robert Graves, who did. According the Strong Interactional Hypothesis, even Graves would not be able to tell that Faulks had not served if Faulks had done the best job possible in terms of gaining interactional expertise in respect of trench life by immersing himself in the discourse of the soldiers who served in the war.[22]

It is true, of course, that, other things being equal, failures to acquire interactional expertise would indicate failures to share experience with contributory experts. Schilhab believes that preliminary Imitation Game tests she has run with midwives and mothers show that the experience of giving birth are not sufficiently well embedded in the language to allow non-mother midwives to acquire the interactional expertise of mothers. If non-mothers cannot acquire the interactional expertise we can be sure they cannot have imaginatively reconstructed the experiences. Again, do the masturbatory fantasies of a virgin – which many of us could try to recollect – match the actual first experience of sex? Probably not – in many cases because the experience is so disappointing, in other cases because it is so unlike what was expected, and in still other cases because the experience was so elevating.

Suppose Schilhab's results are right and suppose also that virgins cannot imaginatively reconstruct sexual experience. These are not decisive proofs that interactional expertise cannot carry experience. Perhaps the problem is that the conditions have not been right for the acquisition of the interactional expertise. Thus my

---

[21] For a discussion in the context of interactional expertise see Schilhab (2007).

[22] Whether this could be achieved as a matter of fact rather than principle is not so clear since serving soldiers' discourse would begin to change as soon as the war ended.

own experience suggests that people do not spend a lot of time in deep conversation about the moment-by-moment physical experience of sex so there is no reason to suspect that one is in a position to gain interactional expertise in respect of sex when one is a virgin, or indeed at any time. The same might be true of Schilhab's midwives. Do women spend hours in deep conversation about the moment-by-moment experience of childbirth? Interactional expertise is not easy to acquire (see below); it may be that a midwife who has attended multiple births has not been much engaged in detailed and skilfully descriptive interchange about moment-to-moment sensation; the birth of a child might be too traumatic for all but the most phenomenological reflexive mothers to engage in such an accounting.

In sum, to know the extent to which interactional expertise carries experience it is necessary to reflect about those experiences prior to which one has good reason to think one already has excellent interactional expertise. It is quite hard to know that one has good interactional expertise, however, as the examples of sex and birth-giving indicate. One might think one knows so much about a topic that one has interactional expertise in it but further reflection might show that this is far from clear. Consider, for example, the case of war. Soldiers say that no-one who has experienced war would ever go to war again. This is in-spite of the fact that huge amounts have been written and spoken about war. If it is true that those without war experience are much more ready to go to war – as is often said about politicians – then it goes against the drift of my argument and it goes against the Strong Interactional Hypothesis. But do those who embrace war more willingly than the experienced soldier really have good interactional expertise? Soldiers are normally portrayed as finding it too painful to talk of the most horrible details of battle while pro-war sentiment is deliberately spread before a war so it may be very hard to gain the real thing because the 'language of war' is subject to continual distortion. Further consideration indicates that in a case like the First World War the discourse is sufficiently strong so that no-one hungers for a repeat of that experience, not even the most bloodthirsty politicians. So the case of war is, at best, ambivalent.

My inclination is, nevertheless, to say that interactional expertise cannot carry every kind of experience. I cannot imagine that talk could ever reproduce the condition for the experience of (a) the rare intense sexual desire for an individual associated with either love or infatuation and (b) the witnessing of the birth of one's own child and the early years of parenthood. The astonishing intensity of these experiences is, in each case, a thing that I cannot imagine being imaginatively reconstructed. But this is not a proof – it may just be a failure of imagination on my part.

Crucially, whether such experiences can be reconstructed through the medium of discourse or not, the Strong Interactional Hypothesis can still be true. The Strong Interactional Hypothesis does not depend on re-experiencing practice even in the imagination. If the Strong Interactional Hypothesis is true, language alone carries a sufficiently good repository of tacit knowledge to enable judgements to be made about practical matters which have not been directly, or even imaginatively, experienced. Perhaps serious drug addiction is an example. This is an experience that I have not had though I would guess it is something like intense sexual infatuation.

But even without imagining it I think discursive experience of it would make it possible to make good judgements in respect of it.

This section has dealt with the question of whether the possession of interactional expertise carries experiences with it which are similar to those of the contributory expert. The visualisation process of athletes, along with the idea of 'mirror neurons,' indicates that some experiences can be reproduced imaginatively in the absence of practice.[23] I have suggested that I find it hard to believe that all experiences can be reproduced in this way. But proofs have not been developed, only invitations for more phenomenological introspection and more experiments on language acquisition.

11.6 How Is Interactional Expertise Acquired?

The easiest way to acquire interactional expertise is to become a contributory expert. While learning to do the things that have to be done in a domain of practice, the language will be learned. Learning to do the thing ensures the most intense immersion in the spoken discourse of the community – the person learning to practice is continually alongside fluent instructors, continually listening to the 'war stories' and hearing the myths of the community being recounted; there is no better way of becoming deeply immersed in the language. Of course, it may turn out that a contributory expert is not very good at language – they are inarticulate or lack interactive ability – but other things being equal, immersion in the practice is the best way of being immersed in the language.

One of the hardest ways to acquire interactional expertise is in the way the author of this chapter acquired the interactional expertise of gravitational wave physics. This was to enter the social domain of the target group and set up specially arranged conversational encounters. Many status barriers have to be jumped, the participants have to be persuaded to expend valuable time, and relatively little time can be spent immersed in the discourse when it is compared to the time spent by a practitioner. It is a long process, taking this author about ten years to gain what he has in the way of interactional expertise in gravitational wave physics. His interactional expertise, by the way, was enough to pass an Imitation Game test and to have his remarks about physics listened to and taken seriously by the practitioners and even, on one or two rare occasion, to win a debating point about the physics itself.[24] Once the initial barriers have been overcome, however, becoming more fluent is an enjoyable process, especially if one enjoys the company of those whose language one is trying to learn.

---

[23] Schilhab (2007).

[24] See also Collins (2004, chapter 23), where I give an account of my participation in a review committee and my feeling that I understood the technology in question better than some of the official reviewers.

These two accounts of acquiring interactional expertise at either end of the spectrum of possibilities obscure a host of subtleties to do with exactly how one mixes with the community. The possibilities have been referred to as 'levels of immersion.'[25] Thus, I claim to have acquired interactional expertise purely from talking to the scientists but actually I spent as much time as I could around the apparatus they were building, looking at it, and 'getting the feel' of it. I suspect that, as a person who likes diagrams, and likes having a visual image of things that are being spoken of, I would have been, at best, much slower at acquiring the expertise without these contacts with the physical correlates of the world of practice. Rodrigo Ribeiro, who is making a special study of levels of immersion and interactional expertise, will almost certainly find that deeper immersion results in better interactional expertise. Whether this is a result of the logistics of deeper immersion, or whether there is some barrier that cannot be surmounted without visual experience of physical settings, is going to be hard to show. My inclination is to say that the arguments and evidence gathered to date indicate that a space has to be left for the possibility of acquiring full interactional expertise in the absence of any visual immersion or physical contact – this is what the experiments on the blind and the argument about the brain-in-the-vat seem to show. But it might also be the case that for many practical purposes there might as well be a barrier between shallow immersion and deep immersion since it is so hard to get a high level of interactional expertise from shallow immersion and it is not something that anyone should choose unless forced.

11.7 Conclusion: Symbols and Language

In the introduction it was argued that language was more than a set of computer-like symbols, it was a repository of understanding. It was then argued that since fluency in the language without ability in the practices to which a language referred, still makes it possible to make good judgements about practical matters pertaining to the domain, language was a repository of tacit knowledge.

How can something that is expressed in symbols be more than the symbols? How can something expressed in the medium of explicit knowledge be a repository of tacit knowledge? The circle is squared because language is itself a practice. It is learned in communities which practice spoken discourse; discoursing is a practical activity. Thus, when one learns a natural language, one learns not only words but how to make well-formed sentences in the language in question and reject some badly-formed sentences or expressions while understanding and generating others that are not well-formed according to any formal rule, or even any rule that can be anticipated, but can be understood as though they did confirm. The rules of sentence or phrase formation are like any 'social rule' – they do not contain the rules for the own application; thus language fluency contains the tacit knowledge of sentence

[25] The term belongs to Rodrigo Ribeiro ex-Cardiff PhD student and professor at the University of Minas Gerais in Belo Horizonte, Brazil.

and phrase formation. In the same way learning the language of a practical domain contains the tacit knowledge needed to make judgements including judgements that would be impossible to anticipate. Perhaps it is contained in the exact nuance or rhythm of words or phrases, perhaps it is in the sentence pattern of paragraphs of the emphases within them – I do not know. Again, the rules of these judgements would not contain the rules of their own application and so knowing how to apply them would be a matter of tacit knowledge. One way or another, it does seem that to become good at the practice of spoken discourse is to come to know how to understand the practical implications of words and the way they capture the life of a community. Remember, it is the way words are bound into practices that would prevent most of us understanding speaking lions, so to speak Lionese fluently – and the idea of interactional expertise and the experiments that correspond to it show it can be done without doing the practices – must be to understand the practices that gave rise to Lionese. When language is taken on in the spirit of a way of being in the world, it contains the tacit knowledge of that way of being in the world.

One can also be bad at spoken discourse, in the way computers are bad, in which case the words are merely frozen representations of momentary states of affairs, not participating in the lived life and lived practices of a community, and not containing the tacit knowledge of the practices. Indeed, computers cannot understand the way to form those bad/good sentences on which so much ordinary speech is composed. The first shock of the new computer user is that every 'word' has to be spelled exactly right for the computer to use it as an instruction whereas humans cope with small mistakes without even noticing. Improved computers are making better guesses at how to rectify mistakes but it is always a matter of statistics not the kind of 'meaning repair' that comes with understanding.

The argument of this chapter begins with the idea that language is a property of social groups and that individuals only share in it. Now that the computer-wars are over, the war damage can be repaired by keeping this constantly in mind.

References

Collins, H. (1990). *Artificial experts: Social knowledge and intelligent machines*. Cambridge Mass.: MIT Press.

Collins, H. (2004). *Gravity's shadow: The search for gravitational waves*. Chicago: University of Chicago Press.

Collins, H. (2008). Response to selinger on dreyfus. *Phenomenology and the Cognitive Sciences, 7*, 309–311.

Collins, H. (2010). *Tacit and explicit knowledge*. Chicago: University of Chicago Press.

Collins, H. (2011). Language and practice. *Social Studies of Science, 41*(2), 271–300. DOI: 10.1177/0306312711399665.

Collins, H., & Evans, R., (2007). *Rethinking expertise*. Chicago: University of Chicago Press.

Collins, H., & Sanders, G. (2007). They give you the keys and say "drive it:" Managers, referred expertise, and other expertises'. *Case Studies of Expertise and Experience: Special Issue of Studies in History and Philosophy of Science, 38*(4), 621–641 [December].

Collins, H. M. (2004a). Interactional expertise as a third kind of knowledge. *Phenomenology and the Cognitive Sciences, 3*(2), 125–143.

Collins, H. M. (2004b). The trouble with madeleine. *Phenomenology and the Cognitive Sciences, 3*(2), 165–170.

Collins, H. M., Evans, R., Ribeiro, R., & Hall, M. (2006). Experiments with interactional expertise. *Studies in History and Philosophy of Science, 37*(A/4), [December] 656–674.

Dreyfus, H. L. (1965). 'Why computers must have bodies in order to be intelligent. *Review of Metaphysics, 21*, 13–32.

Dreyfus, H. L. (1972). *What computers can't do*. New York: Harper and Row

Kuhn, T. (1962). *The structure of scientific revolutions*. Chicago: University of Chicago Press.

Schilhab, T. (2007). Interactional expertise through the looking glass: A peek at mirror neurons. *Case Studies of Expertise and Experience: special issue of Studies in History and Philosophy of Science, 38*(4), 741–747.

Selinger, E., Dreyfus, H., & Collins, H. (2007). Embodiment and interactional expertise. *Case studies of expertise and experience: Special Issue of Studies in History and Philosophy of Science, 38*(4), 722–740 [December].

Wittgenstein, L. (1953). *Philosophical investigations*. Oxford: Blackwell.

Chapter 12
Levels of Immersion and Embodiment

On the Relation Between Different Types of Practice and the Acquisition of Language

Theresa Schilhab

Abstract Contemporary neuroscience seems to suggest that conceptual understanding as in reading and discourse at least in part is perceptually and sensory-somatically corroborated. In other words, conceptual knowledge seems to involve reenacting forms of perceptual experiences. However, in many aspects of life we do not have first hand experiences of the concepts we master to perfection. Who has ever had personal experiences with unicorns, the ice ages or Big Bang? In this chapter, I expand on the relation between symbol use as it applies to the linguistic exchange in professional communities and different levels of immersion in the associated practices to clarify the issue of levels of embodiment from a cognitive point of view.

12.1 Introduction

Knowledge acquisition can be separated according to levels of immersion (see Collins & Evans, 2007). At one end, knowledge seems to involve intense hands on experiences, e.g. to tie shoelaces and, at the other end it involves linguistic 'adroitness' as when one memorises the capitals of the world. Obviously, the degree of bodily involvement differs. Knowing how to tie one's shoelaces is both to know the practice of binding bows and being able to explicate it, whereas hands-on experience seems irrelevant in relation to knowing that Paris is located in France. Tying shoelaces has a practical as well as a theoretical dimension, whereas 'capital knowledge' rests solely on the theoretical dimension (or so it seems). If measured by the degree of bodily involvement, these two examples make up two ends of a 'knowledge axis'.

T. Schilhab (✉)
GNOSIS Research Centre, Danish School of Education, University of Aarhus,
2400 Copenhagen, Denmark
e-mail: tsc@dpu.dk

T. Schilhab et al. (eds.), *The Symbolic Species Evolved*, Biosemiotics 6,
DOI 10.1007/978-94-007-2336-8_12, © Springer Science+Business Media B.V. 2012

Within science and technology studies, versions of these ends have recently been termed 'interactional' and 'contributory expertise',[1] respectively (Collins & Evans, 2002; Collins, 2004; Collins, Evans, Ribeiro, & Hall, 2006). Contributory expertise involves both 'knowing how' and 'knowing that' and applies to full blown immersion in a domain. Ideally, the contributory expert is the physicist who knows his way around the laboratory; a versatile and competent user of equipment who is also theoretically and linguistically proficient.[2] Interactional expertise, on the other hand, applies to extensive linguistic knowledge of things, i.e. objects, phenomena or situations we have never actually encountered before, but which we know of only from conversations and similar indirect means.[3] Interactional expertise depends on a context provided by a linguistic community and is not to be mistaken for symbolic or formal knowledge, which could be picked up by reading a book. As suggested by Collins (2004), interactional expertise could be acquired by the sociologist who has interviewed several physicists and, while being immersed in their culture, has acquired proficiency in speaking about the domain to an extent that makes him indistinguishable, at the linguistic level, from the contributory expert.[4]

Despite the obvious differences between interactional and contributory expertise, they both rely on immersion in a culture that includes embodiment, but to a different extent. Whereas contributory immersion also involves activities concerning objects and actions, interactional immersion exclusively involves linguistic procedures.

The question is; could detailing the components behind the differing degrees of embodiment help us explicate obvious peculiarities, inadequacies and virtues of interactional and contributory expertise? On this basis, would we be able to identify, explicate and eventually anticipate characteristics of knowledge linked to different levels of immersion?

In this chapter, I will address the issue of levels of embodiment in interactional and contributory expertise from a cognitive point of view. To begin answering the questions, we must specify what is meant by 'embodiment' in a cognitive perspective. To that end, we first need to probe the cognitive components and mechanisms behind contributory and interactional expertise.

---

[1] Collins and Evans discuss the taxonomy of expertises, the natural or acquired facility in a specific activity. In this chapter focus is on the characteristics of the knowledge associated to particular expertises.

[2] Physicists (like all specialists) are unique. Lumping them together is therefore bound to result in superficial and inaccurate characterizations. Naturally, some physicists have specialized in experimental designs; others in theoretical approaches.

[3] To count as a genuine interactional expert, knowledge must be obtained away from the concrete context relevant to contributory experts. In reality such clear cut examples are rare. For the sake of argument, I have the serene form in mind.

[4] Bear in mind though that according to the theory, the interactional expert does not contribute to the evolvement of physics as a science.

12.2 Contributory Expertise and Embodiment

Contributory experts acquire knowledge by conversations, writings and hands-on experience. Thus, at some point theory about the domain is or becomes 'coupled' to practice. Simultaneity is of significance to the present discussion, as the pairing of physical activity and exposure to language seems to facilitate embodied knowledge (see Wilson, 2002). To grasp the criticality of concurrency even in elementary settings, think of the acquisition of a common concept, say apples. A child born where apples are a regular part of diet supplement is likely to know apples also by its senses. Possibly, the very first step towards apprehension of the notion of apple happens simultaneously with the first taste and tactile experiences of the fruit (see Glenberg, 2008). On the other hand, children born where apples are rare are more prone to learn about apples 'at a distance' – from fairytales such as 'Snow White and the seven dwarfs' or as a nickname for New York City.

'Apple-knowledge' acquired through first-hand experiences with the referent is likely to rest on a combination of certain linguistic expressions and particular state of affairs; i.e. qualities typical of apples and their context which are usually implicitly perceived.

The fact that knowledge is grounded in non-symbolic processes i.e. perception, is strongly supported by numerous neuropsychological studies (e.g. Barsalou, Simmons, Barbey, & Wilson, 2003; Barsalou, 2008). Studies show activation of 'perceptual symbols' of referents in, for example, the shape of objects when we comprehend sentences such as 'the ranger saw the eagle in the sky' or 'the ranger saw the eagle in its nest' (Zwaan, Stanfield, & Yaxley, 2002) or simulation of perceptual experiences when forming lexical decisions about sentences (e.g. Glenberg & Kaschak, 2002; Barsalou et al., 2003; Holt & Beilock, 2006; Schilhab & Gerlach, 2008a; Schilhab & Gerlach, 2008b). Simulation in these senses is 'reactivation' of the neural circuits that were employed during the original experience. Accordingly, '. . . .perceptual symbols are the residues of perceptual experiences, stored as patterns of activation in the brain' (Zwaan et al., 2002, see p. 168).

Various Reaction Time studies (RT), in which the reaction time of subjects in sensibility judgments is measured, suggest that simulation implies revival of neural traces of past actual experiences. In a study by Glenberg and Kaschak (2002) implied action direction of the sentence was either compatible or contrary to the direction of the response. When asked to assess the sensibility of the sentence 'close the drawer', subjects were to respond 'yes' by either pulling or pushing a handle resulting in a movement towards or away from their body. RT's were significantly lower in the compatibility conditions than in the incompatibility condition. Making sense of the sentence; 'close the drawer', seems to recruit neural connections necessary when executing the actual movement. Also a study by Pecher and colleagues (2003) suggests that knowledge about concepts is partly grounded in perceptual processes (see also Boroditsky & Ramscar, 2002; Markman & Brendl, 2005). In these studies, subjects were exposed to concepts along with an associated property and asked to verify or reject the validity of the association. Subsequently, subjects were exposed to the same concept accompanied by either a property of the same or another modality. For instance 'apple' was accompanied by the visual property 'green' and subsequently 'shiny' (same modality) or 'tart' (different modality). If the second property belonged to the same modality RT's were markedly reduced (see also Marques, 2006).

Imaging studies of the organisation of language show that, in language acquisition, the concurrent stimulation of the senses together with linguistic processes is simulated when we retrieve verbal knowledge (Pulvermüller, 2005). Subsequently, knowledge about and remembrance of e.g. apples seems to be sustained by reactivation of the neural correlate, maintaining the perception *and* linguistic processing. This explains why merely reading words with strong olfactory associations (e.g. 'cinnamon' or 'garlic') recruits primary olfactory cortices normally involved in perceptual processing (González et al., 2006). In fact, mere reading of words recruits neuronal areas which are normally correlated to the actual experience of a smell. (For theories that link words and senso-somatic processing, see Myung, Blumstein, Sedivy, 2006; for causal links between the motor system and the comprehension of language see Glenberg, Sato, & Cattaneo, 2008; Meteyard & Vigliocco, 2008).

12.3 Interactional Expertise and Embodiment

One might argue that immersion in a linguistic culture, constitutive of interactional expertise, is likely to include at least some contact to practical dimensions of the domain. For example an anthropologist who studies Lappish weather rites might not participate directly in the ancient rituals, but observation potentially offers a case of indirect, second hand acquisition of knowledge (for discussion in relation to mirror neurons, see Schilhab, 2007a). The term 'physical contiguity' describes such proximity to practices of a domain that falls short of active involvement or 'hands-on' experience (Ribeiro, 2007a, 2007b).

For the sake of the argument, let us consider what is to become of embodiment, if one obtains knowledge from linguistic socialisation *exclusively*? At least two tracks are open for investigation.

First, to pick up language and speak fluently, one needs ears to hear with (see Steger & Werker, 1997) and something akin to the hyoid bone to manage human speech sounds (2003; Nishimura, Mikami, Suzuki, & Matsuzawa, 2003); inimitable prerequisites that Collins denotes 'minimal embodiment' (Collins, 2004; Selinger, Dreyfus, & Collins, 2007; Schilhab & Gerlach, 2008a). Surely, embodiment in this sense is undisputed under any assertion about knowledge and expertise.

The second track is more radical. This implies accepting that embodiment is a critical prerequisite in all ontogenetic learning (i.e. within the lifetime of the individual). In a linguistic setting, immersion depends on bodies. First, linguistic cultures that sustain interactional expertise depend on contributory experts (i.e. their bodies) that converse and exchange ideas. Naturally, interactional knowledge 'piggybacks' on contributory knowledge, and thus on fully embodied language (Collins, 2004; Schilhab, Juelskjær, Moser, 2008). Secondly, interactional experts participate in the linguistic culture and implicitly perceive regularities in the environment. This pattern recognition makes their participation in the linguistic community similar to the participation of the contributory expert in his community.

Implicit learning, which goes on non-stop, may explain why. Implicit learning presupposes and builds on innate sensitivity and natural assessments which are 'routinely and automatically registered.... .without intention or effort', as described by Kahneman (2002). Implicit learning owes its existence to environmental regularities from which the organism could profit (i.e. Reber, Walkenfeld, & Hernstedt, 1991; Berry & Dienes, 1993; Reber, 1993; Goldstone, 1995; Stadler & Frensch, 1998). Whenever A happens, inevitably B follows, in a one-to-one relation.[5] Such plain stimulus/response relations immanent to the organism originate from prolonged adaptive processes shared by all individuals of the same species. Thus, all that is needed for implicit learning to occur is regularities in the environment (i.e. the world is not random) (Schilhab, 2007b).

One could argue that there is no significant cognitive difference between contributory and interactional expertise. When they pick up phrases, interactional experts immersed in the relevant linguistic community equally expose themselves to contextual stimuli, some of which are linguistic and some non-linguistic. In support of this interpretation, Watkins and colleagues (2003) show that perception of speech, both auditory and visual, result in changes in the excitability of the motor system that allows for speech production. Exposure to conversations primes the listener's ability to speak the words himself.

While contributory experts ground notions and phrases in neural circuits that are also employed for actions and tangible objects, interactional experts ground their notions in circuits predominantly, but not exclusively, concerned with processes of speech (see Ross, Wang et al., 2007). These processes include the above-mentioned auditory and visual clues associated to speech activities and the concrete linguistic context (i.e. the string (sea) of words). The point is that even in the interactional case, contextual regularities corroborate the understanding of the linguistic string as manifested by the neural correlate. For example, fMRI studies in pursuit of the contiguous activation of sensory-motor cortices and linguistic areas seem to emphasise so-called semantic flexibility. It means that 'semantic context determines the degree to which alternative senses and features are processed when a word is heard' (Raposo, Moss, Stamatakis, & Tyler, 2009, p. 394). In this study, subjects were presented with action verbs in isolation such as 'kick' or 'trample', and significant motor activation was found to result. But if action verbs were embedded in metaphoric expressions motor activation was insignificant. The inability of metaphoric phrase reading to activate neural areas corroborating the motor activity of certain effectors, say hands or feet, is also found in Aziz-Zadeh, Wilson, Rizzolatti, and Iacoboni (2006).

---

[5] In the simple version; for example sensitisation and conditioning in Kandel's *Aplysia californica*, a seahare, involves the linking of various external stimuli enabled by the phylogenetically determined, neural connections (Kandel, 2001).

Context matters to interactional experts. Think of the concept of jealousy. Even if it is impossible to literally grasp the referent of the term in the way you grasp an apple, the concept might still appear in conversation under circumstances particular to that feeling which therefore signify it and act as the referent. For instance, it is likely that talk about jealousy is connected to particular grimaces of the talker which affect the listener (in early stages; the learner) (for emotional content and neural underpinning when imitating see Glenberg, Havas, Becker, R., & Rinck, 2005; Bastiaansen, Thioux, & Keysers, 2009). This is the non-verbal (environmental) part of perceivable regularities. However, the notion of jealousy might also appear in particular linguistic connections non-consciously perceived by the listener that is, the linguistic part (linguistic regularities). Sentences like: 'She handed him jealousy' or 'The dog sniffed the jealous tree' are meaningless, since they do not follow conventions governing linguistic use.

So, what are the differences between contributory and interactional expertise after all? While the embodiment of contributory experts involves concrete items in particular, the embodiment of interactional experts is less clear-cut but still involves embodiment in non-trivial ways because of the implicit susceptibility to both environmental and linguistic regularities.

12.4 Characteristics of Knowledge?

12.4.1 Different Kinds of Creativity

So far, I have proposed that embodiment plays a part in both interactional and contributory expertise, but in different ways and to a different extent. I have claimed that both contributory and interactional experts are exposed to linguistic regularities as well as environmental, but to different proportions. Contrary to interactional cultures, contributory cultures consist of vast amounts of bits and pieces; laboratory items and standard actions and practices. In the interactional case, linguistic sources precede more over non-linguistic sources. In the contributory case, however, linguistic and non-linguistic sources jointly compose the conceptual representation. The concreteness of practical life has a higher degree of regularity; laboratories have standard equipment, fume cupboards and plastic gloves. These regularities afford and fixate the senses so that contributory concept formation rests on firm somatosensory perceptions. To interactional experts, environmental regularities are less overt. Actions and regularities are by no means absent in verbal cultures, but the constancy of concrete items and set routines are considerably lower.

For example, interactional experts who gain knowledge of a field would be exposed to standard regularities of a generalised and abstract kind such as sitting on chairs, using keyboards and observing contributory experts giving lectures. In such 'genuine' theoretical conditions, facial muscle contractions, body language and similar concomitant phenomena of contemplative activity is implicitly perceived by the interactional expert. However, these behavioural fragments are generalised in the sense that they are separated from the context of the lecture hall or office since they could easily occur almost anywhere else (see also Schilhab, 2007a, Schilhab, in print).

When they build domain-specific knowledge, interactional experts are then more prone to combine new understanding with already established *theoretical* knowledge (linguistic networks) than contributory experts, who might naturally also incorporate sensations and personal experiences to substantiate their interpretation. While interactional experts create knowledge that is relatively more conceptualised, contributory experts establish knowledge that draws more on recurring somatosensory experiences.

Despite obeying linguistic rules, linguistic understanding is fundamentally abstract, in the sense that interactional understanding is characterised by more degrees of freedom, that is, higher number of 'elements' in the understanding are free to vary. If you only know the exact configuration of items referred to e.g. linguistically; correct completion of procedures or dimensions of the phenomena in question only linguistically (interactional knowledge), you might allow yourself a less strict interpretation of the referent. In contributory knowledge, there is a direct correlation between symbol and referent, which means that changes in the referent will cause changes in the symbol (Barsalou, 1999). Therefore, contributory experts who have formed knowledge by first-hand experience have less flexible interpretations.

As a side-effect of relatively more contextual regularity that feeds the implicit learning capacity, the contributory expert might possess a more rigid understanding of phenomena related to the field. Experts, as defined by the Dreyfus Model of Skill Acquisition (Dreyfus & Dreyfus, 1986) compensate for this effect of practice learning by acquiring many personal learning instances as well as obtaining instructions from more than a single teacher to broaden the reference of experiences.

How could more or less 'degrees of freedom' in conceptualisation be recognized? The linguistic network is the predominate point of reference; one might expect a bolder and more innovative use of phrases and concepts from interactional experts because of their lack of reference to concrete items and actions (for a linguistic analysis of deception, see Newman, Pennebaker, Berry, & Richards, 2003).

However, imagination could be an element of human cognition that might pull interactional conceptual learning significantly nearer to contributory learning. In so far as interactional experts corroborate conceptual learning by self-induced visualization, e.g. by envisioning oneself or a substitute experiencing the situation, they might transcend the alleged boundary of theory and practice, which would eventually blur any distinct signs of differences in experiential quality (Schilhab, 2007a, see also Schilhab, 2011).

If the power of imagination has the compensatory abilities in question, we must eventually explain what it takes to establish imagination. Do imaginative powers rest on versatile somatosensory experiences founded at an early age? And does imagination hinge on the interplay between somatosensory experiences and extensive conscious reflection?

12.4.2 Reduced RT

Could the different levels of immersion produce other traceable effects? A study by Holt and Beilock (2006) in which drawings of domain-specific (or everyday) objects or actions were presented to expert hockey and American football players as well as novices accentuate differences in the pace of neural processing. The task was to judge whether any mentioning of represented objects or actions that followed sensible sentences had occurred. When the respondents assessed the occurrence of pictured items or actions that had also been mentioned in a preceding sentence, experts showed larger reductions in RT (response time) than novices. In the everyday condition, subjects were exposed to sentences like 'the child saw the balloon in the air'. Subsequently, they were exposed to pictures of either an inflated or deflated balloon and had to judge whether the represented object had just been mentioned in the sentence before.

In the case of the inflated balloon the target matched the sentence. In case of the deflated balloon the target mismatched the sentence because deflated balloons are unlikely to appear in the air. Experts and novices were equally competent at responding to everyday objects and situations. However, in the domain-specific condition; when football players and novices were asked to assess a sentence like: 'the trainer saw the offensive lineman protect the quarterback', only experts showed reduced RTs.

In this study, being a non-expert did not exclude the possibility of coming up with the right answer, although response times increased. How are we to explain that?

In the acquisition phase of, say, apples, the perceptual contribution to the neural correlate of knowledge seems to facilitate retrieval processes (Ross, Wang et al., 2007). Apparently, processes in the neural substrate corroborating the concept of apples improve because of the inclusion of somatosensory experiences.

In the study by Holt and Beilock (2006) one obvious explanation of the results with hockey experts (contributory experts) would be that mere reading of the sentence; 'the trainer saw the offensive lineman protect the quarterback' recruits neural areas that are active when we experience that 'the trainer saw the offensive lineman protect the quarterback'. In that case the sentence; 'the trainer saw the offensive lineman protect the quarterback' is, among others things, neurally constructed from different modalities; say sight, smell and texture (due to embodiment in the acquisition phase). The hockey expert has numerous practical experiences of protection of quarterbacks that fit such descriptions. To him, the neural correlate which constitutes the understanding of the sentence consists of input from many different modalities, which, due to concurrent activation, are tightly connected. Another likely explanation is that understanding the sentence induces activity in neural areas that are also active when we make sense of the pictured actions, which ultimately reduces the reaction time.

In contrast, interactional experts on hockey would incorporate the concept of protection of quarterbacks almost entirely in the linguistic framework. The neural representation of the concept would be based on less robust connections, which again would make the recruitment less efficient.

Exposure to (and interest in) concrete items is elementary to the natural development of preverbal children, therefore one might wonder whether concrete contexts are somewhat more familiar and therefore perhaps easier to acquire and incorporate in the knowledge (the neural nets that sustain the learning might be easier to establish, because major parts of it already exist?).

However, this is not to say that the employment of wholly linguistic knowledge is always delayed compared to conceptual knowledge linked to somatosensory experiences. For sure, neural connections that sustain genuine conceptual knowledge could be potentiated by systematic recruitment. One could imagine scenarios in which various conversations that employ (the use of) particular concepts could strengthen and thereby speed up the efficacy of the involved neural connections to an extent which would match or even exceed the efficiency of neural connections that correlate to contributory concepts. This might actually apply to the final stages of becoming an interactional expert.

12.5 Conclusion

To interpret interactional and contributory expertise by levels of embodiment raises questions about the applicability of the cognitive framework. In many scientific domains, students are exposed to theoretical knowledge before they are exposed to practice. In domains where apprenticeship is prevalent, the order might be reversed or theory and practice offered simultaneously during education. No doubt, the order of presentation of theory and practice may have an impact on the underlying neural organization of domain-related knowledge.

Also, we must consider to what extent relevant similarities exist between young children who learn the concept of 'apple' and novices in scientific disciplines who learn the linguistic setting of laboratory practices in the field.

No doubt, the synchronized presentation of referents and concepts is likely to be the 'original' way we obtain our first language (Glenberg, 2008; Glenberg et al., 2008). However, for professional jargons, learning is always second to the well-established first language, so linguistic fundamentals are in place when contributory or interactional expertise is created (e.g. Perani et al., 1998).

Even if learning the language of a new domain is like learning your mother tongue in the sense that it too involves the acquisition of new concepts, being second and therefore feeding on natural language results in an adapted neural organization. Logically, to learn a second language is entirely different from learning the first language, which includes cracking the code of symbol use. The competent language user has long passed this stage by learning the first language. On this account, the comparison of the learning of contributory language with natural language might be flawed.

However, taking different levels of embodiment into account cognitively differentiates contributory and interactional knowledge in fundamentally new ways. Qualifying their differences in cognitive terms provides us with valuable insight into human thinking and stipulates promising subfields that await further research.

References

Aziz-Zadeh, L., Wilson, S. M., Rizzolatti, G., & Iacoboni, M. (2006). Congruent embodied representations for visually presented actions and linguistic phrases describing actions. *Current Biology, 16*, 1818–1823.

Barsalou, L. W. (1999). Perceptual symbol systems. *Behavioral and Brain Sciences, 22*, 577–660.

Barsalou, L. W. (2008). Grounded cognition. *Annual Review of Psychology, 59*, 617–645.

Barsalou, L. W., Simmons, W. K., Barbey, A. K., & Wilson, C. D. (2003). Grounding conceptual knowledge in modality-specific systems. *Trends in Cognitive Sciences, 7*(2), 84–91.

Bastiaansen, J. A. Thioux, C. J. M., & Keysers, C. (2009). Evidence for mirror systems in emotions. *Philosophical Transactions of the Royal Society B, 364*, 2391–2404.

Berry, D. C., & Dienes, Z. (1993). *Implicit learning*. East Sussex: Lawrence Erlbaum Associates Ltd.

Boroditsky, L., & Ramscar, M. (2002). The roles of body and mind in abstract thought. *Psychological Science, 13*(2), 185–189.

Collins, H. (2004). Interactional expertise as a third kind of knowledge. *Phenomenology and the Cognitive Sciences, 3*, 125–143.

Collins, H., & Evans, R. (2007). *Rethinking expertise*. Chicago: University of Chicago Press.

Collins, H., Evans, R., Ribeiro, R., & Hall, M. (2006). Experiments with interactional expertise. *Studies in History and Philosophy of Science, 37*(a), 656–74.

Collins, H. M. & Evans, R. (2002). The third wave of science studies: Studies in expertise and experience. *Social Studies of Science, 32*, 235–296.

Dreyfus, H., & Dreyfus, S. (1986). *Mindover machine*. New York: Free Press.

Glenberg, A. M. (2008). Embodiment for education. Handbook of cognitive science. In P. Calvo & T. Gomila (Eds.), *An embodied approach* (pp. 355–372). Amsterdam: Elsevier.

Glenberg, A., Havas, M. D., Becker, R., & Rinck, R. (2005). Grounding language in bodily states: The case for emotion. In D. Pecher & R. A. Zwaan (Eds.), *Grounding cognition* (pp. 115–128), Cambridge: Cambridge University Press.

Glenberg, A. M., & Kaschak, M. P. (2002). Grounding language in action. *Psychonomic Bulletin and Review, 9*(3), 558–565.

Glenberg, A. M., Sato, M., & Cattaneo, L. (2008). Use-induced motor plasticity affects the processing of abstract and concrete language. *Current Biology, 18*(7), 290–291.

Goldstone, R. L. (1995). Effects of categorization on color perception. *Psychological Science, 6*(5), 298–304.

González, J., Barros-Loscertales, A., Pulvermüller, F., Meseguer, V., Sanjuán, A., Belloch, V., et al. (2006). Reading cinnamon activates olfactory brain regions. *NeuroImage, 32*, 906–912.

Holt, L. E., & Beilock, S. L. (2006). Expertise and its embodiment: Examining the impact of sensorimotor skill expertise on the representation of action-related text. *Psychonomic Bulletin and Review, 13*(4), 694–701.

Kahneman, D. (2002). *Maps of bounded rationality: A perspective on intuitive judgment and choice* (Nobel Prize Lecture). Les Prix Nobel. T. Frangsmyr. Stockholm, Aula Magna, Stockholm University.

Kandel, E. R. (2001). The molecular biology of memory storage: A dialogue between genes and synapses. *Science, 294*, 1030–1038.

Markman, A. B., & Brendl, C. M. (2005). Constraining theories of embodied cognition. *Psychological Science, 16*(1), 6–10.

Marques, J. F. (2006). Specialization and semantic organization: evidence for multiple semantics linked to sensory modalities. *Memory and Cognition, 34*(1), 60–67.

Meteyard, L., & Vigliocco, G. (2008). The role of sensory and motor information in semantic representation: A review. In P. Calvo & T. Gomila (Eds.), *Handbook of cognitive science. An embodied approach* (pp. 293–312). Amsterdam: Elsevier.

Myung, J.-y., Blumstein, S. E., & Sedivy, J. C. (2006). Playing on the typewriter, typing on the piano: Manipulation knowledge of objects. *Cognition, 98*, 223–243.

Newman, M. L., Pennebaker, J. W., Berry; D. S., & Richards, J. M. (2003). Lying words: Predicting deception from linguistic styles. *Personality and Social Psychology Bulletin, 29*(5), 665–675.
Nishimura, T., Mikami, A., Suzuki, J., & Matsuzawa, T. (2003). Descent of the larynx in chimpanzee infants. *Proceedings of the National Academy of Sciences, 100*(12), 6930–6933.
Pecher, D., Zeelenberg, R., & Barsalou, L. W. (2003). Verifying different-modality properties for concepts producesswitching costs. *Psychological Science, 14*(2), 119–124.
Perani, D., Pauesu, E., Galles, N. S., Dupoux, E, Dehaene, S., Bettinardi, V., et al. (1998). The bilingual brain. Proficiency and age of acquisition of the second language. *Brain, 121*, 1841–1852.
Pulvermüller, F. (2005). Brain mechanism linking language and action. *Nature, 6*, 576–582.
Raposo, A., Moss, H. E., Stamatakis, E. A., & Tyler, L. K. (2009). Modulation of motor and premotor cortices by actions, action words and action sentences. *Neuropsychologia, 47*, 388–396.
Reber, A. S. (1993). *Implicit learning and tacit knowledge*. Oxford: Clarendon Press.
Reber, A. S., Walkenfeld, F. F., & Hernstedt, R. (1991). Implicit and explicit learning: individual differences and IQ. *Journal of Experimental Psychology: Learning, Memory and Cognition, 17*(5), 888–896.
Ribeiro, R. (2007a). The role of interactional expertise in interpreting: the case of technology transfer in the steel industry. *Studies In History and Philosophy of Science Part A, 38*(4), 713–721.
Ribeiro, R. (2007b). *Knowledge transfer*. Cardiff: Cardiff University.
Ross, B. H., Wang, R. F. Kramer, A. F. Simons, D. J., & Crowell, J. A. (2007). Action information from classification learning. *Psychonomic Bulletin and Review, 14*(3), 500–504.
Schilhab, T. (2007a). Interactional expertise through the looking glass: a peek at mirro neurons. *Studies In History and Philosophy of Science Part A, 38*(4), 741–747.
Schilhab, T. S. S. (2007b). Knowledge for real – On implicit and explicit representations and education. *Scandinavian Journal of Education, 51*(3), 223–238.
Schilhab, T. S. S. (2011). Neural perspectives on 'interactional expertise': The plasticity of language. *Journal of Consciousness Studies, 18*(7–8), 99–116.
Schilhab, T. S. S., & Gerlach, C. (2008a). Embodiment, corporeality and neuroscience. In T. Schilhab, M. Juelskjær, & T. Moser (Eds.) *Learning bodies* (pp. 19–43). København: Danmarks Pædagogiske Universitetsforlag.
Schilhab, T. S. S., & Gerlach, C. (2008b). Connections in action – Bridging implicit and explicit domains. In J. Hoffmeyer (Ed.), *A legacy for living systems* (pp. 135–144). København: Springer.
Schilhab, T. S. S., Juelskjær, M., & Moser, T. (2008). Post Scriptum – On 'Learning bodies'. In T. Schilhab, M. Juelskjær, & T. Moser (Eds.), *Learning bodies* (pp. 303–315). København: Danmarks Pædagogiske Universitetsforlag.
Selinger, E., Dreyfus, H., & Collins, H. M. (2007). Interactional expertise and embodiment. *Studies in History and Philosophy of Science Part A, 38*(4), 722–740.
Stadler, M. A., & Frensch, P. A. (Eds.). (1998). *Handbook of implicit learning*. Thaousand Oaks: Sage.
Steger, C. L., & Werker, J. F. (1997). Infants listen for more phonetic detail in speech perception than in word learning tasks. *Nature, 388*, 381–382.
Watkins, K. E., Strafella, A. P., & Paus, T. (2003). Seeing and hearing speech excites the motor system involved in speech production. *Neuropsychologia, 41*, 989–994.
Wilson, M. (2002). Six views on embodied cognition. *Psychonomic Bulletin & Review, 9*(4), 625–635.
Zwaan, R. A., Stanfield, R. A., & Yaxley, R. H. (2002). Language comprehenders mentally represent the shapes of objects. *Psychological science, 13*(2), 168–171.

Chapter 13
Emerging Symbols

Stefan Leijnen

Abstract Using a neural network simulation of a series of language training experiments with chimpanzees, the difference between indexical and symbolic interpretation is explored. From the results of the simulation follows a discussion about the systemic requirements for crossing the symbolic threshold and how the primacy of icons applies to computational models.

13.1 The Meanings of *Symbol*

In a study aiming to test the linguistic abilities of chimpanzees, several experiments are devised and conducted to demonstrate how different learning strategies produce different uses of language (Savage-Rumbaugh & Rumbaugh, 1978). The study shows how their learning curves can be understood from the way these chimps acquire language, allowing for a behavioral operationalization of language acquisition. The results are embedded within a larger semiotic theory of symbolic interpretation, distinguishing between three types of signs (icons, indices and symbols) that describe how an object can be related to a referent by an interpreter (Buchler, 1955; Hookway, 1985; Chandler, 2002).

Several other language training studies (Gardner & Gardner, 1977; Premack, 1976; Rumbaugh, 1977) show that apes can acquire large vocabularies. The subject has to point to one or more lexigrams on a board in order to express its thoughts or desires. Researchers stimulate the apes to use the correct lexigrams and apply appropriate grammar rules. However, even though their sentence construction capability can be trained to be more or less flawless, their learning strategy appears to differ from the way humans would approach such a problem. Although they appear to use lexigrams as representations of the objects they stand for (like humans do) their pointing behavior is a trained response to the presented stimulus.

S. Leijnen (✉)
Institute for Computing and Information Sciences, Radboud University Nijmegen, 6525 AJ Nijmegen, Netherlands
e-mail: stefan@leijnen.org

T. Schilhab et al. (eds.), *The Symbolic Species Evolved*, Biosemiotics 6,
DOI 10.1007/978-94-007-2336-8_13, © Springer Science+Business Media B.V. 2012

The often implicit assumption that these apes use lexigrams as representations for something else is not to be easily overlooked. For us to talk about apes using language and having a vocabulary, evidence is required that – indeed – these apes use *linguistic skills* to solve a problem, instead of *associative skills* to merely discover a correlation between stimuli and responses leading to a reward. The difference between these two skills is subtle but crucial, especially considering the principal reason for doing ape language studies is finding out if they are actually capable of learning a language.

So how are we to make this distinction clear? We find two contrasting definitions of symbols in which the difference is expressed (Deacon, 2003):

(S1) A symbol is one of a conventional set of tokens manipulated with respect to certain of its physical characteristics by a set of substitution, elimination, and combination rules, and which is arbitrarily correlated with some referent.

(S2) A symbol is one of a conventional set of tokens that marks a node in a complex web of interdependent referential relationships and specific reference is not obviously discernible from its token features. Its reference is often obscure, abstract, multifaceted, and cryptic, and tends to require considerable experience or training to interpret.

The chimpanzees in the Savage-Rumbaugh and Rumbaugh study are subjected to a training program that causes the disparity between these two kinds of symbols to become salient, demonstrated by a significant difference in performance results. In one experiment, the chimps learn to distinguish lexigrams for four objects (*banana*, *orange*, *coke* and *milk*) and two verbs (*give* and *pour*). The chimpanzees are required to use the correct verb with each noun by arranging them in a sentence. Producing accurate sentences like *give orange* or *pour milk* is rewarded; producing incorrect compounds like *pour banana* or *coke milk* is discouraged.

Once the chimps have learned to associate pairs correctly, a follow-up experiment shows that their symbol use is, in fact, non-symbolic. As the researchers introduce new edibles and liquids to the experiment, the amount of trials needed to learn to embed these words into sentences grows. Instead of using the web of relations to which the lexigrams refer – the chimps know that edibles are given and liquids are poured, but they don't apply this knowledge to the construction of lexigram sentences – they memorize each verb-noun correlation as a rule. The chimps use lexigrams as

> [...] a set of events which come to precede the receipt of a desired action or object. [...] errorless trials, though given in a fashion which closely approximates that of the final choice, do not lead to symbolic learning even in simple tasks such as food names (Savage-Rumbaugh & Rumbaugh, 1978).

The apes have learned to use symbols as defined by *S1*, but not according to the more strict definition *S2*. The relations between the lexigrams are arbitrary as the chimps fail to notice the analogy with the relations between objects and actions. *S1* is a rather shallow, computational definition of symbols that doesn't capture the way humans use symbols as expressed in *S2*. Hence, phrased in semiotic terms, the chimpanzees have learned to use lexigrams as indices. An index pairs two things together based on their co-appearance, like a thermometer (number and temperature) or a windsock (position and wind direction). In this case, a noun lexigram is paired with a verb lexigram.

For the ape subjects to use the lexigrams as symbols (according to *S2*) a reference is required to the network of relations for which the lexigrams stand. Evoking such a reference is exactly the goal of the next experiment in the chimp language training program. It is set up in almost the same way as in the previous ones, but this time the apes' attention is drawn towards the food and drink dispensers by increasing their saliency with light and sounds signals. The apes now notice the dispensers opening, also when they're empty. This causes some of the apes to pair their understanding of objects and actions with their understanding of lexigrams, and transfer knowledge between these networks. Instead of memorizing each and every lexigram combination as an index these chimps have created a symbolic link, which offers them a more efficient way of storing information in the long run.

13.2 Simulated Learning

The chimp language training research supports the claim that symbolism is not intrinsic to a word, lexigram or object, but is dependent on the interpretation itself. Interpreters can be iconic, indexical and symbolic, and some of the apes where capable of all three of these skills while others could only reach the indexical level. In order to explain this gap, it would be insightful to take a peek inside a chimp's head, study how signals travel between neurons and how eventually a lexigram sentence comes about. In a meticulous study of the chimp's interpretation process, the differences that cause the symbolic shift could be unveiled. Of course, the sheer complexity and size of the brain would result in far too many parameters for us to make sense of. As an alternative, computer simulated models of smaller, less complex brains can be used in order to discover the systemic requirements for symbolic interpretation.

For our experiments we will use an artificial neural network: a three-layer perceptron (McCulloch & Pitts, 1943) with full connectivity (Fig. 13.1). The nodes in the hidden and output layer are implemented with a step activation threshold function (1) (cf. Table. 13.1).

$$y_j = \begin{cases} 1 \; if \; \sum_{i=1}^{n} (w_i x_i) \geq \theta \\ 0 \; if \; \sum_{i=1}^{n} (w_i x_i) < \theta \end{cases} \tag{13.1}$$

By varying the connection weights between neurons different network architectures are generated, each with a potentially different behavior (i.e. returning a specific output in response to a certain input). After a set of random weight

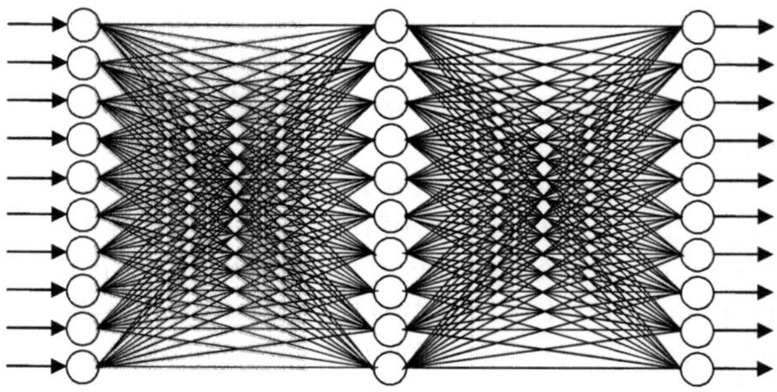

Fig. 13.1 Three-layer perceptron with full-connectivity

Table 13.1 Step activation threshold parameters

y_j	Base value for output connection j
x_i	Base value for input connection i
w_i	Weight of input connection i
n	Number of input connections
θ	Threshold parameter (0.85)

Table 13.2 Parameters of the genetic algorithm

# Children per generation	50
# Elites per generation	10
# Maximum generations	30000
# Learning runs	100
P (mutation) per bit	0.01

configurations has been selected, each of their input layers is activated with trial data and propagated as an activation wave through the network. Weight configurations are stored in a binary array. A score is awarded to each network based on the percentage of desired output values in a series of training sessions. The highest scoring networks (the elites) are then recombined using cross-over and mutation to form a new generation of network configurations, and so on. Due to the similarity with biological evolution and the storage of information in gene-like data arrays, this method is formally known as a *genetic algorithm* (Holland, 1975). The parameters of this particular GA are given in Table 13.2.

13.3 Experiments

Using the computational tools described above, the difference between indexical and symbolic interpretation is shown in a series of experiments. The two types of chimps (symbolic and non-symbolic) of the original language training research are modeled as neural networks. Objects, actions and lexigrams are replaced by binary strings of input and output data. The genetic algorithm acts as a training program, forwarding input data into the networks and evaluating the results.

For the indexical learning model, the objects, actions and lexigrams are coded according to the method displayed in Table 13.3. There are a couple of things

Table 13.3 Binary encoding examples for the indexical experiment

Network input	Binary string	Correct output	Binary string
banana + bias	1000000001	banana lexigram + give lexigram	1000000010
coke + bias	0100000001	coke lexigram + pour lexigram	0100000001
orange + bias	0010000001	orange lexigram + give lexigram	0010000010

that should be noted about this encoding. First, it disregards iconic interpretation processes by translating multifaceted entities into easily discernable icons. The chimpanzees are required to make distinctions between, bananas, yellow lexigrams, cans of coke and acts of pouring, but the neural network simply uses a ten bit binary string as input and output of the indexical process. This ensures that the neural network learns to create indexical associations, instead of a mixture of icons and indices: marginalizing the role of iconic interpretation isolates the indexical interpretation process which facilitates the study of its features. Also, in order to allow for a fair comparison with the symbolic network, a bias unit is added to the input vector.

The neural network is trained by the genetic algorithm to output the correct binary string, given a certain input string. For the input string, the leading eight bits indicate the presence of a particular edible or liquid, the ninth bit is always zero and the tenth bit is always one. The output string uses the leading eight bits to signify the use of a food or drink lexigram. The trailing two bits denote the use of an action lexigram.

Once the first pairing has been learned (i.e. *banana* with *give banana*), a second pair is added to the dataset. The learning continues with the same network and a training set of two possible input strings. This process is repeated until all eight objects have been associated with correct output sentences. The time it takes the network to learn each additional object is displayed in Fig. 13.2a.

The chimps that learn to manipulate lexigrams as symbols are induced to adopt a new learning strategy by the food and drink dispensers. These dispensers make them reconsider the relation between the lexigram buttons and obtaining a reward. They notice a systemic similarity between the system of lexigrams and the system of objects and actions (Deacon, 1997) and use their existing knowledge of the object domain to produce correct lexigram sentences.

For the symbolic learning model we use the same approach as for the indexical simulation, with the exception of the domain knowledge being available in the input string. In other words, the subject already knows that a banana is given (not poured) and takes this knowledge into account when it constructs a sentence. The additional information helps to predict the correct outcome, as actions and action lexigrams are correlated. The training data is shown in Table 13.4, the resulting learning curve in Fig. 13.2b.

A comparison between the learning curves of the indexical and symbolic models is somewhat biased. Just as the chimpanzees were at some point required to learn that bananas are given and milk is poured, so should the symbolic network, one

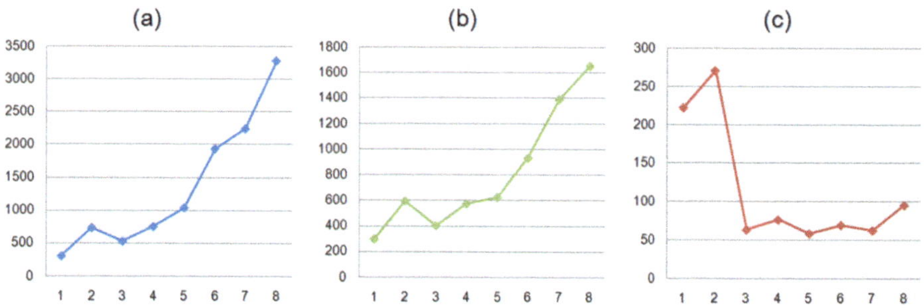

Fig. 13.2 Learning curves for the indexical task (**a**), the symbolic task (**b**) and the domain task (**c**). The y-axis indicates the number of generations it takes for each additional object (x-axis) to be learned

Table 13.4 Binary encoding examples for the symbolic experiment

Network input	Binary string	Correct output	Binary string
banana + give	1000000010	banana lexigram + give lexigram	1000000010
coke + pour	0100000001	coke lexigram + pour lexigram	0100000001
orange + give	0010000010	orange lexigram + give lexigram	0010000010

Table 13.5 Binary encoding examples for the domain experiment

Network input	Binary string	Correct output	Binary string
banana + bias	1000000001	give + bias	1000000001
coke + bias	0100000001	pour + bias	0100000001
orange + bias	0010000001	give + bias	0010000001

could argue. The goal of these experiments is to test the difference between indexical and symbolic learning; to exclude learning the domain knowledge would be a bias. Therefore, a third experiment is carried out. A neural network learns to associate objects with corresponding actions, using the same method as in the previous experiments. Table 13.5 contains the training data, the resulting learning curve is displayed in Fig. 13.2c.

13.4 Conclusion

A neural network model is used to simulate two different learning strategies in a series of three experiments. A genetic algorithm operates on a population of networks to train them in producing the desired output string. To generate a training dataset with input and output patterns, eight objects, two actions and ten lexigrams that were also used in the chimpanzee trainings tasks are encoded into binary patterns. For each of the experiments this results in a learning curve, showing the average number of generations needed by the genetic algorithm to find a working network configuration when a new object is inserted into the training dataset. The first experiment (indexical task) simulates how much learning time is required to map objects to lexigram sentences. In the second experiment (symbolic task) both the object and the action are part of the input. Finally, a third experiment (domain task) is added to avoid a possible bias. In comparing the indexical and symbolic task the learning time required for the domain knowledge task is added to the learning time for the symbolic task. This gives four learning curves, as shown in Fig. 13.3.

Several conclusions can be drawn from these curves. The domain knowledge task takes considerably less time than the other tasks, which can be attributed to the required output containing only one variable (either give or pour) instead of two. Also, there is an overall decrease in learning time after the third object is added. Once the two possible output patterns have been learned, the network has created a tendency to produce the right kinds of output patterns in the future. This holds for the indexical and symbolic tasks as well as for the domain task; however, due the steep learning curves of the former two this effect is not as significant.

The chimpanzee experiment claims that the apes that adopted a symbolic approach required more training time and made more errors during training, but once they had crossed the symbolic threshold they were able to produce better sentences and learn new symbols faster. Figure 13.3 shows that this also holds for the simulated interpreters. Requiring less time to learn the first objects, the indexical learning curve grows steeper than the symbolic learning curve in the long run.

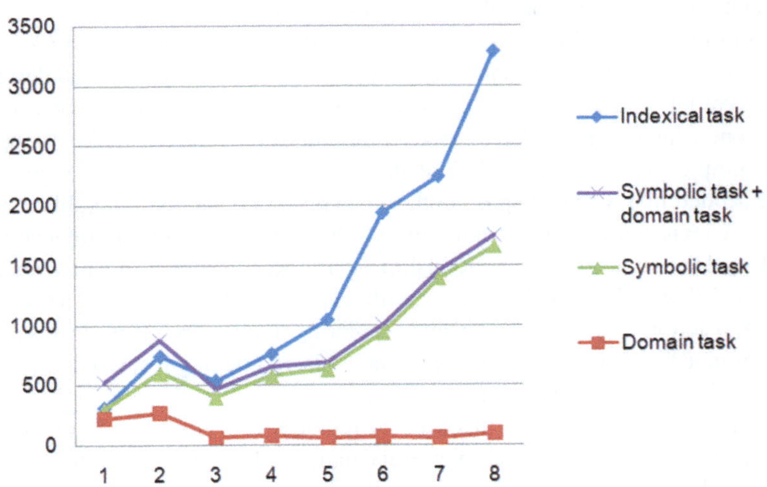

Fig. 13.3 Learning curves for all three tasks compared. The y-axis indicates the number of generations it takes for each additional object (x-axis) to be learned

13.5 Discussion

We have set up the neural network experiments in order to investigate the differences between indexical and symbolic learning. Although such a difference can be shown to exist in our models, the experimental findings do not prove the accuracy of the models used nor do they validate the conversion from the chimpanzee language training program to the simulation. It should be noted that too many simplifications and assumptions had to be made to call these networks either indexical or symbolic interpreters. In order to reduce the complexity and tractability of the learning task, a relatively straightforward neural network is used. Also, even though a bias is avoided by adding the domain task, it is unclear how exactly the learning curve of the domain task and the symbolic curve ought to compare to the results of the indexical task. One should therefore be prudent with generalizing the particular model and approach used in these experiments.

However, when the results are projected onto the semiotic theory (similar to the approach taken by Savage-Rumbaugh and Rumbaugh), they do allow for interesting conclusions to be drawn. The learning curves help to identify the mechanisms that underlie the shift to symbolism. The findings show that this shift serves a practical purpose as it allows the subject to off-load memory from one domain to another, thereby avoiding duplication of information. With selection pressure favoring language use, this gives an advantage to symbolic over non-symbolic systems. The findings also indicate that for a symbolic shift to take place, the different domains (e.g. the domain of objects and actions and the domain of lexigram relations) are required to be mapped onto each other by the interpreter. Understanding how this mapping takes place is an important step towards a more accurate simulation of the interpretation process and the role of symbols herein.

Recall our two definitions of symbols, *S1* and *S2*. In the case of *S1* a lexigram would point directly to a referent (i.e. an index). According to the second definition *S2*, the symbol would also have a pointing relation to its referent, albeit a more obscure one which is embedded in a web of interdependent referential relationships. In the chimp experiments, the relations that exist among objects and lexigrams are also embedded in a web that spans both the lexigram domain and the object-action domain. A lexigram can be an index for another lexigram: their simultaneous use will likely lead to a pointing relationship from one to another (*banana lexigram* is usually followed by *give lexigram*, hardly ever by *pour lexigram*). The realm of objects and actions has a similar system of pointing relations (coke is always poured and never given). Therefore, a symbolic relation is, as one might say, a *higher-order pointing relation* from one domain to another.

For the interpreter to create this kind of relation, it needs to find domains that can be mapped onto one another. Not every pairing of indexical systems is viable, there has to be a correlation between them that makes linking them purposeful. The input data presented in the symbolic task has some redundancy in it, so it makes sense for the interpreter to correlate the system of lexigrams with the system of actions and objects (cf. Table 13.4). It is exactly this redundancy or *system iconicity* (redundancy implies a lack of difference) in the topology of the systems that makes a symbolic relation advantageous (Deacon, 1997). A symbol, therefore, is a triadic relation that requires two systems of indices with topological redundancy, resulting in a higher-order index between two loci in those systems. The recognition of this redundancy, the *insight* that two domains are alike, is prerequisite for the symbolic shift to occur in an individual.

We can take this deconstruction of the sign one step further and consider what an index, being the constituent of symbols, is itself composed of. A pointing relation always points from one thing A to another B, which may in turn point to a third C and so on. The index from A to B is activated by the recognition of A (which is an iconic process). By virtue of their indexical relationship, A causes B to become active (as though B has been recognized). Suppose for example that A is smoke and B is a fire. The thought of a fire may cause a new thought C, no matter whether the fire was perceived directly (icon) or thought of after perceiving smoke (index). Consequently, what is caused by an index is also an icon.

The pointing relation itself is caused by a recurring appearance of signal and referent, being in close proximity to each other in one or more dimensions (i.e. spatial or temporal). Recognizing B frequently after recognizing A causes the interpreter to make a prediction about the future occurrences of B after A. The commonality of these situations is the simultaneous occurrence of signal and referent. Once the signal appears again, the interpreter recognizes the state as one of those situations where both signal and referent occur together. This *recognition* is itself a higher-order icon, because it classifies the signal-referent relation as one of many that have occurred before. Hence, an index is a relation between two icons that exists by virtue of a higher-order icon: their regular co-occurrence.

As an index is solely composed of icons, and a symbol is a particular configuration of indices, it follows that icons are the primary building blocks for all three types of interpretation. This conclusion does not imply that every iconic interpreter is also an indexical or symbolic interpreter. As the ape language training tasks as well as the simulation experiments show, a specific configuration is required for symbolic interpretation. Some apes were clearly unable to do symbolic interpretation even though they had indexical capacities. The neural networks that were trained to learn indices clearly show a behavior that differs from symbolic networks. Likewise, indexical interpretation requires a specific setup of iconic skills in order to induce the formation of a higher-order icon.

This conclusion *does* imply that iconic interpretation is a fundamental skill for interpretation. The firstness of icons is argued for in semiotics (Peirce, 1894) but also by the proficiency of simple neural network models in classification tasks, where their robustness allows them to deal with distorted data (Kohonen, 1982; Harnad, 1990). The potential of these computational models for recognition and classification tasks makes them a good starting point for further investigations into associative and symbolic models of interpretation.

References

Buchler, J. (1955). *Philosophical writings of Peirce*. New York: Dover.
Chandler, D. (2002). *Semiotics: The basics*. London: Routledge.

Deacon, T. W. (1997). *The symbolic species: The co-evolution of language and the brain*. New York: W. W. Norton & Co.

Deacon, T. W. (2003). Universal grammar and semiotic constraints. *Language Evolution, 7*, 111–139.

Gardner, R. A., & Gardner, B. T. (1977). Comparative psychology and language acquisition. *Psychology: The state of the art, 309*, 37–76

Harnad, S. (1990). The symbol grounding problem. *Physica, D42*, 335–346.

Holland, J. H. (1975). *Adaptation in natural and artificial systems*. Ann Arbor, MI: University of Michigan Press.

Hookway, C. (1985). *Peirce*. London: Routledge & Kegan Paul.

Kohonen, T. (1982). Self-organized formation of topologically correct feature maps. *Biological Cybernetics, 43*, 59–69.

McCulloch, W., & Pitts, W. (1943). A logical calculus of the ideas immanent in nervous activity. *Bulletin of Mathematical Biophysics, 7*, 115–133.

Peirce, C. S. (1894). What is a sign? *The essential peirce: Selected philosophical writings, 2*, 4–10. Indiana University Press.

Premack, D. (1976). *Intelligence in ape and man*. Mahwah, NJ: Lawrence Erlbaum Associates.

Rumbaugh, D. (1977). *Language learning by a Chimpanzee: The Lana project*. New York: Academic.

Savage-Rumbaugh, S., & Rumbaugh, D. M. (1978). Symbolization, language and chimpanzees: A theoretical reevaluation on initial language acquisition processes in four Young Pan Troglodytes. *Brain and Language, 6*, 265–300.

Chapter 14
Gender in Innovative Techno Fantasies

Cathrine Hasse

Abstract Human beings are a symbolic species with a special capacity for fantasy. It has been argued by Terrence Deacon that the prefrontal cortex has developed alongside with language and tool-making and that this part of the brain is also connected with making plans (Deacon, 1997). This could mean that human agents have developed a particular capacity for creating their habitats according to their fantasies about how they would like to live in the future. However, this general argument does not allow for a deeper understanding of difference in human fantasies and how these differences might relate to gendered experiences. Human fantasies about future developments might differ with gendered human experiences. In feminist studies it has been a recurring theme whether we can argue for a gendered status of epistemic agency. In this article I shall inspect the claim of gendered epistemologies from the angle of gendered connectionism in relation to robotics and physics. The argument I want to make is that there might be no detectable difference in how female and male researchers envision scientific innovation, but there are differences in feminist and masculine techno fantasies and this might influence how we plan our future with technological tools.

To take off his filthy garments & clothe him with Imagination.
William Blake "Milton"

Science and technology are two interdependent realms of creativity which shape our futures in particular ways. In understanding the nature of scientific practice, feminists have asked 'whose knowledge' scientists are creating (Harding, 1991). They have answered by a harsh critique of what they have seen as a western, white male scientific rationality. Feminists have claimed that science and technology is basically developed by white, masculine fantasies, which by and large excludes marginalized people from non-western countries and women, while at the same time upholding an imaginary idea of a value-free, culture-free, objective scientific practice. Feminists furthermore claim that the masculine gender is inherent not only in how science is practiced, but also in how it is developed and how it is implemented in new technologies.

C. Hasse (✉)
Department of Learning, Institute of Learning, University of Aarhus, Copenhagen, Denmark
e-mail: caha@dpu.dk

T. Schilhab et al. (eds.), *The Symbolic Species Evolved*, Biosemiotics 6,
DOI 10.1007/978-94-007-2336-8_14,

Feminist epistemology has been engaged in this deconstruction of how masculinity has reconfigured science in its own picture of 'normal subjectivity', which is not (only) masculine, but 'phallogocentric' (nostalgia for the presence of the One true Word) (Haraway, 1991). Science will, in this perspective come to understand itself as being universal, rational and capable of transcendence through a denial of the importance of the body.

Would female epistemology open up for an entirely different approach to science than the masculinity approach? This is frequently discussed by feminist epistemologists – but just as often rejected partly because what is underlined is the situated nature of all subjectivities. The critique of the detached knower, which feminist epistemology share with the general community of Science and Technology studies (STS), has lead to an acknowledgement of *situated knowledges* (Haraway, 1991). This is the true feminist objectivity.

Scientists' agency matters in the creation of technologies. Scientists' engineers and their techno fantasies literally reconfigure our worlds through their inventions. In the words of the feminist physicist Karen Barad: 'Agency is not an attribute but the ongoing reconfigurings of the world' (Barad, 2003, p. 818). Often the new developments are invented and carried out by scientists in collaboration with engineers. Both of these fields have a lack of females researchers involved in the development and innovations of science and technology. It is however an open question whether more female scientists would make a difference in science.

If gender matters to physicists' agency and if it is a particular kind of fantasies guiding the agency of especially male scientists and engineers (e.g. science fiction inspired fantasies), it is no innocent claim. As science has been utopian and visionary from the start, the question is not only 'whose knowledge', but 'whose visions and techno fantasies' (Haraway, 1991, p. 186) will make up our future man-made worlds?

14.1 Emotional Human-Robot Interaction

One area where science and technology recently have contributed with innovation which almost certainly radically will change human interrelations is in the area of robotics. One of the salient examples of science fiction-like techno fantasies infiltrating with the real life of people is found in the so-called 'artificial emotional' robots already being put to use as devices to calm down senile people in old-age homes e.g. in Japan and Denmark.

Often these robotic techno fantasies of Human-Robot Interaction (HRI) have been created to solve problems, which have gender dimensions. In 2050 demographics show that one third of the population in Japan will be over 65. In Denmark the picture is more or less the same. In both countries it is traditionally the role of women to look after elderly people, and there is no doubt that the increasingly aging population will put a pressure on the demand for old-age home workers. This can lead to a demand for a more educated workforce, who can drive innovation and develop innovative new thinking on-the-spot.

Another solution has been suggested from robot scientists: create robots which can take some of the pressure from the demanding tasks done by the primarily female old homes staff. One such robotic invention is Paro, a creature of aluminium, nuts and bolts covered with soft, white, antibacterial fur formed in the shape of a baby harp seal and with huge dark eyes. Paro was originally designed in 1993 by a male scientist, Takanori Shibata, from Intelligent Systems Research Institute at the National Institute of Advanced Industrial Science and Technology (AIST) in Japan. The creature weighs three kilo and is designed to create emotions in old people through physical interaction. It has inbuilt tactile sensors and adaptors, which makes it possible for the seal to 'learn' from human contact and it can react to touch, sound etc. In clinical tests it has been shown to relax and comfort elderly senile people and even increase their brain activity. In a small study of the effect Paro had on elderly people the robot apparently made old people more communicative than usual and even showed effects on dementia (Wada, Shibata, Saito, & Tanie, 2004; Tamura et al., 2004, p. 609).

It has already been tested and put to use in a number of old peoples homes in Denmark, where it apparently stands out as a huge success.

At for example the old people's home, Bakkeager, in the municipality of Vejle, Paro was met with enthusiam. The home has 50 inhabitants from the age 66–100 years and 75 % of the old people are senile.

It is underlined by the manager that Paro is not a substitute for human contact with the staff, but should be seen as a supplement. The staff follows courses where they are certified in the use of Paro, so they can make professional use of the robot in their daily work. Employees at old peoples' homes thus learn how to deal with the innovative device developed in Japan rather than develop innovative solutions themselves. One area where Paro can help the staff is when conflicts arise at the centre. Here Paro can be used as a 'diversion'. On a homepage connected to the old-age home it is described in these words: 'The lively and affectionate baby seal has proven to have a positive and stimulating effect on the people with senile dementia. It awakens positive sentiments, curbs aggressive behaviour and can in some cases increase brain activity in the residents with senile dementia.'[1]

Could a female scientist have developed a similar creature? Would the predominantly female staff have imagined this solution to their everyday problems themselves? Is Paro the result of a particularly male scientists' techno fantasy or are new assistive robot technologies a sign, that women bring experiences from health care environments to the sciences and are opening for a new development of assistive technologies building on women's fantasies?

1 www.teknologisk.dk/paro

14.2 Lack of Female Scientists

There are three arguments for why women generally could be expected to drive the development of techno fantasies in science to a lesser degree than their male colleagues:

1. Women's' work receive less acknowledgement than males in natural science and engineering and often they do not receive full credit for their contributions or support for development of their ideas.
2. Women in general care less about developments in science and technology than men. They prefer to study human subjects, such as psychology and languages, and are underrepresented in studies of science and engineering.
3. Women who set out on a career path in science and engineering are to a comparatively easier pushed out through subtle cultural mechanism than their male colleagues and thus excluded from developing science education and practice.

It is well documented, that female scientists are not receiving the same attention and support and that their work in science and engineering is considered less important than the work of their male colleagues – even when they were working side by side. The place of women in science has been in the margins. This has been documented in close-up studies of women's possible career paths within a wide array of scientific fields such as biology (e.g. Haraway, 1989; Keller, 1983), engineers (e.g. Meilwee & Robinson, 1992), physics (e.g. Schiebinger, 1989; Keller, 1977; Wertheim, 1995), mathematicians (e.g. Henrion, 1997). It has even been argued, that the few women entering a career in science, try to hide their womanhood (Schiebinger, 1989) and deny the experiences and preferences of being female. Even though women have entered the sciences in increasing numbers at student level the dismal picture has not changed much over the past thirty years (Chimba & Kitzinger, 2010).

The lack of women in science and engineering has increasingly been enunciated by politicians as a problem in line with universities having turned into 'mass universities' – not least due to a massive intake of female students. Higher education in science and engineering have not benefitted from the development. Women have chosen to study human subjects such as psychology, design, anthropology and other areas which involve human factors. Girls already in secondary school deselect science issues and show much less motivation than boys for learning scientific and technological subjects (Sjöberg & Schreiner, 2005). This has led to discussions of whether the diversity in interests mirrors basic differences between boys and girls (Stadler, Duit, & Benke, 2000). The lack of women in science and engineering has also been documented statistically in a number of studies, Osborn et al. (2000), Bebbington and Glover (2000), Rees (2002), European Commission (2009).

The same studies show that women enrolling in science and engineer studies do not advance to the same degree as their male colleagues. Even though we find cultural differences (Barinaga, 1994; Rees, 2002; Hasse & Trentemøller, 2008) it is a general trend in all countries that women do not advance in science and engineering to the same extent as their male counterpart – no matter how equal their numbers are when enrolling in higher education to study science or engineering (TERSTI, 2003, p. 263; Osborn et al., 2000). One explanation for this leaking from science careers (Alper, 1993) or rather the free choice of not wishing for at career in science (Svinth, 2008) has been argued to be tied to a male dominated research environment which through subtle everyday exclusion mechanisms exclude women from advancing in their careers and eventually make them want to leave (Hasse & Trentemøller, 2008). There is a certain 'policing' of the borders of science (Rouse, 1991) which seems to exclude women. But women also in their own right deselect natural science and engineering at an early stage in their educational patterns and prefer other kinds of more humanistic oriented studies. (e.g. Etzkowitz, Kemelgor, & Uzzi, 2000). As a new trend women enrol in technological and scientific studies when they have been connected to softer skills than just mechanical and technical skills – for example new combinations of design and engineering. All though this cultural pattern of in- and exclusion is not directly linked to knowledge production, male and female scientists might work from different gendered experiences which are more or less acknowledged by the scientific community and in the end also affect the scientific knowledge itself (Hasse & Trentemøller, 2008). Would science change if more women held top-positions in science and women's experiences were allowed to influence the creation of what was considered valuable scientific knowledge and technological innovations? Would we see fewer or more robots like Paro? Do women prefer science with a more humanistic aspect to it?

The answer to the latter question is a cautious 'yes': not only do we find more women in the humanities than in natural science (European Commission, 2009); within natural science many women prefer to work in areas with direct implications for human beings like e.g. in medical physics rather than in the more 'aloof' areas of theoretical physics (Hasse, Trentemøller, & Sinding, 2008, pp. 150–153). And an even more cautious 'yes' to the first question: maybe science would change if more women had science careers. But does that necessarily mean that gender *matters* in scientific re-configuration of our common world? I shall argue, though more research is needed to make the argument stronger, that we would see more emotional robots like Paro if more women entered science. Not because they are women, but because the new type of emotional robots represent a turn to soft skills more connectable to many women's experiences – skills which are also increasingly shared by men.

The development of emotional robots can be seen as a post-Fordist development, which has come to trump hard skills. Hard skills were during the Industrial Revolution and through the Fordist era, associated with manual and mechanical operations and led to scientists and engineers development of industrial robots. Soft skills are conceptualized as being parts of ones self such as social sensitivity, emotionality and modes of thinking, communication and conflict handling social skills – and it is these capacities which today are wanted, even dictated by an inventory of national, social, and market needs (Urciuoli, 2008). These skills have been the ones sought-after in the typical woman occupations – such as the staff at old people's homes.[2] Making robots like Paro illustrates the change from a focus on hard skills in the Industrial society coupled with a focus on hard technology towards a new soft skill society dominated by values traditionally and stereotypically ascribed to womanhood.

14.3 From Sci-Fiction to Sci-Facts

Industrial robots are an integrated aspect of human worklife today as well as occupy an important position in the science fiction fantasies of tomorrow consumed (and sometimes also written) by male physicists and engineers. Physicists and engineers share a passion for what has been called 'hard science fiction'. Science fiction motivates many physicist students to study physics – but mainly the male students. Even though there is no clear connection between the actual doings of science, technology and science fiction many possible links have been documented.

In my own work as an anthropologist at the Niels Bohr Institute I enrolled as a first year student of physics with the aim of studying gender in relation to what could be analysed as a cultural world of physics as seen from a position of a newcomer (a woman and anthropologist – with all the problems and possibilities following from these positions). More specifically I studied how newcomers became old-timers (Lave & Wenger, 1991) through processes of embodied learning connections between systems of meaning and physical objects (Hasse, 2008a, 2008b). This is in line with the feminist methodology in science studies. 'If natural sciences and their preoccupations in reporting on nature are embedded in and complicituos with social projects, then a causal, scientific grasp of nature and how to study it must be embedded in – be a special area of – causal, scientific studies of social relations and how to study them' (Harding, 1991, p. ix).

As also noted by the anthropologist Mary Douglas, inspired by Ludwig Fleck, science is fundamentally a 'thought world' (Douglas, 1987). To come to know this thought world you must engage in the everyday practice, which opens up your own phenomenological learning process (Hasse, 2008a, 2008b). This is not least true if you want to learn about the fantasy and imaginaries connected to the thought world. Anthropologists have formerly studied apparently more exotic 'imagined thought worlds' than the what can be found at the Niels Bohr Institute. Even so I was also in the apparently familiar environment introduced to strange traditions and behaviours. I agree with one of the founding fathers of interpretative anthropology, Clifford Geertz, when he states that: 'Anthropology is only apparently the study of customs, beliefs, or institutions. Fundamentally it is the study of thought' (Geertz, 1973, p. 352). It is, however, also a study of materiality and the entanglement between thoughts (and fantasies) and physical objects and spaces.

[2] It is a general trend in Western Countries that labour markets are gender segregated and that women take jobs in areas connected with care and soft skills. See e.g. International Labour Organization, http://www.ilo.org.

In my study I used the method of participant observation and took it literally that participation in everyday activities of all kinds is the road to learning (Lave, 1997) also in educational institutions (Billett, 2004). I followed introductory courses in physics, math, astronomy and I also followed the students in other everyday activities connected to physics study. I went to FREBAR (Friday Bar hang-out-time), parties, and participated in the yearly student theatre show 'Fysikrevy'. What I learned was that, in addition to what others have had to say about learning physics as a thought world – most notably the science historian Thomas Kuhn (1977) – this thought world consisted of much more than learning symbolic reading of nature through learning equations combined with learning an arsenal of 'best exemplars' and other elements of a disciplinary matrix in classes such as these on the class schema.

In the study of thought worlds what matters for anthropology is to bring 'connections to light' to quote another anthropologist Marc Augé (1999). Following a cultural-historical approach these connections can be perceived as connections between meaning-systems and materiality (Cole, 1996), but we could add to this the connection between manifest expressions of fantasies of the future (e.g. in books, movies and talk) and dreams of how to make this future come true. This is what I have elsewhere termed the 'relational zone of proximal development' of physics (Hasse, 2001). What became the biggest surprise for me learning among the physicists students were the new unexpected connection I learned to make between the scientific 'core' of textbook physics and a lot of seemingly extracurricular and therefore seemingly unconnected other fantastic aspects of a physicist students everyday life (Hasse, 2008a).

In my further analysis one aspect stood out: the frequent reference to science fiction (primarily among the students). It was a surprise that although male students actively engaged in conversations about the connections between science fiction and physics – almost no women participated neither in everyday conversations, nor in the 'Fysikrevy' where these issues were treated in an ironic fashion by the students themselves. In this particular thought world I learned that material objects like magazines on humanistic subjects were 'out' and science fiction literature was 'in'. In the institutional cultural logic reading science fiction was considered a serious occupation because it contained the seeds of tomorrow's future physics. Old timers among the students discussed science fiction literature, and science fiction related subjects openly. Discussions in the student room and other places often brought in science fiction topics like 'warp speed', antimatter as propellant fuel, space creatures and cultures like 'the Klingons'. Through science fiction discussions the students formed questions that are also important for part of physics science today: is there life in space? Is travel to other solar systems possible? The students would take as a point of departure for questions of time and curving space the episode of 'Starwars' where the pilot Han Solo makes his starship 'Millenium Falcon' speed up. From this point of departure they would discuss how it would be possible to develop what in the movies are called hyperdrive propulsion system that propel a starship through an alternate dimension of hyperspace and thereby make travel between star systems possible. They would also ridicule Han Solo for saying that his space ship is '[t]he one that made the Kessel run in [x] parsecs!' because parsecs are a measure of distance, not of time. But they did not question the extremely stereotyped presentations of gender and the almost primitive psychological landscape in these visions.[3]

In a survey among the newcomer students I asked about their reasons for enrolling in the physics studies. Among 14 possible answers only two other reasons for enrolling got higher scores than 'science fiction as contributory cause'. Almost one third of the male students, 32%, and 7% of the female students, gave this explanation for their motivation to study physics, which made it the third most popular answer only surpassed by 'Reading books on great physics theories' and 'Engaged teaching in physics in high school' (Hasse, 1998, pp. 16–18).

Others have noticed the connection between the development of science and hard science fiction as an inspirational source for the physicists-to-be. Hard science fiction is technically oriented and in physics it is fictions like W.C. Well's time machines, Arthur C. Clark's universe, the Star Wars movies and the television series Star Trek which have received the most response from physicists (e.g. Nahin, 1993; Krauss & Hawking, 1996, Kaku, 1994). At the annual American Association for the Advancement of Science (AAAS) hard science fiction has also been discussed and Leroy Dubeck has presented his theory on learning science through science fiction building on the book Fantastic Voyages: Learning Science through Science Fiction Films. And several American universities have advertised courses with titles like 'The Physics of Star Trek' and 'Cosmology: Science Fact to Science Fiction' (see e.g. Dubeck, Moshier, Bruce, & Boss, 1993). Hard science fiction is even sometimes explicitly 'helping science' to get ideas as when science fictions writers are invited to help envision the future paths of science.[4] Many natural scientists have also turned into writers of science fiction novels. Science fiction is connected to gender in so far hard science fiction is connected with male science fiction writers, whereas soft science fiction – which includes fantasies of transformed bodies and social skills – has more female writers. Gender is also connected to science fiction in general as this genre as a whole is connected to maleness. Boys and science fiction are for example often connected in literature on science learning, and science fiction is used for learning purposes to spur boys' interests for science as well as reading (Lie, Linnakylä, & Roe, 2003, p. 52)

[3] The gendered stereotypes in science fiction can also be found in robotics – see for example the many chatterbots presented as 'sexy women' (like Amythechatterbot) or the Starfleet officer Captain Kirk from Star Trek.

[4] (One example of these encounters took place in 2001 when science fiction novelist Ben Bova was an invited speaker at NASA's 'Turning Goals into Reality' conference on aerospace transportation in May 18–19). http://tgir.msfc.nasa.gov

14.4 From Hard Science to Female Fantasies?

On the one hand science has generally proven to be hostile to the claims of imagination in science (Daston, 1998). Science has been argued to have an internalist self-understanding resting on the central assumption that the success of science is insured by its internal features (Harding, 1998, p. 2). Science sees itself as being 'outside of culture' and can in the internalist self-understanding be perceived as an objective scientific endeavour whose practitioners develop science through hard work and natural skills and where 'temperament, gender, nationalism or other sources of disorder' are of secondary importance (Traweek, 1988, p. 162).

Feminist epistemologies cover a number of diverse themes countering this perspective using gender as en eye-opener for new kinds of epistemic analysis. Many feminist studies of science and technology have demonstrated that scientific practices build on what we with Helen Verran can call 'ontic-epistemic fantasies' connected to the specific co-configuring practice which connects engagement, metaphors and visions with our daily interactions with the material world (Verran, 1998). Verran names these fantasies and tropes 'imaginaries' – specific fantasies tied up with knowledge systems, which have been simultaneously used and denied in the western world (ibid., 250).

Scientific knowledge is in the feminist and STS-perspective 'location, partial embodiment and partial perspective' (Haraway, 1991, p. 191). In feminist science studies the situated nature of knowledge has consequences. When science is considered a human, embodied and material practice (Haraway, 1991, 1988), boundaries between subjects and objects are never fixed and observation of scientific objects is also a practice. Therefore science needs to accept the value of multiple perspectives on how to construct science to ensure values and taken-for-granted criteria of scientific practices to be called into question (Longino, 1990). Sandra Harding even claim that

> [a]dequate social studies of the sciences turn out to be the necessary foundations upon which more comprehensive and less distorted descriptions and explanations of nature can be built (Harding, 1991, p. 15).

Feminist science studies is not denying some kind of reality but it is a reality bound to be, in the words of Barad, an 'agential realism' (1999) including materiality and specific situated, and never abstractable, knowers as agents. The agential realist framework changes the question from how discourse comes to matter to how matter comes to matter (Barad, 1998, pp. 89–90).

Gender is not shut out of these processes. Gender is rather emerging when performed in the everyday practice of science. Or in the words of the physicist and feminist Barad in our 'intra-actions', which create mattering matter and gender as well. Barad (1998, p. 108) describes the process as: '(M)aterialization is an iteratively intra-active process whereby material-discursive bodies are sedimented out of the intra-action of multiple material-discursive apparatuses through which these phenomena (bodies) becomes intelligible'. Gender is not something we have or are in a biological sense, but something we do (Fenstermaker & West, 2002,

Butler, 1990). This approach dissolves the stereotypical notions of culture and gender found in mainstream science fiction. If scientific 'doing' involves reading fiction we have to consider how it can 'be the opportunity for the individual's imagination and memory to experience the existence of other imaginations and other imaginary worlds' (Augé, 1999, p. 99).

For many the idea of robots like Paro is an important step towards the humanoid robot worlds depicted in science fiction literature. Paro is not alone out there either. In Denmark cleaning robots is already a natural part of everyday life in public institutions and private homes. They might, like Paro, be made of aluminum, bolts and nuts but they have anthropoid names like 'Skupido' and 'Roberto'. Techno fantasies at Japan's University of Tsukuba has reportedly also created robots like the 2009-launch of the baby-robot Yotaro, a baby simulator which train parents to deal with babies shedding robotic tears before they give birth to their own child. Robots of this kind have one thing in common. What is special about this new type of robotics creating Paro and other artificial emotional robots is, that they are designed to do tasks which have traditionally primarily been done by women.

Is the future of science and technology shaped through a phallogocentric and masculine thought world feeding on science fiction fantasies, dreaming about a world where female soft skills workers are replaced by robots?[5]

The empirical material can, when combined, can be argued to support a thesis of techno fantasies connected to gendered embodiment and experiences (which again could be related to the argument put forward by Theresa Schilhab) (Chapter 12, this volume). This is, from the perspective of a feminist epistemology, not necessarily the same as an argument for a gendered epistemology based on biological difference.

14.5 Female or Feminist Epistemologies

In the natural sciences there has for long been a struggle for trying to get more women to become physicists. Not to improve or change science, but because it is seen as unjust that women do not have the same opportunities as men. The idea that this should make physics become more feminist has been seen, by male and female physicists alike, as a kind of heresy (Auchincloss, 1998). Gender has been seen as neutral in relation to how science develops.

Most feminists have countered the claim that science is gender neutral. In the book Reflections on Gender and Science Evelyn Fox Keller argued for the link between masculinity, notions of objectivity and the exclusion of women from the development of science (Keller, 1985). Rosi Braidotti and her colleagues and Sandra Harding have discussed the relation between the first-world scientific agenda, excluding local knowledge traditions and gender (in relation to protection of the environment vs. Exploitation) (Braidotti, Lazaroms, & Vonk, 2001; Harding, 1998).

[5] In South Korea BBC News could recently report, that a "Robot Ethics Charter" has been drawn which discuss the rights of robots versus people.

Many studies have shown how sex and gender have been culturally constructed from a masculinity perspective in biology and how culture in general provide science with perspective, metaphors and stereotypes reproduced and reinforced by scientific practices and results (e.g. Daston & Park, 1998; Hird, 2002; Franklin, 1995). It has even been argued that there is a possibility for masculinity to influence even the most objective sciences such as physics (Rolin, 2001) and that physics is embedded in national cultures and their respective more or less masculine dominated gender patterns (Traweek, 1988; Hasse & Trentemøller, 2008).

Some have insisted that *women* for example would focus less on 'explosions' and solid matter and more on bodies and fluids (e.g. Bleier, 1984; Irigaray, 1985) thereby placing a direct link between biological sex and scientific research preferences. So far there has not been much serious research connected to this claim – and it is often rejected by female scientists themselves (e.g. Stengers, 1997, 2000) as well as feminists who see it as a return to an essentialist point of view (Wajcman, 1991). In a feminist epistemology it is pointed out those notions of fixed biologically determined 'sex' is as 'situated' as knowledge (Haraway, 1989). Therefore a feminist epistemology does not in general support the idea of biologically fixed gender differences.

In most of the feminists studies of today we find the notion that biological sex as well as gendered performances are constructed along side with the construction of the phallogocentric 'view from nowhere' (Haraway, 1991) approach. Donna Haraway has attacked biological determinism by showing how constructions of the female sex in biological research are inherently cultural and historical (1989).

The gendered body is never fixed and bounded in stereotypes, but is a process of reconfigurations, which might seem as bounded and fixed, but which in reality is moving when we are.

> [O]f materialization that stabilizes over time to produce the effect of boundary, fixity, and surface we call matter. . .Crucially, then, [the construction of bodies] is neither a single act nor a causal process initiated by a subject and culminating in a set of fixed effects (Butler, 1993, pp. 9–10).

It seems therefore problematic to claim that a science governed by *women* should be more carried by different scientific techno fantasies than science made by men and that these biologically fixed categories of gender is what produces differences in techno fantasies.

14.6 Female Techno Fantasies

It should come as no surprise from this feminist perspective that one of the worlds leading robot developers is a woman Cynthia. The work of her research group at MIT was named one of the 50 Best Innovations of 2008 by TIME magazine. Her special approach is to build soft skills such as sociability and capacities for learning into robots.

> Social (or sociable) robots are designed to interact with people in a natural, interpersonal manner – often to achieve social-emotional goals in diverse applications such as education, health, quality of life, entertainment, communication, and collaboration. The long-term goal of creating social robots that are competent and capable partners for people is quite a challenging task. They will need to be able to communicate naturally with people using both verbal and nonverbal signals. They will need to engage us not only on a cognitive level, but on an emotional level as well. They will need a wide range of social-cognitive skills and a theory of other minds to understand human behavior, and to be intuitively understood by people. A deep understanding of human intelligence and behavior across multiple dimensions (i.e., cognitive, affective, physical, social, etc.) is necessary in order to design robots that can successfully play a beneficial role in the daily lives of people. This requires a multidisciplinary approach where the design of social robot technologies and methodologies are informed by robotics, artificial intelligence, psychology, neuroscience, human factors, design, anthropology, and more. (Breazeal,Takanishi, & Kobayashi, 2008).

In taking this approach to robot design Breazeal integrates particular branches of humanistic sciences which have for long been dominated by women. She is, like the feminists, breaking stereotypes of how natural and human sciences should be kept apart. But she is also reproducing the stereotype of women connected with soft skills interests.

In most universities in the western world women constitute the majority in the humanistic studies, whereas male students dominate in the natural sciences (European Commission, 2009). Even within the narrow discipline of physics many women seek more down-to-earth subjects like geophysics and physics used for developing hospital equipment rather than theoretical physics (Hasse, and Trentemøller, 2008, p. 213).

Like Paro the robots created by Breazeal (e.g. Cog and Kismet, Huggable) are 'artificial emotional' robots which open up for a new understanding of what robots can be used for. Instead of the focus on technical and mechanical robots prevalent in the Fordist era, these post-Fordist robots focus on the soft skills. That these skills primarily were understood as connected with being female might be what is really the major paradigmatic change in science as well as society.

As noted by Lucy Suchman the dream of machines (and the robotic visions connected to it) as the new servant class have been changing from the industrial age to the age of the service economy. We find this vision presented and discussed in many critical feminist contributions – most notably by Suchman herself (e.g. Suchman, 2007).

Paradoxically lots of funding has been given to make it possible for the scientists in natural science and to developers of technology to reconfigure our material worlds with robots whereas the question of whether gender matters for the actual development of scientific endeavours remain under researched. Very little empirical work has been invested in looking into the obvious diversity on male and female access to influence scientific reconfigurations – and how these reconfigurations are connected to techno fantasies. From the argument presented above we should not look for biological sex for answers. Feminism has changed science (Schiebinger, 1999; Alcoff & Potter, 1993), because we from this perspectives can ask new political questions on how situated human beings, with partial perspectives (Strathern,

1991) create each others worlds. 'Feminist research orients us not only to relations and symmetries among persons and things, but also to the politics of difference' (Suchman, 2005, p. 6).

14.7 Cultured Bodies?

Feminist epistemologies on the one hand criticise masculine scientific techno fantasies, but deny they are connected to the sexed body. Different politics of science fantasies might be embodied, but bodies are not fixed gendered stereotypical entities. When feminist epistemology stress the situatedness of knowledge – including bodily situatedness (Haraway, 1991) they come close to other STS theories which also acknowledge the importance of embodied experiences in science. Postphenomenology for example also oppose the 'classical' notion of epistemology where the identity of the knowing subject is not important. In this move from decontextualised to situated knowledge both post phenomenology and feminist theory reject knowledge understood as decontextualized and stable. Knowledge is situated knowing in action. The postphenomenological understanding of perception/knowledge as it appears in the work of Don Ihde operates with an analytical distinction of the body into two:

> Body one: Micro-perception (bodily-sensory dimensions)
> Body two: Macro-perception (cultural-hermeneutic dimensions) (Ihde, 1993, p. 74).

Culture is in this perspective not a superficial layer placed as a circle outside an equally encircled individual. Agent and culture are intertwined in a manner which makes it senseless to speak of where the body begins and the culture stop. Sense, emotion and culture form each other.

In cultural-historical theory we find the same assumption, which has led to many studies of how cultural learning processes form our self-evident understandings of how to engage with the world, which is never separate from us but emerge as we interact (Hasse, 2008a, 2008b). Not least the practice-based approach by Jean Lave and Etienne Wenger (1991), Yrjö Engeström (1987) and Michael Cole (1996) have thought us that knowledge is embedded in materiality, materializes historically and culturally and is situated in relations. We might even speak of a cultural epistemology which makes us understand numbers in culturally informed ways (e.g. Verran, 1998). Gendered epistemology can also be argued to be cultured as cultural diversity is formed in gendered embodied experiences. Also scientists embodied experiences are mattering matter.

In some arguments it has been argued that being a *woman* gave access to different life experiences and thus different morals, senses of self (Gilligan, 1982). In the cultural-historical perspective experiences are in a pragmatic and post phenomenological way more than just senses, but the whole of the body engaging with the surrounding world (Ihde, 2002). It is not about being a stereotype (woman), but about the cultural-historical changing possibilities for engaging with the world.

Cultural embodied knowledge is thus to be understood as embedded in subjects. The learning body might stabilize over time to produce the effect of some kind of boundary. The boundary formed by ongoing experiences is the gendered epistemology building on the politics of difference. This is normally taken for granted, but feminist epistemology has opened our eyes for how differences, also in techno fantasies in science and the values they propose, are culturally created and installed in bodies. Therefore it is likely to the point when Harding is ascertaining that

> [s]ocial values cannot remain undetected in studies of the abstract laws that govern the movements of the physical universe. It is far more likely the researchers import their social values into studies of other humans than into the study of stars, rocks, rats, or trees. (Harding, 1991, p. 79).

14.8 Towards a New Vision of Science

Paro and the emotional robots may not be the materialized results of male techno fantasies. They rather point to a change in techno fantasies in general. If science is changing it is likely because science is entangled in an emerging Post-Fordist soft skill society. We see changes in the increasing number of women in science, the increasing number of students studying humanistic subjects as well as in the artificial-emotional robot designs in particular. In the new politics of difference new scientific fantasies might appear and they might challenge what we have considered to be innovative in the past. Technological inventions have in a naturalized way guided scientific fantasies.

All cognitive functions including thinking in mataphors and symbols can be argued to be tied to fantasy (Raudla, 2008). In relation to the arguments put forth by Terrence Deacon in *The Symbolic Species*, it could be argued that humans can change their environment through symbolic means and that this process is somehow connected with human fantasies. Humans can, in other words, fantasise about future technologies which can act as tools to change their habitat. The science fiction books and movies I encountered at the Niels Bohr Institute did, I shall argue, just that. Fantasies act as directional forces on physicists and engineers constructions of mattering matter, which in subtle ways led scientific endeavours forward towards changes in the human habitat inspired by the fictional Star War or Star Trek societies.[6] But will these kinds of fantasies also in the future be our source of inspiration? Here a new interesting discussion about how symbolic meaning of technological tools change over time is opened up. What is *considered* innovative fantasies might be relational and tied to the diversity of human gendered experiences. With the words of Suchman:

---

[6] Even Breazeal has told reporters that she was inspired by science fiction and not least the Star War series.

> We need to ask how projects to reclaim
> creativity, invention and the like might themselves be reproductive of a, specific, cultural and historical preoccupation with these particular values. Must those not presently identified as creative be shown in fact to be inventors in order to be fully recognized? This question suggests that we need to pay close attention to the tensions and contradictions that arises when we adopt a strategy that distributes practices previously identified exclusively with certain people and places (for example, with privileged white men working in elite institutions of science and technology) across a wider landscape (one thatincludes women). (Suchman, 2005, p. 2).

The political reason for giving female perspectives more space in science is not that they are women but that women may be carriers of different embodied experiences in body one as well as body two and that these experiences might point to the development of a more human and less humanoid robot vision of our future. Many women have a passion for reading literature very different from the stereotypical one praised in science fiction circles. What Susan Star and Lee Bowker questions in informatics might concern science in general:

'Why should computer scientists read African-American poets? What does information science have to do with race-critical or feminist methods and metaphysics? The collective wisdom in those domains is one of the richest places from which to understand these core problems in information systems design: how to preserve the integrity of information without a priori standardization and its often attendant violence? In turn, if those lessons can be taken seriously within the emerging cyber world, there may yet be a chance to strengthen its democratic ethical aspects. It is easy to be ethnocentric in virtual space; more difficult to avoid stereotypes. The lessons of those who have lived with such stereotypes are important, perhaps now more than ever' (Bowker & Star, 1999, p. 302).

Would science develop differently if it was practiced by other kinds of people, who for example have experienced the stereotypes of being women or black on their own bodies? Will new symbols be created which can feed in to new fantasies acting as directional forces on what tools to use and how humans should construct their habitas? If science can be said to be limited in its scope by such stereotypes how do they connect to the development of science? If personal experiences, like reading poetry, might open up for new spaces of scientific inquiry – what kind of driving forces develop science today? And are these present driving forces inherently connected to male experiences? These questions are not just relevant from a feminist point of view, but concern the very nature of scientific knowledge and its development. When society move to value other kinds of soft skill experiences we might find conflicts – not between industrial robots being outnumbered by artificial emotional robots but a war between techno fantasies and poetic fantasies.

In our contemporary world our problems develop when images build upon images and when images are lost. This is 'the risk that we run today with the war of dreams' (Augé, 1999, p. 56).

A counter project is to question the very force with which techno fantasies are supported in society and the way innovation itself is understood as tied up with technology. The moral and political aspects of science (Haraway, 1991; Alcoff, 1996) also concerns what kind of embodied experiences can develop science. Access of the marginalized to scientific practices will change the kind of questions asked in science and thus the scientific endeavour as such (Hartsock, 1989, Harding, 1991, 1998). Maybe soft skills in the future politics of science will be connected (by male and female scientists alike) to more poetic versions of the future than innovations build into robots of aluminium and antibacterial fur. Maybe the innovation should concern the staff at old people's homes together with the old people and their possibilities to invent and create new fantasies not already embedded in ready-to-use technologies. Feminism directs our attention to the importance of 'the human aspect' of science and towards research in our basic lack of understanding of how culturally formed fantasies and dreams can be built into technologies shaping our lives.

References

Alcoff, L. M. (1996). *Real knowing*. Ithaca, NY: Cornell University Press.

Alcoff, L. M., & Potter, E. (Eds.). (1993). *Feminist epistemologies*. New York: Routledge.

Alper, J. (1993). The pipeline is leaking women all the way along. *Science, 260*, 409–411.

Auchincloss, P. (1998). Physics and Feminism. In *The Back Page - APS News*, May/1998. Retrieved January 5th 2010, from http://www.aps.org/apsnews/0598/059815.html.

Augé, M. (1999). *The war of dreams: Studies in ethno fiction* (L. Heron, Trans.). London: Pluto Press.

Barad, K. (1998) 'Getting real: Technoscientific practices and the materialization of reality' differences. *A Journal of Feminist Cultural Studies, 10*(2), 87–128.

Barad, K. (1999). Agential realism: Feminist interventions in understanding scientific practices. In M. Biagioli (Ed.), *The science studies reader* (pp. 1–11). New York: Routledge.

Barad, K. (2003) Posthumanist performativity: Toward an understanding of how matter comes to matter. *Signs: Journal of Women in Culture and Society, 28*, 801–831.

Barinaga, M. (1994). Surprises across the cultural divide. *Science, 263*, 1468–1472.

Bebbington, D., & Glover, J. (2000) *Women and scientific employment: Mapping the European data*. Brussels: Commission of the European Communities, Research Directorate General.

Billett, S. (2004). Workplace participatory practices. *The Journal of Workplace Learning, 16*(6), 3123–324.

Bleier, R. (1984). Science and gender: A critique of biology and its theories on women. In G. Bowle & R. Duelli-Kleine (Eds.), *Athene series*. New York: Pergamon Press.

Bowker, G. C., & Star, S. L. (1999). *Sorting things out. Classification and its consequences*. Cambridge, MA: The MIT Press.

Braidotti, R., Lazaroms, I., & Vonk, E. (Eds.). (2001). *The making of european women's studies* (Vol. III). Utrecht: Athena/Utrecht University Press.

Breazeal, C. Takanishi, A., & T. Kobayashi (2008). Realism of the unreal: the Japanese robot and the performance of representation. *Visual Communication, 7*, 345–362.

Butler, J. (1990). *Gender trouble*. New York: Routledge.

Butler, J. (1993). *Bodies that matter*. New York: Routledge.

Chimba, M., & J. Kitzinger (2010) Bimbo or boffin? Women in science: an analysis of media representations and how female scientists negotiate cultural contradictions. *Public Understanding of Science, 19*, 609–625.

Cole, M. (1996). *Cultural psychology: A once and future discipline*. Cambridge, MA: The Belknap Press of Harvard University Press.

Daston, L. (1998). Fear and loathing of the imagination in science. *Dædalus, 127*, 73–95.

Daston, L., & Park, K. (1998). *Wonders and the order of nature*. New York: Zone Books.
Deacon, Terrence. (1997). *The symbolic species: The co-evolution of language and the brain*. New York: Norton.
Douglas, M. (1987). *How institutions think*. London: Routledge & Kegan Paul.
Dubeck, L. W., Moshier, S. E., Bruce, M. H., & Boss, J. E. (1993). Finding the facts in science fiction films. *Science Teacher, 60*(4), 46–48.
Engeström, Y. (1987). *Learning by expanding. An activity-theoretical approach to developmental research*. Helsinki: Orienta-Konsultit.
Etzkowitz, H., Kemelgor, C., & Uzzi, B. (2000) *Athena unbound: The advancement of women in science and technology*. Cambridge: Cambridge University Press.
European Commission. (2009). *She figures 2009: Statistics and indicators on gender equality in science*. Brussels: Publication Office of the European Union.
Fenstermaker, S., & West, C. (Eds.). (2002). *Doing gender, doing difference – Inequality, power, and institutional change*. New York and London: Routledge.
Franklin, S. (1995). Romancing the helix: Nature and scientific discovery. In J. Stacey & L. Pearce (Eds.), *Romance revisited* (pp. 63–77). London: Falmer Press.
Geertz, C. (1973/1993). *The interpretation of cultures*. London: Fontana Press.
Gilligan, C. (1982). *In a difference voice: Psychological theory and women's development*. Massachusetts: Harvard University Press.
Haraway, D. (1988). Situated knowledges: The science question in feminism and the privilege of partial perspective. *Feminist Studies, 14*, 575–99.
Haraway, D. (1989). *Primate visions: Gender, race, and nature in the world of modern science*. New York: Routledge.
Haraway, D. (1991). *Simians, cyborgs, and women: The reinvention of nature*. New York: Routledge.
Harding, S. (1991). *Whose science? Whose knowledge? Thinking from women's lives*. New York: Cornell University Press.
Harding, S. (1998). *Is science multicultural? Postcolonializms, feminists, and epistemologies*. Bloomington, IN: Indiana University Press.
Hartsock, N. (1989). *the feminist standpoint revisited and other essays*. Boulder, CO: Westview Press.
Hasse, C. (1998). Kulturelle forestillinger og køn i videnskabens samfund [Cultural conceptions and gender in the scientific society]. In I. Henningsen (Ed.), *Køn i den akademiske organisation* (Working paper nr. 4). Copenhagen: Københavns Universitet.
Hasse, C. (2001). Institutional Creativity – The relational zone of proximal development. *Culture & Psychology, 7*(2), 199–221.
Hasse, C. (2008a). Cultural body learning – The social designation of institutional code-curricula. In T. Schilhab, M. Juelskjær, & T. Moser (Eds.), *Body and learning* (pp. 193–215). København: The Danish School of Education Press.
Hasse, C. (2008b). Postphenomenology – Learning cultural perception in science. In *Human Studies* (pp. 43–61). Hamburg: Springer.
Hasse, C. & Trentemøller, S. (2008). *Break the pattern! A critical enquiry into three scientific workplace cultures: Hercules, caretakers and worker bees*. Tartu: Tartu University Press. See also cordis.europa.eu/documents/documentlibrary/116810421EN6.pdf
Hasse, C., Trentemøller, S., & Sinding, A. B. (Eds.). (2008). *Draw the line!: International conference, Copenhagen 2008. Papers, proceedings and recommendations*. Tartu: Tartu University Press. See also cordis.europa.eu/documents/documentlibrary/116810421EN6.pdf
Henrion, C. (1997). *Women in mathematics: The addition of difference*. Bloomington, MI: Indiana University Press.
Hird, M. (2002). Re(pro)ducing sexual difference. *Parallax, 8*(4), 94–107.
Ihde, D. (1993). *Postphenomenology—essays in the postmodern context*. Evanston, IL: Northwestern University Press.

Ihde, D. (2002): *Bodies in technology (Electronic mediations) (Vol. 5)*. Minneapolis: University of Minnesota Press.
Irigaray, L. (1985). *This sex which is not one*. Ithaca, NY: Cornell University Press.
Kaku, M. (1994) *Hyperspace: A scientific odyssey through parallel universes, time warps, and the 10th dimension*. Oxford: Oxford University Press.
Keller, E. F. (1977). The Anomaly of a Woman in Physics. In S. Ruddick & P. Daniels (Eds.), *Working it out: 23 women writers, artists, scientists, and scholars talk about their lives and work* (pp. 77–91). New York: Pantheon.
Keller, E. F. (1983). *A feeling for the organism*. New York: W. H. Freeman.
Keller, E. F. (1985). *Reflections on gender and science*. New Haven, CT: Yale University Press.
Krauss, L., & Hawking, S. (1996). *The physics of startrek*. New York: Harperperennial Library.
Kuhn, T. (1977) *The essential tension*. Chicago: The University of Chicago Press.
Lave, J. (1997). Learning, apprenticeship, social practice. *Nordisk Pedagogik, 17*(3), 140–151.
Lave, J. & Wenger, E. (1991/1996). *Situated learning, legitimate peripheral participation*. Cambridge: Cambridge University Press.
Lie, S., Linnakylä, P., & Roe, A. (2003) *Northern lights on PISA. Unity and diversity in the Nordic countries in PISA 2000*. Oslo: University of Oslo Press.
Longino, H. (1990). *Science as social knowledge*. Princeton: Princeton University Press.
Meilwee, J. & Robinson, J. (1992). *Women in engineering: gender, power, and workplace culture*. New York: SUNY Press.
Nahin, P. J. (1993). *Time machines: Time travel in physics, metaphysics, and science fiction*. New York: American Institute of Physics.
Osborn, M., Rees, T., Bosch, M., Hermann, C., Hilden, J., McLaren, A., et al. (2000). *Science policies in the European Union: Promoting excellence through mainstreaming gender equality. A Report from the ETAN Network on Women and Science*. Luxembourg: Office for Official Publications of the European Communities.
Raudla, T. (2008). Vico and Lotman: poetic meaning creation and primary modelling. *Sign Systems Studies, 1*, 137–165.
Rees, T. (2002). First Results from the Helsinki Group on Women and Science Conference: Policy review. In L. Maxwell, K. Slavin, & K. Young (Eds.), *Gender & research* (pp. 53–56). Luxembourg: Office for Official Publications of the European Communities.
Rolin, K. (2001). Gender and physics: A theoretical analysis. *Journal of Women and Minorities in Science and Engeneering, 7*, 1–8.
Rouse, J. (1991). Policing knowledge: Disembodied policy for embodied knowledge. *Inquiry, 34*, 353–364.
Schiebinger, L. (1989). *The mind has no sex? Women in the origins of modern science*. Cambridge, MA: Harvard University Press.
Schiebinger, L. (1999). *Has feminism changed science?* Cambridge, MA: Harvard University Press.
Sjöberg, S., & Schreiner, C. (2005). How do learners in different countries relate to science and technology? Results and perspectives from the project ROSE. *Asia Pacific forum on science learning and teaching, 6*(2), 1–17. See also http://www.ied.edu.hk/apfslt/
Stadler, H., Duit, R., & Benke, G. (2000). Do boys and girls understand physics differently? *Physics Education, 35*, 417–422.
Stengers, I. (1997). *Power and invention: Situating science*. Minneapolis: University of Minnesota Press.
Stengers, I. (2000). *The invention of modern science*. Minneapolis: University of Minnesota Press.
Strathern, M. (1991). *Partial connections*. Lanham, MD: Rowan and Littlefield.
Suchman, L. (2005). Agencies in technology design: Feminist reconfigurations. *Paper presented at the workshop on gendered innovations in science and engineering, Stanford University, 15–16 April*.
Suchman, L. A. (2007). *Human-machine reconfigurations: Plans and situated actions* (2nd ed.). Cambridge: Cambridge University Press.

Svinth, L. (2008). Women in physical science. In C. Hasse, S. Trentemøller, & d A. B. Sinding (Eds.), *Draw the line! International conference, Copenhagen 2008. Papers, proceedings and recommendations* (pp. 19–43). Tartu: Tartu University Press.

Tamura, T., Yonemitsu, S., Itoh, A., Oikawa, D., Kawakami, A., Higashi, Y., et al. (2004). Is an entertainment robot useful in the care of elderly people with severe dementia? *Journals of Gerontology Series A: Biological Sciences and Medical Sciences., 59A*, 83–85.

TERSTI. (2003). *Third European report on science & technology indicators 2003. Towards a knowledge-based economy*. Brussels: European Commission, Directorate-General for Research. Retrieved February 8th 2008 and May 2008, from http://www.dife.de/~mristow/2003EU_3rd_report.pdf

Traweek, S. (1988). *Beamtimes and lifetimes. The world of high energy physicists*. Cambridge: Harvard University Press.

Urciuoli, B. (2008). Skills and selves in the new workplace, *American Ethnologist, 35*(2), 211–228.

Verran, H. (1998). Re-imagining land ownership in Australia. *Postcolonial Studies, 1*(2), 237–254.

Wada, K., Shibata, T., Saito, T., & Tanie, K. (2004). Effects of Robot-Assisted activity for elderly people and nurses at a day service center. *Proceedings of the IEEE, 92*(11), 1780–1788.

Wajcman, J. (1991). *Feminism confronts technology*. Cambridge: Polity Press.

Wertheim, M. (1995). *Pythagoras' trousers: God, physics, and the gender wars*. New York: Times Books/Random House.

Index

T. Schilhab et al. (eds.), *The Symbolic Species Evolved*, Biosemiotics 6,
DOI 10.1007/978-94-007-2336-8, © Springer Science+Business Media B.V. 2012

W

Y

Z

MIX
Papier aus verantwortungsvollen Quellen
Paper from responsible sources
FSC® C105338

If you have any concerns about our products,
you can contact us on
ProductSafety@springernature.com

In case Publisher is established outside the EU,
the EU authorized representative is:
Springer Nature Customer Service Center GmbH
Europaplatz 3, 69115 Heidelberg, Germany

Printed by Libri Plureos GmbH
in Hamburg, Germany